OVERTRAINING in SPORT

Richard B. Kreider, PhD
Andrew C. Fry, PhD
Mary L. O'Toole, PhD

EDITORS

Human Kinetics

Library of Congress Cataloging-in-Publication Data

Overtraining in sport / edited by Richard B. Kreider, Andrew C. Fry,
   Mary L. O'Toole.
         p.   cm.
      Includes bibliographical references and index.
      ISBN 0-88011-563-7
      1. Sports--Physiological aspects.   2. Physical education and
   training.   3. Stress (Physiology)   4. Burn out (Psychology)
   I. Kreider, Richard B., 1962-   .   II. Fry, Andrew C., 1956-   .
   III. O'Toole, Mary Louise.
   RC1235.094   1997
   617.1'027--dc21   97-31205
                                                                            CIP

ISBN: 0-88011-563-7

Permissions notices for material reprinted in this book from other sources can be found on
page xi.

Managing Editor: Alesha G. Thompson; **Assistant Editor:** Erin Sprague; **Copyeditor:**
Judy Peterson; **Proofreader:** Erin Cler; **Indexer:** Mary E. Coe; **Graphic Designers:** Judy
Henderson and Nancy Rasmus; **Graphic Artist:** Francine Hamerski; **Cover Designer:**
Robert Reuther; **Illustrator:** Sara Wolfsmith; **Printer:** Braun-Brumfield

Printed in the United States of America   10   9   8   7   6   5   4   3

Human Kinetics
Web site: http://www.humankinetics.com/

United States: Human Kinetics, P.O. Box 5076, Champaign, IL 61825-5076
1-800-747-4457
e-mail: humank@hkusa.com

Canada: Human Kinetics, 475 Devonshire Road, Unit 100, Windsor, ON N8Y 2L5
1-800-465-7301 (in Canada only)
e-mail: humank@hkcanada.com

Europe: Human Kinetics, P.O. Box IW14, Leeds LS16 6TR, United Kingdom
+44 (0)113-278 1708
e-mail: humank@hkeurope.com

Australia: Human Kinetics, 57A Price Avenue, Lower Mitcham, South Australia 5062
(08) 82771555

This book is dedicated to athletes, coaches, and sports scientists who strive to optimize human performance through sport and sport science research.

Contents

Overtraining in Sport: Terms, Definitions, and Prevalence

Richard B. Kreider, PhD, Andrew C. Fry, PhD, Mary L. O'Toole, PhD

In order to optimize athletic performance, athletes must be optimally trained. Athletes who under train may not perform to their potential, whereas athletes who train too often or too intensely may experience negative training adaptations and decreased performance capacity (see table 1.1). The challenge for the athlete and coach is to determine the appropriate type and volume of training, that optimizes performance yet does not elicit negative training adaptations (see figure 3.1). Unfortunately, determining the optimal amount of training for athletes is difficult in that the volume of training, which may optimize performance in some athletes, may under train or over train others. In addition, the internal and external psychological pressures of being a competitive athlete may add to the stress of training and promote negative adaptations. Whether a consequence of physical or psychological stress, once maladaptations to training are elicited, performance may deteriorate for several weeks or months despite rest or reductions in training volume. In addition to the potential adverse medical consequences of overtraining, a prolonged deterioration in performance may interrupt the season or ruin the career of the athlete. Consequently, overtraining is a major problem among competitive athletes.

The negative physiological and psychological effects of training too often or too intensely have been a significant focus of research in exercise science over the last 20 years. In addition, significant research has been conducted on ways to monitor and prevent overreaching and overtraining. The purpose of the *International Conference on Overtraining in Sport,* which was held at The University of Memphis preceding the 1996 Olympic Games in Atlanta, was to bring many of the world's leading researchers, physicians, and coaches together to discuss the issues of overtraining in sport. An additional purpose of the gathering was to collaborate on the first comprehensive book analyzing the physiological, biomedical, and psychological aspects of overtraining in sport.

Many different terms have been used in the literature to describe the phenomenon of overtraining: *overwork, overtraining, overreaching, overstraining, staleness, burnout, overstress, overfatigue,* and so on. The different terminology used to describe this phenomenon has created much confusion when interpreting literature in this area. Therefore, for the purposes of interpreting the literature discussed in this book, we have defined this phenomenon as follows:

Overreaching. An accumulation of training and nontraining stress resulting in a short-term decrement in performance capacity with or without related physiological and psychological signs and symptoms of overtraining in which restoration of performance capacity may take from several days to several weeks.

Overtraining. An accumulation of training or nontraining stress resulting in long-term decrement in performance capacity with or without related physiological and psychological signs and symptoms of overtraining in which restoration of performance capacity may take several weeks or months.

The following should be noted from the above definitions. First, we propose that the critical factor in describing the overreaching and overtraining phenomenon is training that results in decreased performance capacity, not simply manifestations of reported signs and symptoms of the overtraining syndrome (see table 1.1). In this regard, some athletes who experience decrements in performance capacity exhibit no overt signs and symptoms of the overtraining syndrome, while other athletes who may present with signs and symptoms of overtraining may not experience decrements in performance. Moreover, different types of training appear to elicit different signs and symptoms to overreaching or overtraining. Second, we feel it is important to differentiate between training that results in short-term decrements in performance (overreaching), and training that results in more prolonged decrements in performance (overtraining). While we understand that it may be difficult to differentiate between overreaching and overtraining and that a continuum from under training, to optimal training, to overreaching, to overtraining may exist, we feel that this differentiation is important for interpreting and designing research in this area.

Unlike other thematic proceedings, this book was carefully designed to comprehensively cover the issues of overtraining without excessive overlap among chapters. In addition, since the issues related to overtraining are of interest to the athlete, coach, and researcher, special care was taken to present the information in an easy-to-understand manner. Further, since the research on overtraining is multidisciplinary and readers may not have comprehensive backgrounds in each area, we have attempted to provide enough background information so that the interested reader can understand each area without having extensive knowledge in the area.

This book is divided into seven main sections. In the first section, the prevalence, physiological responses, and methods of monitoring and preventing overreaching and overtraining in endurance athletes are discussed. In the second section, factors contributing to overreaching and overtraining in strength/power athletes as well as physiological responses to alterations in resistance volume and intensity are discussed. The third section of the book discusses the medical consequences of overreaching and overtraining including cardiovascular and hematological responses; neuroendocrine responses; and, musculoskeletal/orthopedic considerations. The fourth section discusses the effects of overreaching and overtraining on the immune system and possible interventions to prevent immunosuppression in athletes. In the fifth section, nutritional factors that may contribute to overreaching, overtraining, and central fatigue are discussed. The sixth section discusses the psychological aspects of overreaching and overtraining as well as potential treatment and preventative methodologies. The last section of the book summarizes the status of overtraining research and provides an analysis of future research needs and directions.

It is our hope that this book will serve as an outstanding resource for athletes, coaches, and research scientists interested in understanding the potential contributing factors and interventions related to overreaching and overtraining. While many questions remain to be answered, we hope that this book serves as a valuable contribution to the field of exercise science and a stimulus for additional research on overreaching and overtraining in sport.

Acknowledgments

The editors would like to express their appreciation to the Department of Human Movement Sciences and Education at The University of Memphis and Human Kinetics for co-sponsoring the *International Conference on Overtraining in Sport* held at The University of Memphis, July 14-17, 1996. This conference served to bring many of the world's leading researchers, physicians, and coaches together prior to the 1996 Olympic Games in Atlanta to discuss the issues of overtraining in sport detailed in this book.

The editors would also like to thank the researchers who presented papers at the *International Conference on Overtraining in Sport* and contributed chapters to this book. We would also like to express our appreciation to the students and staffs at The University of Memphis and Human Kinetics who assisted in the organization and preparation of this book.

Credits

Table 1.1: Reprinted, by permission, from R.W Fry, A.R. Morton, and D. Keast, 1991, "Overtraining in athletes: an update," *Sports Medicine* 12 (1): 34.

Figure 3.3: Reprinted, by permission, from M.E. Levine, A.N. Milliron, and L.K. Duffy, 1994, "Diurnal and seasonal rhythms of melatonin, cortisol and testesterone in interior Alaska," *Arctic Medical Research* 53: 25-34.

Figure 3.4: Reprinted, by permission, from R.W. Fry, A.R. Morton, and D. Keast, 1991, "Overtraining in athletes: an update," *Sports Medicine* 12 (1): 37.

Figure 4.3: Adapted, by permission, from R.U. Newton, W.J. Kraemer, K. Häkkinen, B.J. Humphries, and A.J. Murphy, 1996, "Kinematics, kinetics, muscle activation during explosive upper body movements," *Journal of Applied Biomechanics* 12: 31-43.

Figure 7.1: Reprinted, by permission, from P.B. Raven and G.J.J. Stevens, 1988, Cardiovascular Function and Prolonged Exercise. In *Perspectives in Exercise Science and Sports Medicine, Volume 1 Prolonged Exercise*, edited by D.R. Lamb and R. Murray (Carmel, IN: Cooper Publishing Group), 48.

Figure 7.2: Reprinted by permission of the publisher from "Effect of brief and prolonged exercise on left ventricular function," M.T. Upton, *American Journal of Cardiology*, Vol. 45, p. 1155. Copyright 1980 by Excerpta Medica Inc.

Figure 7.3: Reprinted, by permission, from K.O. Niemela, 1984, "Evidence of impaired left ventricular performance after an uninterrupted competitive 24-hour run," *Circulation* 70: 352 (©1984 American Heart Association).

Figure 8.1: Reprinted, by permission, from A. Weltman et al., 1991, "Endurance training amplifies the pulsatile release of growth hormone: effects of training intensity," *Journal of Applied Physiology* 72: 2188-2196.

Tables 9.1 and 9.2: Reprinted, by permission, from W.B. Kibler, 1990, "Clinical aspects of muscle injury," *Medicine and Science in Sports and Exercise* 22: 450-452.

Figure 11.3: Reprinted, by permission, from M. Gleeson et al., 1995, "The effect on immunity of long-term intensive training in elite swimmers," *Clin. Exp. Immunol.* 102: 214. (London: Blackwell Science Ltd.)

Figure 16.3: Used by permission from R.S. Lazarus and S. Folkman, 1984, *Stress, appraisal, and coping* (© Springer Publishing Company, Inc., New York 10012).

Figure 16.5: Reprinted, by permission, from J.P. Whelan, A.W. Meyers, and C. Donovan, 1995, Competitive recreational athletes. In *Sport psychology interventions*, edited by S.M. Murphy (Champaign, IL: Human Kinetics Publishers, Inc.), 85.

Figure 17.1: Adapted, by permission, from D.L. Costill, R. Thomas, R.A. Robergs, D. Pascoe, C. Lambert, S. Barr, and W.J. Fink, 1991, "Adaptations to swim training: influence of training volume," *Medicine and Science in Sports and Exercise* 23: 371-377.

Figure 17.2: Reprinted, by permission, from M.G. Flynn, K.K. Carroll, H.L. Hall, B.A. Kooiker, C.A. Weideman, C.M. Kasper, and P.G. Brolinson, 1994, "Cross training: indices of training stress and performance," *Medicine and Science in Sports and Exercise* 26: S153.

Physiology of Overtraining in Endurance Athletes

CHAPTER 1

Overreaching and Overtraining in Endurance Athletes

Mary L. O'Toole, PhD

Introduction

An endurance athlete's formula for success, hard training, may also, unfortunately, be the formula for demise. The apparent limitation to endurance exercise performance is the ability of the athlete to endure strenuous training in order to elicit positive adaptations without breakdown or maladaptation in any physiological system. Unfortunately, no model exists that can be used so that the overload training stimulus results in optimal improved performance while minimizing the potential to develop overreaching or the overtraining syndromes. These maladaptations are essentially an imbalance between stimulus and recovery. The universal finding in overreached or overtrained athletes is a decrease in performance ability. Accompanying signs and symptoms may be diffuse in nature resulting in generalized fatigue with or without a variety of specific physiological symptoms. Alternatively, a specific physiological system may break down, e.g., the breakdown of some aspect of the musculoskeletal system may result in an overuse syndrome such as a stress fracture.

The boundary between hard training and overtraining is not clear. Individual variability in response to training is such that an appropriate training load for one athlete may cause the overtraining syndrome in another. Additionally, not all aspects of performance are affected simultaneously nor to the same degree. Identification of physiological markers prodromal to the overtraining syndrome in endurance athletes is likewise difficult. Some markers are seemingly contradictory, e.g., both increased and decreased resting heart rates have been reported in overtrained runners. Other overtrained athletes may lack specific symptoms, yet experience performance decrements and an inability to train at customary levels.

Prevalence, Signs, and Symptoms in Endurance Athletes

Overtraining has been recognized as being detrimental to endurance performance since the early 1920s. Much of the early published information about overtraining was limited to anecdotal accounts making quantification of prevalence difficult. Multiple signs and symptoms have been associated with overtraining. R.W. Fry et al. (10) have provided a comprehensive list of major symptoms reported in the literature and have categorized them according to physiological/ performance, psychological/information processing, immunological, and biochemical manifestations (see table 1.1).

It appears that any one or multiple symptoms in any combination may be present in an athlete suffering a performance decrement. More recently, cross-sectional studies and studies of short duration designed to deliberately evoke an overtraining response have been reported (4). Many of these studies have focused on one specific category of overtraining responses, e.g., immunology. Because of the short but very intense training, most of these studies produce overreaching rather than true overtraining. Only a few studies are available in which a group of athletes has been monitored for a prolonged period, such as a full season. These prolonged studies, however, provide the most information about the true overtraining problem, including its prevalence as well as the major signs and symptoms. Information is thus available about the scope and severity of the problem for specific endurance athlete groups.

Runners

The prevalence of the overtraining syndrome appears to be quite high in middle to long distance runners. Morgan et al. (23) reported that 65% of elite runners experienced staleness during some point in their competitive careers. The overtraining syndrome is, however, not unique to elite runners (5, 33). The most susceptible runners seem to be the highly driven athletes who fail to include enough recovery time in their training (10). The prevalence of overtraining in any group of runners during a competitive season is variable with exact numbers depending on the interaction of multiple exogenous and endogenous factors. Exogenous factors include length of the season, events being trained for, competitive schedule, etc. Endogenous factors include training intensity and volume, individual biomechanics, length of recovery periods, etc.

Because of the insidious nature of the onset of overtraining, few studies have truly evaluated the development of overtraining during the course of a runner's season or career. Most studies have been short-term, acute overload studies aimed at creating overtraining (more accurately, overreaching) in all subjects. R.W. Fry et al. (9) had five trained runners perform two intensive interval training sessions for 10 days in a row. At the end of the 10 days of training, run performance was decreased in all; fatigue and immune system responses were not back to baseline. Verde et al. (35) increased the training of 10 highly trained distance runners by 38% for three weeks. Running performance was not affected, but six of the 10 runners reported sustained fatigue and decreased vigor. No useful changes in resting heart rate, perceived exertion during submaximal exercise, or sleep patterns were observed. These programmed overloads more appropriately reflect a state of

Table 1.1 The Major Symptoms of Overtraining as Indicated by Their Prevalence in the Literature

Physiological performance

Decreased performance

Inability to meet previously attained performance standards or criteria

Recovery prolonged

Reduced toleration of loading

Decreased muscular strength

Decreased maximum work capacity

Loss of coordination

Decreased efficiency or decreased amplitude of movement

Reappearance of mistakes already corrected

Reduced capacity of differentiation and correcting technical faults

Increased difference between lying and standing heart rate

Abnormal T wave pattern in ECG

Heart discomfort on slight exertion

Changes in blood pressure

Changes in heart rate at rest, exercise, and recovery

Increased frequency of respiration

Perfuse respiration

Decreased body fat

Increased oxygen consumption at submaximal workloads

Increased ventilation and heart rate at submaximal workloads

Shift of the lactate curve towards the x axis

Decreased evening postworkout weight

Elevated basal metabolic rate

Chronic fatigue

Insomnia with and without night sweats

Feels thirsty

Anorexia nervosa

Loss of appetite

Bulimia

(continued)

Table 1.1 *continued*

Amenorrhea or oligomenorrhea

Headaches

Nausea

Increased aches and pains

Gastrointestinal disturbances

Muscle soreness or tenderness

Tendonostic complaints

Periosteal complaints

Muscle damage

Elevated C-reactive protein

Rhabdomyolysis

Psychological/information processing

Feelings of depression

General apathy

Decreased self-esteem or worsening feelings of self

Emotional instability

Difficulty in concentrating at work and training

Sensitive to environmental and emotional stress

Fear of competition

Changes in personality

Decreased ability to narrow concentration

Increased internal and external distractibility

Decreased capacity to deal with large amounts of information

Gives up when the going gets tough

Immunological

Increased susceptibility to and severity of illnesses, colds, and allergies

Flu-like illnesses

Unconfirmed glandular fever

Minor scratches heal slowly

Table 1.1 *continued*

Swelling of the lymph glands

One-day colds

Decreased functional activity of neutrophils

Decreased total lymphocyte counts

Reduced response to mitogens

Increased blood eosinophil count

Decreased proportion of null (non-T, non-B) lymphocytes

Bacterial infection

Reactivation of herpes viral infection

Significant variations in CD4:CD8 lymphocytes

Biochemical

Negative nitrogen balance

Hypothalamic dysfunction

Flat glucose tolerance curves

Depressed muscle glycogen concentration

Decreased bone mineral content

Delayed menarche

Decreased hemoglobin

Decreased serum iron

Decreased serum ferritin

Lowered TIBC

Mineral depletion (Zn, Co, Al, Mn, Se, Cu, etc.)

Increased urea concentrations

Elevated cortisol levels

Elevated ketosteroids in urine

Low free testosterone

Increased serum hormone binding globulin

Decreased ratio of free testosterone to cortisol of more than 30%

Increased uric acid production

Reprinted from Fry, Morton, and Keast 1991.

overreaching rather than overtraining. Others have proposed using even shorter inappropriate overloads to study overtraining outcomes in runners. Marinelli et al. (22) measured free testosterone/cortisol ratios (FTCR) before and after a marathon at 4000 m altitude. They suggest that the FTCR can be useful in monitoring fitness, overstrain, and overtraining since acute changes can be seen with a single bout of strenuous exercise. Lehmann et al. (19) compared neuromuscular excitability in runners who had either doubled their training mileage (from 86 to 175 km/wk) or had a 152% increase in tempo-pace and interval runs during four weeks. The neuromuscular excitability of the vastus medialis and rectus femoris deteriorated following increased volume, but not increased intensity.

Swimmers

Of all endurance athletes, swimmers have been the most studied group in terms of overtraining. Nineteen elite swimmers were monitored, without interfering with their training, during six-month's preparation for national team selections (12, 13). Five swimmers did not complete the six-month training period, mainly because of viral infections. At the end of six months, three of the 14 (21%) remaining swimmers were classified as stale, an incidence comparable to previous estimates. Additionally, some of the five dropouts may have been more susceptible to infection because of heavy training. There were no significant differences in the training programs of the stale and nonstale swimmers. Changes in the usual physiological parameters (e.g., HR, BP, VO_2 and blood lactate) did not differentiate between the stale and nonstale swimmers. Higher neutrophil and plasma norepinephrine levels during the taper period did, however, differentiate between the two groups. Interestingly, athlete ratings of sleep and fatigue at midseason predicted staleness before performance deteriorated several weeks later. Flynn et al. (8) studied five swimmers who trained overly hard for two weeks. Compared with the initial nine weeks of training, testosterone levels (total and free) were lower, suggesting perhaps an early sign of overreaching.

Cyclists

As with runners and swimmers, most of the studies examining responses to extreme overload in cyclists have been overreaching studies in which training was intensified for a short period of time, usually two to six weeks. After only two weeks of intensive interval training (2-3 h sessions/d), Jeukendrup et al. (15) reported decreased maximal power output as well as decreased time trial performance. Cyclists exhibited conflicting symptoms. Sleeping heart rate was increased, suggesting, according to Kuipers and Keizer (17), a sympathetic type of overreaching. However, maximal heart rate and heart rates during the time trials were both decreased. Paradoxically, submaximal lactates were decreased with the work rate at 4 mmol/L lactate increased. Maximal lactates decreased by as much as 50%. After two weeks of reduced training, lactate responses were increased. Decreased training restored performance, suggesting that these athletes were overreached rather than overtrained. These investigators subsequently reported decreased maximal and submaximal lactates in a national-class cyclist in whom overtraining was identified by decreased performance, irritability, and sleeplessness. Snyder et al. (31) suggested that a lactate/perceived exertion ratio (HLa/RPE) might be an early sign of overreaching in cyclists. Seven well-trained cyclists trained normally for two weeks, followed by two weeks of extreme overload. The maximal HLa/RPE was decreased 29% after one week and 49% after two weeks. They suggested that athletes (cyclists) are overreached when the HLa/RPE is less than 100. To induce an overreached state, Snyder et al. (32) increased training of eight cyclists for a two-week period in order to determine the role of appropriate glycogen repletion in the prevention of overtraining. All participants met at least three of five criteria for clinical classification of overtraining despite maintenance of appropriate resting muscle glycogen levels. This finding is in confirmation of the findings of Bruin et al. (2) who demonstrated overtraining in horses whose resting muscle glycogen levels were adequately maintained.

Other Endurance Athletes

Both speed skaters (1) and rowers (37) have been monitored over the course of their seasons in an attempt to identify an early sign of overtraining. The FTCR was monitored in eight elite speed skaters during the course of an eight-month season. Data suggested that a decrease of 30% or more in this ratio was an indication of incomplete recovery (1). Veroon et al. (37) studied a group of rowers during nine months of pre-Olympic training. They noted that the FTCR decreased with heavy training and that decreases of 30% or more in the FTCR were related to incomplete recovery, but not necessarily indicative of overreaching or overtraining. Neither power at 4 mmol/L lactate nor maximal power was related to changes in the FTCR (37). Urhausen et al. (34) also reported significantly decreased testosterone/cortisol ratios as a normal response to heavy training in rowers.

Inconsistent Signs

In an attempt to induce overreaching or the overtraining syndrome, Bruin et al. (2) excessively trained race horses. One of the first signs that adaptation was compromised was an inability of the overreached horses to complete the high-intensity interval training workouts. This was in the absence of other characteristic warning signs for overtraining. For example, neither plasma urea concentration nor plasma creatine kinase activity was a good marker for the imbalance between training and recovery. Additionally, insufficient time for recovery of something other than muscle glycogen appeared to be responsible for the overreached state since resting muscle glycogen levels remained normal throughout the study.

Although overreaching and overtraining have anecdotally been reported in almost all endurance sports, very little information is available regarding prevalence or specific signs and symptoms indicative of overreaching or overtraining. Several groups of endurance athletes regularly engage in extreme amounts of exercise training. While some of these athletes undoubtedly become overtrained, the majority do adapt to these extremes. For example, triathletes training for the Hawaii Ironman Triathlon have been reported to average approximately 280 miles a week of swimming, cycling, and running (25). Much more information is needed to quantify and better understand the magnitude of the problem of overtraining for endurance athletes.

Overload: The Training Response and Improved Performance

Progressive overload is the foundation of all successful training. According to the general adaptation syndrome first proposed by Selye (28), stress causes a temporary decrease in function followed by an adaptation that improves function. In the training response, overload is the stress that causes fatigue (temporary decrease in exercise ability), and improved performance (following recovery from fatigue) is the adaptation (see figure 1.1). During muscular exercise, all fatigue is ultimately the result of an inability to generate energy at a rate sufficient to meet the needs of the performance. The specific energy pathways responsible for fatigue depend on the type of activity. For endurance athletes, the physiological systems necessary for providing oxygen and fuel delivery to the muscles as well as the systems necessary for extracting and generating energy are the systems that must be stressed and forced to adapt to a higher functional demand. A good example of the stress, fatigue, adaptation cycle in endurance athletes is the procedure for glycogen supercompensation. This well-known procedure involves reducing muscle glycogen stores to minimal levels with strenuous exercise (the stress phase). Exercise capacity is thus severely limited. During the recovery phase, increased carbohydrate dietary intake results in one and one-half times more glycogen than usual to be stored in the muscles. This adaptation of the muscle to store more glycogen will

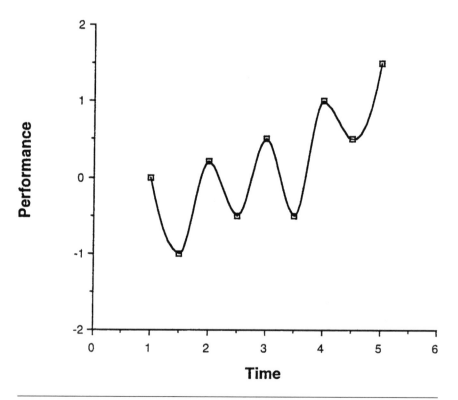

Figure 1.1   Schematic of training overload resulting in improved performance.

result in improved performance in events in which availability of carbohydrate as substrate can be a limiting factor, e.g., marathon running. Similarly, other aspects of endurance performance undergo adaptations in response to the training overload that result in improved rates of aerobic metabolism. Discussion of the many well-known adaptations, such as increased number and density of mitochondrial enzymes, as well as improved cardiovascular and thermoregulatory function, is beyond the scope of this chapter.

For the endurance athlete, overload is accomplished through some combination of increased volume and intensity. In general fitness programs, these factors are reciprocally related. That is, for the same fitness gains, high intensity, short duration exercise can be made to equate with lower intensity, longer duration exercise. For the athlete, however, volume and intensity are adjusted in accord with the specific demands of competition and in particular, by the requirement for the rate of energy generation necessary for a specific performance.

For example, the marathoner who wants to run under a 2 h 30 min marathon must run at a pace slightly faster than 5 min 45 sec per mile. If minor changes in running efficiency are disregarded, the calculated rate of energy generation to allow this pace is approximately 56 ml/kg/min. The overload of the training program should be aimed at adaptations that make this pace reasonable. Obviously, methods of aerobic training that have been shown to improve VO_2max need to be undertaken first. But, to realize his/her full potential as a marathoner, one needs to be able to use as much of his/her aerobic capacity as possible. Therefore, training must also be aimed at raising the anaerobic threshold as close as possible to maximal capacity. Elite marathoners have been reported to race at an average pace equivalent to 86% VO_2max, a pace that represents approximately 93% of their 4 mmol/L lactate threshold (30). Slower runners have been reported to run only at a pace equivalent to 65% VO_2max.

Training Volume

Although typical training volumes for most endurance sports can be found in the literature, the volume of training that is optimal for performance improvement is difficult to define. Although weekly training volumes are usually reported, daily training volumes or even single session training volumes are important for optimizing the training.

The ideal adaptation results from being able to manipulate the combination of volume and intensity in the correct ratio of work and rest. Additionally, individual capabilities may differ, causing the same training volume to be excessive overload for one individual, but less than that necessary to cause sufficient adaptation in another. Many athletes and coaches have tended to equate large volumes with success. There are numerous anecdotal accounts of elite athletes performing extreme volumes of training. For example, the amount of training done by triathlete Dave Scott was legendary, including 400 miles of cycling per week. Similarly, it was not unusual in the mid-1960s for runners to put in 200 miles of running per week (21). Although cause and effect can not be definitively established for many musculoskeletal injuries, many of these overuse syndromes (a musculoskeletal manifestation of overtraining) have been associated with high training volumes.

More recently, scientific evidence is accumulating that training volumes may be reduced by as much as one half in some sports with no detrimental effects on performance. For example, changes in swim performance of a 100-yard swim were compared over a four-year period in two groups of swimmers of similar ability. One group trained more than 10,000 meters per day; the other no more than 5000 meters per day. The average improvement of 0.8% per year was identical in the two groups (7). The amount that training volume contributes to performance improvement is likely related to competition distances. When training was examined in relation to the components of the Ironman Triathlon (2.4-mile swim, 112-mile bike, 26.2-mile run), the effects of training volume were related to component length. Average weekly training included 10 miles of swimming, 200 miles of bicycling, and 45 miles of running, a volume that would likely be considered excessive by most standards. The fastest swimmers trained greater distances and at a faster pace than slower swimmers, but the only differences between faster and slower cyclists and runners were greater training distances (25).

Training Intensity

In addition to volume overload, manipulation of intensity functions to overload specific metabolic pathways. The timing and length of recovery intervals, either during sets of a single exercise session or from session to session, should be governed by the specifics of the systems that have been stressed. For example, short, all-out bursts of activity overload the high energy phosphate systems. For optimal adaptation, complete recovery should occur before a second stimulus is given. On the other hand, to build up tolerance to high lactate levels, initiating a second stimulus before complete recovery may be desirable. Between workouts, however, optimal response is thought to occur during the overshoot following recovery from the previous stimulus. If the time to the next training session is too long, the overcompensation will regress to the original functional state and progressive improvement will not occur. Conversely, if the training stimulus is given too frequently, such that it interrupts the recovery/overcompensation phase, adaptation will not occur. If the overload has been properly designed, a progressive improvement in performance will result.

Overload: Maladaptation and No Improvement in Performance

Just as progressive overload is the foundation of successful training, it also has the potential to be the cause of overreaching and overtraining. For example, with multiple hard training sessions, the recovery/regeneration phase may be compromised. If this continues for an appreciable period of time, overreaching occurs and performance declines. Unfortunately, the decline in performance frequently causes an athlete to further increase his/her training. Although athletes may deliberately provoke overreaching in order to stimulate full supercompensation, most times overreaching has an insidious onset and, if not identified, even the most experienced athletes will progress to the overtraining syndrome. Continuation of this overwork will ultimately cause the quality of performance to be severely hampered and the athlete to become progressively overtrained (see figure 1.2).

The dilemma for the endurance athlete, then, is how to know when overload is enough, too much, or too little. For endurance athletes, there is general consensus

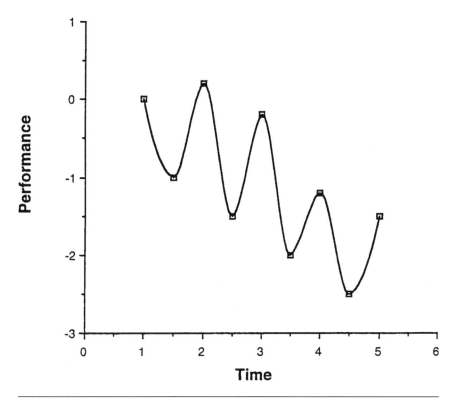

Figure 1.2   Schematic of training overload resulting in decreased performance.

that undertraining is better than overtraining, but is unusual in seriously training athletes (27). In athletes getting an appropriate amount of training, performance continually improves. Conversely, underperformance is the hallmark of maladaptation. Although fatigue, even at rest, is a common symptom of maladaptation, no physical signs have consistently been associated with the overtraining syndrome. Additionally, some of the physical signs may be contradictory (e.g., increased or decreased resting heart rate or decreased lactate response to submaximal exercise) and are not well understood. Increased susceptibility to musculoskeletal injury or infections such as head colds may be indicators of a state of overreaching or overtraining, but may be misinterpreted as isolated, local problems rather than manifestations of the overtraining syndrome.

Several studies have been designed to purposefully provoke maladaptation in order to better characterize the early or intermediate stages in the development of overtraining. Both excessive volume overload and intensity overload as well as reduced time for recovery have been used as stimuli to provoke an overreached state. Jeukendrup et al. (15) used two weeks of intensified training to provoke overreaching in well-trained amateur male cyclists. The intensified training consisted of 2-3 h sessions/d of extremely heavy interval training. This type of high intensity overload with probable incomplete recovery between work bouts resulted in decreased submaximal lactate concentrations and a decrease of up to 50% in maximal lactate production during a graded exercise test. These findings occurred despite adequate muscle glycogen. The investigators suggest that a reduced sympathetic drive or a reduced sensitivity to catecholamines may be responsible for these findings. The relationship between these maladaptations in the lactate response and decrements in actual race performance or in progression to over-training is not clear.

The role that excessive volume plays in the development of the overtraining syndrome is likewise unclear. Lehmann et al. (18, 20) purposefully overloaded distance runners by either doubling their weekly training volume in three weeks or doubling the amount of interval and speed workouts. Increased training volume caused endurance performance to plateau and decreased maximal performance, while increased intensity resulted in performance improvements. Morgan et al. (23) note that little epidemiologic evidence exists dealing with the incidence of the overtraining syndrome in swimmers. They suggest that this is, in part, the result of the varying volume loads of various programs. They further suggest that overtraining is unlikely in programs in which peak volumes are limited to 3000-5000 yds/d, but a potential problem in programs that incorporate 10,000-15,000 yds/d.

Susceptibility of Endurance Athletes to the Overtraining Syndrome

The imbalance between training and recovery that may lead to overreaching and ultimately to the overtraining syndrome in endurance athletes may be the result of increased training or decreased recovery. Highly motivated athletes who are experiencing either a plateau or slight decrement in performance are likely to increase their training and are particularly susceptible to overreaching or overtraining.

Increased Training

Increased training can be in the form of either increased intensity or increased volume. Studies attempting to provoke overreaching or overtraining have increased one or both of these factors. Using a horse model, Bruin et al. (2) increased training initially by increasing the total amount of work done during each interval of interval training. When the expected decrement in performance did not occur within 2.5 months, the severity of training was further increased by increasing the intensity of 20 minutes of endurance running. So, in the first case, the volume of high-intensity exercise was increased without creating a performance decrement and in the second case, the intensity of the volume training was increased causing the desired endpoints. Both Costill et al. (6) and Kirwan et al. (16) reported on a group of swimmers whose training volume was doubled for 10 days. Performance did not decrease, suggesting that short-term excessive training may not be harmful, thus making detection of early signs of overreaching or overtraining difficult. Hooper et al. (12, 13) suggest that volume of training, rather than intensity, may be the major contributing factor in the development of the overtraining syndrome in 21% of a sample of elite swimmers. They further suggest that volume may have been a major factor in 33% of the overtraining syndrome demonstrated in the Indian National basketball team (36).

Decreased Recovery

R.W. Fry et al. (10) suggest that failure to allow adequate recovery is the leading training-related factor in the progression to overreaching or overtraining. In support of this hypothesis, Bruin et al. (2) demonstrated that even markedly increased training loads could be tolerated as long as intensive exercise days were separated by days of moderate intensity endurance running. When the intensity of the endurance running was increased, thereby reducing recovery time, overreaching could be achieved. Presumably, low-intensity exercise or rest allows the organism to recover and adjust to the physical stress, while harder training prevents complete restoration of homeostasis.

Although recovery practices of athletes have not been studied in detail, the concept of periodicity of training is well accepted (10). R.W. Fry et al. (11) have hypothesized that cyclic training may be essential to elicit optimal performance and to avoid the overtraining syndrome. In its simplest form, athletes have been pursuing a hard/easy training pattern for years. In more complete form, the training year is divided into macrocycles in which a particular type of training is emphasized in relation to a major competition. The macrocycles are further subdivided into mezo- and microcycles. Microcycles with very low training loads allow recovery and regeneration to occur periodically throughout the training year. Additionally, the technique of tapering to optimize performance by allowing adequate recovery from hard training before competition is universally accepted. During a taper, some combination of training frequency, intensity, and volume is altered to reduce the training stimulus. Most evidence indicates that a rather drastic reduction in volume (up to 85-90% of usual training volume) in combination with short intense workouts gives the best result (14, 29). This type of taper has been shown to result in a performance improvement of approximately 3% in swimmers and distance runners. This taper pattern provides indirect evidence for the reduction of training volume and increase in recovery time to avoid the detrimental effects of excessive training, including mood disturbances. Following a taper, mood-state, including increased vigor and decreased anxiety, has been shown to be improved (23, 24, 26).

Other Contributing Factors

Factors other than those of heavy training may lead to a training/recovery imbalance and may lead to overreaching in response to a previously well-tolerated training program (3). Poor diet, and in particular inadequate carbohydrate and fluid intakes, as well as transient infections such as the common cold may make it difficult to tolerate usual training loads. There is also anecdotal evidence that social, academic, and economic factors may also contribute to an athlete's susceptibility to overreaching or overtraining (3).

References

1. Banfi, G., M. Marinelli, G.S. Roi, V. Agape. 1993. Usefulness of free testosterone/cortisol ratio during a season of elite speed skating athletes. *International Journal of Sports Medicine* 14: 373-379.

2. Bruin, G., H. Kuipers, H.A. Keizer, G.J. Vander Vusse. 1994. Adaptation and overtraining in horses subjected to increasing training loads. *Journal of Applied Physiology* 76: 1908-1913.

3. Budgett, R. 1990. Overtraining syndrome. *British Journal of Sports Medicine* 24: 231-236.

4. Callister, R., R.J. Callister, S.J. Fleck, G.A. Dudley. 1990. Physiological and performance responses to overtraining in elite judo athletes. *Medicine and Science in Sports and Exercise* 22: 816-824.

5. Costill, D.L. 1986. *Inside running: basics of sports physiology.* Indianapolis: Benchmark Press.

6. Costill, D.L., M.G. Flynn, J.P. Kirwan, J.A. Houmard, J.B. Mitchell, R. Thomas, S.H. Park. 1988. Effects of repeated days of intensified training on muscle glycogen and swimming performance. *Medicine and Science in Sports and Exercise* 20: 249-254.

7. Costill, D.L., R. Thomas, R.A. Roberts, D.D. Pascoe, C.P. Lambert, S.I. Barr, W.J. Fink. 1991. Adaptations to swimming training: influence of training volume. *Medicine and Science in Sports and Exercise* 23: 371-377.

8. Flynn, M.G., F.X. Pizza, J.B. Boone Jr., F.F. Andres, T.A. Michaud, J.R. Rodriguez-Zayas. 1994. Indices of training stress during competitive running and swimming seasons. *International Journal of Sports Medicine* 15: 21-26.

9. Fry, R.W., J.R. Grove, A.R. Morton, P.M. Zeroni, S. Gaudieri, D. Keast. 1994. Psychological and immunological correlates of acute overtraining. *British Journal of Sports Medicine* 28: 241-246.

10. Fry, R.W., A.R. Morton, D. Keast. 1991. Overtraining in athletes: an update. *Sports Medicine* 12: 32-65.

11. Fry, R.W., A.R. Morton, D. Keast. 1992. Periodisation and the prevention of overtraining. *Canadian Journal of Sports Science* 17: 241-248.

12. Hooper, S.L., L.T. Mackinnon, R.D. Gordon, A.W. Bachmann. 1993. Hormonal responses of elite swimmers to overtraining. *Medicine and Science in Sports and Exercise* 25: 741-747.

13. Hooper, S.L., L.T. Mackinnon, A. Howard, R.D. Gordon, A.W. Bachmann. 1995. Markers for monitoring overtraining and recovery. *Medicine and Science in Sports and Exercise* 27: 106-112.

14. Houmard, J.A., R.A. Johns. 1994. Effects of taper on swim performance: practical implications. *Sports Medicine* 17: 224-232.

15. Jeukendrup, A.E., M.K. Hesselink, A.C. Snyder, H. Kuipers, H.A. Keizer. 1992. Physiological changes in male competitive cyclists after two weeks of intensified training. *International Journal of Sports Medicine* 13: 534-541.

16. Kirwan, J.P., D.L. Costill, M.G. Flynn, J.B. Mitchell, W.J. Fink, P.D. Neufer, J.A. Houmard. 1988. Physiological responses to successive days of intense training in competitive swimmers. *Medicine and Science in Sports and Exercise* 20: 255-259.

17. Kuipers, H., H.A. Keizer. 1988. Overtraining in elite athletes: review and directions for the future. *Sports Medicine* 6: 79-92.

18. Lehmann, M., P. Baumgartl, C. Wiesenack, A. Seidel, H. Baumann, S. Fischer, U. Spori, G. Gendrisch, R. Kaminski, J. Keul. 1992. Training-overtraining: influence of a defined increase in training volume vs training intensity on performance, catecholamines and some metabolic parameters in experienced middle- and long-distance runners. *European Journal of Applied Physiology* 64: 169-177.

19. Lehmann, M., E. Jakob, U. Gastmann, J.M. Steinacker, J. Keul. 1995. Unaccustomed high mileage compared to intensity training-related neuromuscu-

lar excitability in distance runners. *European Journal of Applied Physiology* 70: 457-461.

20. Lehmann, M., U. Gastmann, K.G. Petersen, N. Bachl, A. Seidel, A.N. Khalaf, S. Fischer, J. Keul. 1992. Training-overtraining: performance, and hormone levels, after a defined increase in training volume versus intensity in experienced middle- and long-distance runners. *British Journal of Sports Medicine* 26: 233-242.

21. Lucas, J. 1977. A brief history of modern trends in marathon training. In *The marathon: physiological, medical, epidemiological, and psychological studies*, ed. P. Milvy, 858-861. New York: New York Academy of Sciences.

22. Marinelli, M., G.S. Rio, M. Giacometti, P. Bonini, G. Banfi. 1994. Cortisol, testosterone, and free testosterone in athletes performing a marathon at 4,000 m altitude. *Hormone Research* 41: 225-229.

23. Morgan, W.P., D.R. Brown, J.S. Raglin, P.J. O'Connor, K.A. Ellickson. 1987. Physiological monitoring of overtraining and staleness. *British Journal of Sports Medicine* 21: 107-114.

24. Morgan, W.P., D.L. Costill, M.G. Flynn, J.S. Raglin, P.J. O'Connor. 1988. Mood disturbances following increased training in swimmers. *Medicine and Science in Sports and Exercise* 20: 408-414.

25. O'Toole, M.L. 1989. Training for ultraendurance triathlons. *Medicine and Science in Sports and Exercise* 21: S209-213.

26. Raglin, J.S., W.P. Morgan, P.J. O'Connor. 1991. Changes in mood states during training in female and male college swimmers. *International Journal of Sports Medicine* 12: 585-589.

27. Ryan, A.J., R.L. Brown, E.C. Frederick, H.L. Falseti, E.R. Burke. 1983. Overtraining in athletes: a round table. *Physician and Sportsmedicine* 11: 93-110.

28. Selye, H. 1957. *The stress of life*. London: Longmans Green.

29. Shepley, B., J.D. MacDougall, N. Cipriano, J.R. Sutton. 1992. Physiological effects of tapering in highly trained athletes. *Journal of Applied Physiology* 72: 706-711.

30. Sjodin, B., J. Svedenhag. 1985. Applied physiology of marathon running. *Sports Medicine* 2: 83-99.

31. Snyder, A.C., A.E. Jeukendrup, M.K.C. Hesselink, H. Kuipers, C. Foster. 1993. A physiological/psychological indicator of over-reaching during intensive training. *International Journal of Sports Medicine* 14: 29-32.

32. Snyder, A.C., H. Kuipers, B.O. Cheng, R. Servais, E. Fransen. 1995. Overtraining following intensified training with normal muscle glycogen. *Medicine and Science in Sports and Exercise* 27: 1063-1070.

33. Town, G.P. 1985. *Science of triathlon training and competition*. Champaign, IL: Human Kinetics.

34. Urhausen, A., T. Kullmer, W. Kinderman. 1987. A 7-week follow-up study of the behaviour of testosterone and cortisol during the competition period in rowers. *European Journal of Applied Physiology* 56: 528-533.

35. Verde, T., S. Thomas, R.J. Shephard. 1992. Potential markers of heavy training in highly trained distance runners. *British Journal of Sports Medicine* 26: 167-175.

36. Verma, S.K., S.R. Mahindroo, D.K. Kansal. 1978. Effect of four weeks of hard physical training on certain physiological and morphological parameters of basketball players. *Journal of Sports Medicine* 18: 379-384.

37. Veroon, C., A.M. Quist, L.J.M. Vermulst, W.B.M. Erich, W.R. deVries, J.H.H. Thijssen. 1991. The behaviour of the plasma free testosterone/cortisol ratio during a season of elite rowing training. *International Journal of Sports Medicine* 12: 257-263.

CHAPTER 2

Physiological Responses to Short- and Long-Term Overtraining in Endurance Athletes

Manfred Lehmann, MD, PhD, Carl Foster, PhD, Nikolaus Netzer, MD, PhD, Werner Lormes, PhD, Jürgen M. Steinacker, MD, PhD, Yufei Liu, MD, PhD, Alexandra Opitz-Gress, MD, and Uwe Gastmann, MD, PhD

Introduction

During the 1996 European Championships, a German gymnast shook his head almost imperceptibly, closed his eyes briefly and left the arena without looking up; he was fatigue personified. "Suddenly, I just couldn't do any more. I just wanted to rest." A look at his schedule showed why: two international competitions in March, World Championships in April, European Championships in mid-May, Olympic selection trials at the end of May, national championships and additional selection trials in June 1996. (*Süddeutsche Zeitung*, 13 May 1996). On the other hand, we also receive a clear message when a tennis professional says after an ATP final, "My recent successes are due to less tennis, more regeneration, and the enforced break (due to injuries); I am less exhausted and burnt out than the other players." (*Süddeutsche Zeitung*, 27 October 1995). Or as a professional cyclist states after his success in the Giro d`Italia 1995, "I am better this year, because I train less; in other years, I was already tired before the race." (*Neue Zürcher Zeitung*, 28 May 1995). And when a former soccer professional notes, "I always asked myself why the others don't learn from us; it is not necessary to train so much to be good." (*Süddeutsche Zeitung*, 2 September 1995). Or a tennis professional states after losing the opening match in Indian Wells, "I have to take a break, otherwise, I can't keep up the pace." (*Stuttgarter Zeitung*, 14 March 1996).

The message seems clear, but is the message understood? We have to say no, since the consequences of overtraining are experienced by more than 60% of distance runners at least once during a career (60), by 21% of athletes of the Australian swimming team during a half-year season (34, 35), by 33% of players of the Indian basketball team during a six-week training period (75), and by more than 50% of soccer players after a four-month competition season (46).

Therefore, a chronic problem in high performance and particularly in professional sports remains the continual risk of an imbalance in the training, competition, recovery cycle as the cause for developing the overtraining syndrome called burnout or staleness (8, 11, 18, 19, 36, 41, 49, 51, 54, 60, 65, 68, 73). Glycogen deficit (9, 10), autonomic imbalance (36, 39), amino acid imbalance (53, 62, 66, 67, 78), neuroendocrine imbalance (1, 3, 31, 33, 76, 77), and catabolic/anabolic imbalance (1, 28-30) are all attributed to too much training and too many competitions with too little time for regeneration. Nontraining stressors, such as social, educational, occupational, economical, nutritional, travel factors (18, 49, 51, 54), and monotony of training (7, 18) exacerbate the problem in a completely individual manner.

For prognostic reasons, overreaching (short-term overtraining) has to be distinguished from long-term overtraining, which can lead to a clinical state described as overtraining syndrome, also called burnout or staleness (11, 18, 19, 32, 41, 49, 65, 73). From a retrospective standpoint we can state, if supercompensation (for definition, see reference 20) occurs after a cycle of overtraining and an adequate tapering or recovery period, only short-term overtraining or overreaching has occurred (13, 18, 20, 41, 49). However, if performance incompetence persists after intensified training and an appropriate regeneration period, and a disease has been excluded, the affected athlete most likely suffers from an overtraining syndrome (18, 36, 41, 49). But this trivial definition can not really help affected athletes and coaches to avoid an overtraining syndrome.

Key symptoms of overtrained endurance athletes have been found to be (1) persistent performance incompetence, (2) persistent high complaint ratings, (3) altered mood-state, and (4) suppressed reproductive function (8, 11, 18, 29, 35, 36, 41, 60, 65). High fatigue rating, suppressed mood-state, and individually differing muscle complaints are usually well correlated with training load (18, 35, 36, 49, 60). Disease must be ruled out in any case of suspected overtraining syndrome (18, 36, 41, 49, 73). In endurance sports, overtraining syndrome is characterized by persistent high fatigue ratings and apathy, that has been described as the parasympathetic type of overtraining syndrome (36) or staleness (11, 66), and differentiated from the sympathetic type (see table 2.1); these terms refer to a possible autonomic imbalance (36).

Since a graph of the relationship between training load and performance is an inverted U-shaped curve enclosing only a small target area for optimal training (18), ineffective undertraining and ineffective overtraining are common even among experienced athletes such as Dr. Ron Hill, the dominant international long-distance runner of the late '60s and early '70s. Dr. Hill's optimal training load appears to have been about 160-170 km/wk (63), since his marathon performance deteriorated above 180 km/wk (18, 63). Over a nearly 10-year period there was no day on which Ron Hill did not run, even before major events (63). Accordingly, it may be suggested that he was not able to fully utilize his potential at major competitions due to a chronic

Table 2.1 Some Findings in Sympathetic and Parasympathetic-Type Overtraining Syndromes

Overtraining syndrome	
Sympathetic type	**Parasympathetic type**
Impaired performance	Impaired performance
Lack of supercompensation	Lack of supercompensation
Restlessness, irritability	Fatigue, depression, apathia
Disturbed sleep	Not sleep disturbed
Weight loss	Constant weight
Increased resting heart rate	Low resting heart rate
Increased resting blood pressure	Low resting blood pressure
Retarded recovery after exercise	Suppressed heart rate-exercise profile
	Suppressed glucose-exercise profile
	Suppressed lactate-exercise profile
	Suppressed neuromuscular excitability
	Suppressed sympathetic intrinsic activity
	Suppressed catecholamine sensitivity
	Altered hypothalamo/pituitary, adrenal/gonadal function

Based on data from references 1-3, 8, 11, 18, 19, 29, 34-38, 41, 45-53, 59, 60, 64, 72.

imbalance in the training, competition, recovery cycle, with the result that Frank Shorter took the marathon gold metal in Munich in 1972.

Increasing Training Volume Versus Increasing Training Intensity

Experienced Athletes

On a commonsense basis, overtraining syndrome would be expected to be more a consequence of increase in training intensity than training volume (39, 65, 75). This concept was examined in two prospective experimental studies with nine experienced distance runners (47, 48). In the first study, increase in training volume (ITV), the training volume was increased from baseline by about 100% within four weeks; 93-98% of the training consisted of monotonous long-distance runs. Each day was a hard day! A year later in the companion study, increase in training intensity (ITI), the total training volume was increased only about 37%, but the proportion of intensive training measures, that is of tempo-pace and interval runs, increased by 150% following the concept of alternating hard and easy training days (18).

During ITV, persistent performance incompetence (see figure 2.1) and persistent high complaint ratings (47) were observed. All athletes failed to equal or improve their personal records during the subsequent three months. A typical statement by an athlete was, "You step on the gas, but nothing happens!"

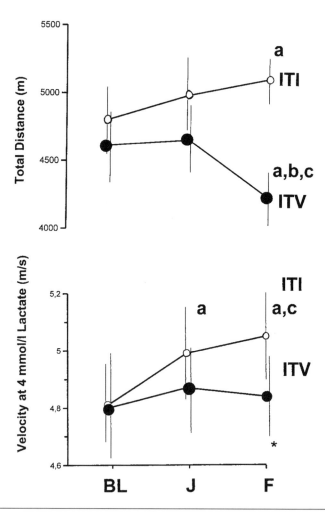

Figure 2.1   Scheme of two prospective studies with 9 runners. Maximum (total distance during the maximum incremental test) and submaximal running velocity (velocity at 4 mmol/L lactate) increased continuously during the increase in training intensity (ITI) study, but decreased (maximum) or showed no progression (submaximal running velocity) during the increase in training volume (ITV) study. BL = baseline; J = interim; F = final examination; index a, b: significant vs. BL; index c: significant between ITV and ITI.

Contrary to the ITV study, the ITI study represented less stressful training, resulting in improved submaximal and maximal power output (see figure 2.1) and a complaint index staying mostly at a moderate level (47). The athletes recorded personal best performances during the three-month follow-up, and athlete M.B. won the cross-country championship in Baden-Württemberg. Despite a 150% × 4 wk increase in intensive training measures from an average 9 to 23 km/wk tempo-pace and interval runs, the athletes tolerated this increasing training better, most likely because of (1) a restricted total training volume, (2) a one-day training ban in each training week, and (3) clearly reduced training monotony attributable to frequent regeneration days on alternating hard and easy training days.

In agreement with other overtraining studies (21, 34, 35, 37, 71, 72), physiological parameters like body mass, resting heart rate, blood pressure, etc. did not have any diagnosis-relevant importance (47, 48). By contrast, during ITV compared to ITI hematological or blood-chemical parameters such as decrease in serum albumin, summed serum amino acid concentration, leukocyte counts, serum ferritin and iron concentrations or levels of energy-rich substrates, etc. (47, 48) make a diagnosis easier based on a multifactorial approach.

Recreational Athletes

Contrary to expectations and the ITV study, it was not possible to induce persistent decrease in performance and persistent high complaint ratings with ITI, so another prospective experimental approach, an intensive ergometer training (IET) study, was performed with six less-adapted recreational athletes (50). They participated in a 6 wk × 6 d/wk intensive steady state (90-96% of 4 mmol/L lactate threshold (4 LT)) and interval cycle ergometer training (117-127% of 4 LT) of 40-60 min/d (including short warm-up and cool-down periods). The training stopped each day on muscular exhaustion. On day seven (regeneration day), training was low intensity for about 30-40 min (see figure 2.2). The athletes trained three to six times as much as before without a regeneration microcycle in weeks three and four. Submaximal and maximum power output were significantly increased after three weeks, but stopped improving between three and six weeks or decreased (power output at 2 mmol/L lactate threshold (2 LT); see figure 2.3). According to the ITV study, at submaximal workload, there was a clear suppression of blood glucose-, free fatty acids-, and blood lactate-performance curves after six weeks (see figure 2.4), which may reflect decreased substrate stores and a decreased sensitivity of ß2-adrenoreceptors to catecholamines, so an overestimation of submaximal power output and in fact a decrease must also be assumed at this point. On the other hand, glycogen depletion is not obligatory in the overtraining process (72). In addition, the slightly higher maximal power output at final examination during IET suggests higher motivation, as reflected by higher maximal lactic acid concentrations (13.27 vs. 12.47 mmol/L), rather than higher performance ability.

In the IET study, even after two weeks of regeneration (training of about 2 h/wk) no supercompensation was evident, but instead a decrease in maximum power output to below the baseline level was observed (see figure 2.3). Therefore, in this study the findings may be interpreted as indicative of developing overtraining syndrome. These findings reflect the difficulties in establishing a definitive diagnosis of overtraining syndrome even based on longitudinally controlled

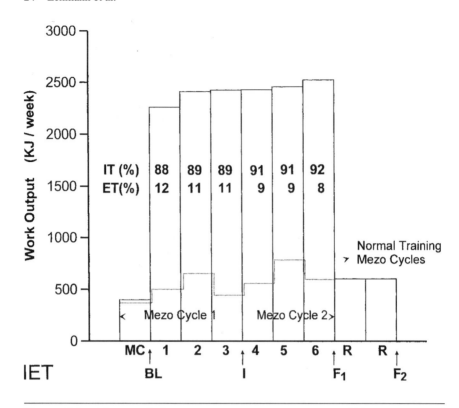

Figure 2.2 Scheme of the intensive ergometer training (IET) study with 6 recreational athletes. Also shown are two subsequent normal training mezo cycles. During microcycles 1-6, training was about 6 times their usual weekly activity (microcycle 0) and consisted of 88-92% of intensive tempo-pace and interval training (IT) and 12-8% extensive (regenerative) training (ET). MC = microcycle; BL = baseline; I = interim; F = final examination; R = recovery.

laboratory experiments, and underscore the complexity of this problem for athletes and coaches in daily training practice.

Fry et al. (21), Jeukendrup et al. (37), and Snyder et al. (71, 72) observed a decrease in performance of special soldiers after 10 days (21) and of experienced athletes after 14 days of intensified intensive training (37, 71, 72), but with 2-3 h/d of training that was accompanied by supercompensation after 1-2 weeks of recovery (or tapering). A decrease in performance was also reported by Dressendorfer et al. (14) after seven successive days of unaccustomed prolonged training (mean seven day distance 129.2 km) in fitness joggers (previous mean distance 17.1 km/wk), and by Costill et al. (10) after 10 days of intensified swimming training in some elite swimmers. Summarizing these results as well as the results of the ITV and IET studies, an increased risk of an overtraining syndrome has to be considered after about three weeks of intensified (intensive ergometer training) or prolonged monotonous endurance training (increase in training volume).

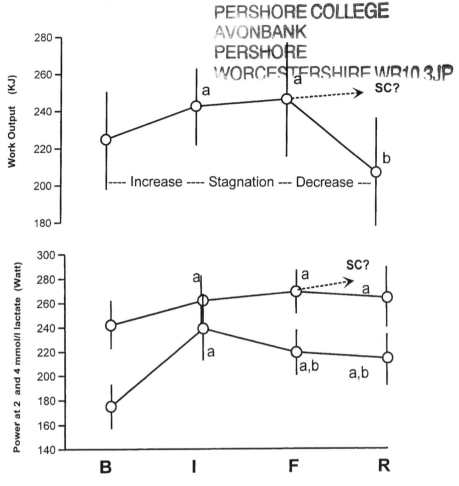

Figure 2.3   During IET, maximum power output (during maximum incremental testing) and submaximal power output (power at 2 and 4 mmol/L lactate) were increased in week 3 (interim = I), showed no further progression (final examination = F) or decreased in week 6 (power at 2 mmol/L lactate), and showed no further progression or decreased (maximum power output) after two weeks of recovery (R). a = significantly different from baseline, b = significant from interim examination, SC = points to supercompensation which was not observed during this study.

Borderline Between Short-Term (Overreaching) and Long-Term Overtraining

The decisive question of how long overreaching can be extended in adapted athletes was examined additionally by Lormes et al. (56) in elite rowers during the preparation period for the junior world championships in a rowing

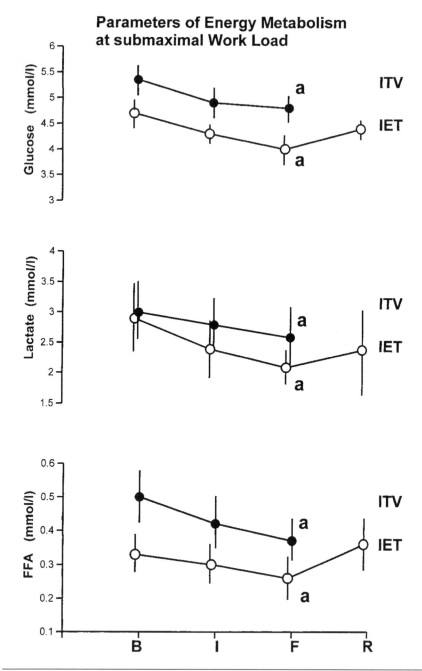

Figure 2.4 During the ITV and IET overtraining studies, blood glucose, lactate, and free fatty acid levels were significantly lower at a fixed submaximal work load (200 watts and 16 km/h, respectively) during final examination (F) compared to baseline (a: $p < .05$), and increased again during the recovery period (R). B = baseline; I = interim.

overreaching study (ROS). This team was examined over five periods each of 7-10 days between national and world championships (see figure 2.5). The mean training volume amounted to 2.2 h/d (period 0, baseline), 2.9 and 3.2 h/d (periods 1 and 2, 19 days including two rest days), and approximately 2.2 h/d (recovery, periods 3 and 4). A maximum graded rowing test revealed a decrease in power output at 4 mmol/L lactate (except for the coxed eight) and a stagnation in maximum power output after period 2 compared to period 0 (see figure 2.6). However, since the competition times during the world championships were mostly better than the times during the national championships, sufficient regeneration may be assumed. This is also confirmed by the four gold medals and one silver medal won.

The ROS confirms the stated opinion that the critical borderline to long-term overtraining in adapted endurance athletes may be around three weeks of intensified or prolonged training of about 3 h/d. If this period is completely utilized, two recovery microcycles (2 weeks) may be required for complete regeneration in order to avoid an overtraining syndrome. A rest day per week and the training principle of alternating hard and easy training days clearly reduce the risk of an overtraining syndrome.

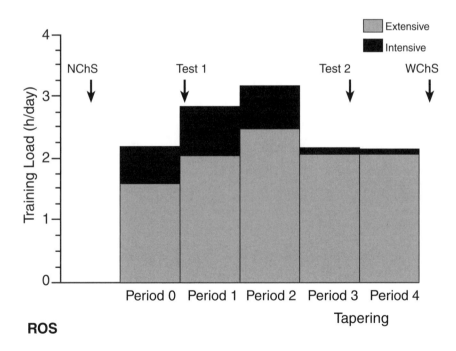

Figure 2.5   Scheme of the rowing overreaching study (ROS) which took place between national championships (NChS) and world championships (WChS). Standardized rowing tests took place after baseline period (period 0) and late in period 3 (tapering period). Each period consisted of 7-10 days.

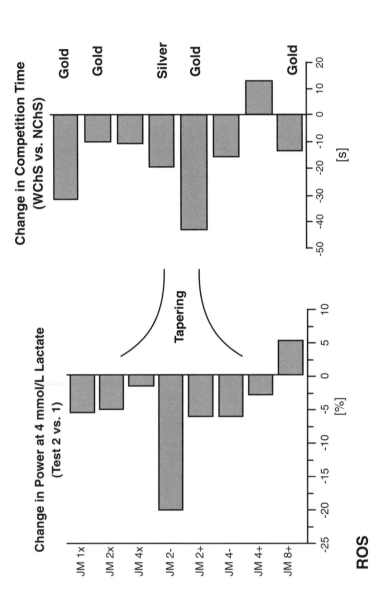

Figure 2.6   Scheme of the change in power and competition time in the eight rowing teams in the ROS. Power output at 4 mmol/L lactate was decreased in most of the rowing teams after two overreaching training periods (see figure 2.5), except for "coxed eight" (JM 8: Junior Male Eight). After tapering, however, during WChS, the competitive results were very favorable compared to NChS demonstrating that the overreaching in training did not produce a persistent performance incompetence, i.e., an overtraining syndrome.

Foster's Monotony Hypothesis

Bruin et al. (7) demonstrated that the constancy (monotony) of training at a high training load may be just as critical in race horses in the generation of overtraining syndrome as the total training load. This is to say that race horses are more tolerant of alternating hard days and easy days of training. Even when the severity of training on hard days was progressively increased the horses did quite well. Additional increases in the training load on recovery days, however, resulted in rapid decompensation of the horses into a state resembling overtraining syndrome in humans. This observation has been confirmed in human athletes by Foster and Lehmann (18). In order to quantify monotony, Foster et al. (15-18) suggest a rating of perceived exertion (RPE) times duration method to estimate training load in athletes (daily training duration × global RPE on Borg's 10-point scale).

Training monotony may be estimated on a weekly basis using the quotient of mean daily training load divided by standard deviation of mean (SD) daily training load. A small SD and a high quotient indicate small day-to-day variation and great monotony, a large SD and low quotient confirm great variation and low monotony of training (18). In this scheme of monitoring training, the daily details of training do not seem to matter. For example, alternating days of very extensive or very intensive training would still lead to every day being a day with high training load and could be considered monotonous. In several athletes, high training load and monotony were related to poor performance and an increased frequency of banal infections which may be markers of emerging overtraining syndrome; whereas reductions in total training load and particularly in the monotony of training were related to personal best performances (18). Reports of Miguel Indurain's training when preparing for the 1995 world championships clearly demonstrate how an athlete can avoid monotony of training by following the concept of alternating hard and easy training days (18).

Selected Physiologic Responses to Overtraining

Autonomic Balance

From a clinical standpoint, Israel (36) suggests that prolonged overtraining produces a sympathetic-parasympathetic or autonomic imbalance. Fatigue and apathy dominate in the parasympathetic type of overtraining, which is typical for endurance sports. Restlessness and hyperexcitability dominate in the sympathetic type of overtraining, which is more typical in explosive sports or related to additional significant nontraining stress factors. In agreement with this hypothesis, a markedly elevated nocturnal (basal) catecholamine excretion was observed in a middle-distance runner, H.P.F., who presented with performance incompetence, hyperexcitability, and restlessness before the 1984 Olympic Games in Los Angeles, for which he was qualified but in which he did not participate for personal reasons (nontraining stressors). By contrast, a German tennis professional presented with burnout syndrome (parasympathetic type of overtraining syndrome) and markedly decreased basal catecholamine excretion after three successive ATP finals on three different continents within approximately four weeks (see figure 2.7).

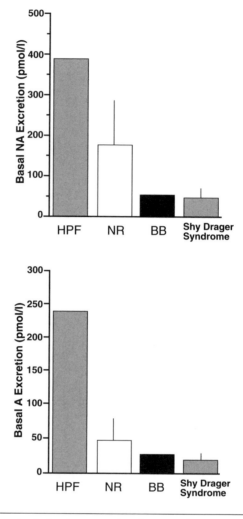

Figure 2.7   Urinary catecholamine excretion during night rest in a middle-distance runner (HPF) showing the so-called sympathetic-type overtraining syndrome, compared to well-trained controls; 8 distance runners before the ITV study (NR: normal range) and to a top tennis player (BB) suffering from a burnout syndrome with an excretion rate in the range of Shy Drager syndrome patients.
Based on data from Lehmann, Petersen, and Khalaf 1991.

Intrinsic Sympathetic Activity.   A significant overload-related decrease in basal catecholamine excretion was also observed in road cyclists during intensified training during the weeks prior to the 1988 Games in Seoul with return to a higher excretion rate during the regeneration period immediately prior to the Games (46). Unexpectedly, this team won two medals. By contrast, the basal catecholamine excretion in the simultaneously-examined track cyclists did not attain the minimum until directly prior to the Games, possibly due to incomplete regeneration (46). These athletes were unexpectedly eliminated in the qualification rounds during the 1988 Games. Since such low catecholamine excretion rates have been observed otherwise only in patients suffering from sympathetic insufficiency (Shy Drager or Bradbury Eggleston syndrome) (45), the very low basal catecholamine excretion in overtrained athletes may indicate a transient reduced catecholamine metabolism and a decrease in intrinsic sympathetic activity indicative of a parasympathetic-type overtraining syndrome. This opinion was confirmed by the ITV study (46-48) showing a negative correlation between basal urinary noradrenaline excretion and trivial complaint index of which fatigue was a main factor (see figure 2.8). The marked decrease in basal catecholamine excretion, however, may be a late finding, and not an early indicator of a developing overtraining syndrome. A decreasing, and after four months clearly reduced, basal excretion (see figure 2.9) was also observed in overloaded and stale semiprofessionals of a soccer team (46), as recently was confirmed by Naessens et al. (61). A decreasing basal excretion could not be confirmed by the IET study (50), however, probably due to the clearly lower total training and nontraining load compared to the road cyclists (46), soccer players (46), and the ITV study (47, 48) (approx. 1 vs. 3 h/d).

Besides inhibitory afferent signals from the overloaded muscles, the reduced intrinsic sympathetic activity may be related to an overtraining-dependent amino acid imbalance (23, 24, 62, 66, 67) as a linkage between muscle metabolism and the brain neurotransmitter metabolism. This imbalance is characterized by elevated

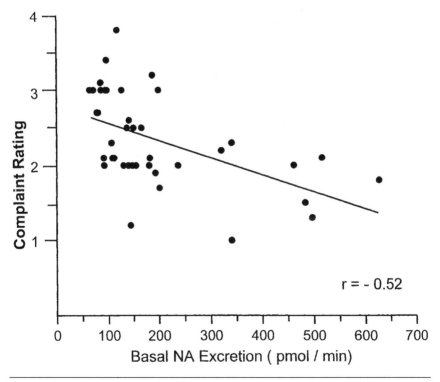

Figure 2.8 During the increase in training volume (ITV) overtraining study, a negative correlation was observed between complaint rating (on a trivial 4-point scale) and basal urinary noradrenaline excretion.

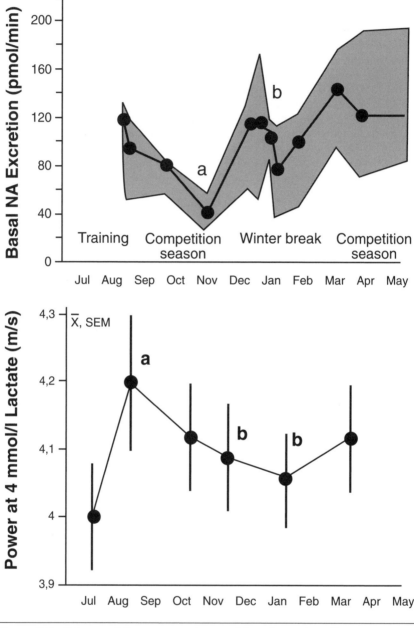

Figure 2.9 Scheme of the power output and basal noradrenaline excretion of stale soccer players. A decrease in basal noradrenaline and power output at 4 mmol/l lactate was also observed in stale semiprofessional soccer players between August and November, the first part of competition season, which they finished at league rank 7. The basal excretion increased again during the winter break, and remained on the average in a normal range during the second part of the competition season from February to May which they finished at league rank 2. a = significantly different from baseline, b = significantly different from the preceding.

brain phenylalanine, tyrosine, and tryptophan uptake, the precursors of cat-echolamines (phenylalanine, tyrosine) and serotonin (tryptophan). A central inhibition-excitation imbalance is hypothesized as a consequence of the in-creased brain aromatic amino acid uptake, particularly of tryptophan (62, 66, 67). The relationship of this hypothesis to overtraining is discussed in detail in chapter 15.

Catecholamine Sensitivity.   The decrease in basal catecholamine excretion was accompanied during the ITV study (47, 48) by a significantly increased plasma noradrenaline concentration before exercise and by the response at identical submaximal workload. That is, despite reduced basal catecholamine metabolism, plasma noradrenaline stress response was increased compared to baseline. Elevated noradrenaline plasma levels were recently confirmed by Hooper et al. (34, 35) in overtrained elite swimmers. However, elevated noradrenaline plasma levels were not found at lower training loads during the IET study (50). The elevated submaximal plasma noradrenaline stress response is accompanied, however, by lower heart rate, blood glucose, lactate, and free fatty acid responses (see figure 2.10), which suggests a reduced sensitivity of the organism to catecholamines. This suggestion is confirmed by Jost et al. (38), who found a reduced beta-adrenoreceptor density, and isoproterenol-stimulated cycle-AMP activity of lymphocytes in distance run-ners and swimmers during high-volume training compared to tapering (38), in accordance with our results in distance runners who showed an average decrease in β-2-adrenoreceptors of 400 receptors on intact polymorphonuclear leucocytes from an elevated density back to the normal range of untrained subjects (43, 44, 58) subsequent to high-volume training (unpublished results). The reduced, that is, again-normalized adrenoreceptor density and sensitivity to catecholamines com-pared to a well-trained state (43, 44, 58), is most likely the result of a daily over-loading of several hours' duration of receptors and post-receptor mechanisms, as was experimentally observed by Tohmeh and Cryer (74). The observed elevated plasma noradrenaline stress response can be interpreted as an attempt to compen-sate for the reduced catecholamine sensitivity. The reduced catecholamine sensi-tivity mimics in the athletes the effects of taking beta-blockers (42) and can be taken as a parameter of "peripheral fatigue." The signal transmission of ergotropic (catabolic) signals to target organs becomes impaired and may explain statements such as "You step on the gas and nothing happens."

In the parasympathetic type of overtraining syndrome, therefore, a predomi-nance of vagal tone is based on a transient decrease in both intrinsic sympathetic activity and a decrease in sensitivity of the organism to catecholamines as part of a complex protective mechanism of target organs against overload-dependent irre-versible cellular damage.

Neuromuscular Excitability

Well-trained endurance athletes show increased neuromuscular excitability of trained, nonfatigued reference muscles (5). Reduced neuromuscular excitability (NME) of stressed muscles is found after prolonged exercise (70). In a one-legged stress test, this fatigue can only be demonstrated in the stressed leg (70). This indicates that the neuromuscular fatigue reaction affects less central mecha-nisms. Reduced NME was also observed after night rest in the overtrained ath-letes following the increase in training volume study (52), and also after the in-tensive ergometer training study (54) (see figure 2.11). Comparable to the observed

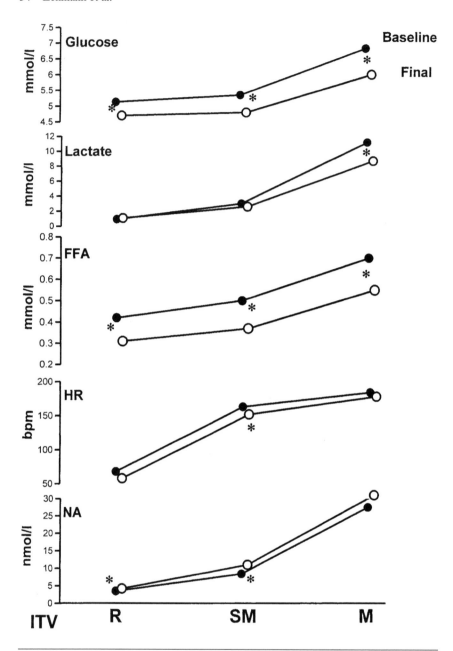

Figure 2.10   Submaximal (SM) and maximum (M) blood glucose, lactate, and free fatty acid responses, and heart rate response were significantly lower at final examination (*; p < .05) compared to baseline during the ITV overtraining study using an incremental cycling test. However, the decrease in these substrates was related to significantly higher resting plasma NA levels and a higher submaximal plasma NA response. R = rest; FFA = free fatty acids; HR = heart rate; NA = noradrenaline.

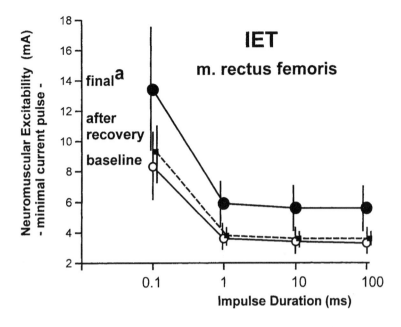

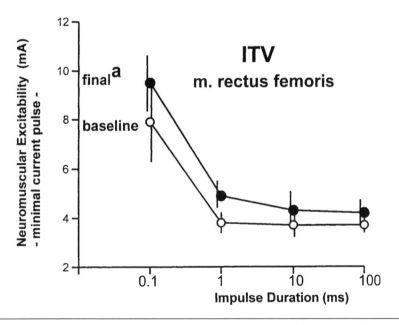

Figure 2.11   As an index of impaired neuromuscular excitability, minimal current pulses to induce a single contraction of fibers of the reference muscle, rectus femoris, showed a significant increase (deterioration) at final examination in relation to different impulse durations, both during the intensive ergometer training (IET) study and the increase in training volume (ITV) study, that went along with an impaired performance.

decrease in adrenoreceptor density discussed earlier, this is primarily to be interpreted as an overload-dependent loss in adaptation (5, 22, 52, 54). A decrease in neuromuscular excitability can also be interpreted as an impaired signal transmission to target organs and as a parameter of peripheral fatigue.

Hormonal Balance

The plasma levels of free testosterone and cortisol should reflect the anabolic and catabolic activity in the tissues (1, 2, 12, 27-30, 33, 59, 76). A catabolic-anabolic imbalance should be indicated by an increase in the cortisol/free testosterone ratio and serum urea concentration reflecting increased proteolysis. This ratio should be a useful parameter in the early detection of overtraining syndrome (1). An elevated basal cortisol level and impaired GH, ACTH, cortisol, and PRL responses to insulin-induced hypoglycemia should indicate hypothalamic dysfunction in overtrained athletes (3). On the other hand, power at 4 mmol/L lactate and maximum power did not show a relation to the hormonal parameters in athletes during a season of elite rowing training (76).

Available overtraining studies (21, 34, 35, 40, 47, 48, 50) were not able to confirm either a convincing increase in basal cortisol and urea or a decrease in free testosterone in already-adapted athletes. An endurance training-related decrease in serum testosterone levels in less-adapted subjects without change in leutinizing hormone (LH) pulsatile release must be differentiated from this as an early regular training-dependent adaptation (77).

The ITV, IET, and ROS studies also show neither elevated basal cortisol concentrations, nor a decrease in free testosterone, except for the less-adapted athletes after the recovery period (see table 2.2), according to Wheeler et al. (77). An increase in serum urea was observed only during the ROS, not, however, during overreaching periods but in the later recovery period along with increasing hematocrit and serum aldosterone levels. This can be explained by the fact that training predominantly took place on the water during the tapering phase and the athletes drank less. The suitability of serum urea levels for training monitoring is contested by the observation that there was no increase in serum urea levels during prolonged strenuous exercise over two hours even with additional supplementation of branched-chain amino acids and increased amino acid metabolism; but rather there was an increase in glutamine concentration (23, 24). It may therefore be concluded that under these conditions, ammonia detoxification did not proceed via urea synthesis but via glutamine synthesis.

Adrenal Sensitivity to Adrenocorticotropin Hormone. The basal and exercise-induced cortisol levels decreased slightly during overtraining (21, 34, 35, 40, 47, 48), likely because of overloading of the adrenal cortex, and comparable overloading of adrenoreceptors and neuromuscular structures, as discussed earlier. Thus, after prolonged overtraining, corticotropin releasing hormone (CRH)-induced pituitary adrenocorticotropic hormone (ACTH) release was about 81% higher combined with about 25% lower adrenal cortisol release (see figure 2.12) (50), according to Barron et al (3). They also observed an impaired cortisol release in overtrained marathon runners, however, combined with elevated basal plasma cortisol levels, which is in contrast to prospective overtraining studies and may indicate a more advanced stage (21, 34, 35, 40, 47, 48).

Table 2.2 Baseline (B) and Final (F) Concentrations of Serum Urea, Hematocrit Value (Hct), Serum Cortisol, and Gonadotropins (LH and FSH) During Different Training, or Overtraining Studies

		ITV		ITI		IET		ROS		
		B	F	B	F	B	F	B	F	R
Urea (mmol/L)	x	6.1	6.0	6.0	5.8	5.1	4.3	6.5	6.0	8.5*
	SD	1.3	1.6	1.1	1.3	1.6	1.9	1.2	1.3	1.8
Hct (%)	x	42.7	40.7*	46.1	42.1*	43.7	42.9	46.5	45.5	48.1*
	SD	2.7	2.5	1.5	1.6	3.7	2.7	1.4	1.5	1.9
Cortisol (mmol/L)	x	442	320	265	320	458	414	450	463	469
	SD	179	94	55	113	163	88	86	110	88
Free testosterone (pmol/L)	x	66	66	69	73	105	102[a]	87	83	94
	SD	21	17	20	21	34	49	16	15	23
LH (mU/mL)	x	2.8	3.0	2.8	3.0	2.3	2.0	3.6	3.6	3.0
	SD	0.9	1.3	0.8	1.2	0.4	0.5	1.2	1.0	0.8
FSH (mU/mL)	x	4.2	4.1	5.2	5.2	2.7	2.7	6.1	5.3	5.5
	SD	1.8	1.7	1.8	2.3	0.7	0.7	2.3	2.0	2.1

[a] 85 ± 20 pmol/L (p < .05) after 2 weeks of recovery (R).
* p < .05
ITI, ITV (see "Experienced Athletes"), IET (see "Recreational Athletes"), ROS

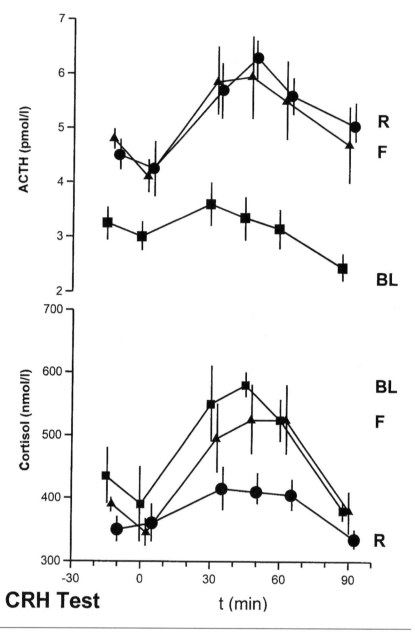

Figure 2.12   After 6 weeks of monotonous intensive ergometer training (IET), a pituitary stimulation test (50) revealed an approximate 80% higher pituitary ACTH response, and a 25% lower adrenal cortisol response, which may indicate a decreased, overtraining-dependent, adrenal sensitivity to ACTH. BL = baseline; F = final examination; R = after recovery.

Reprinted, by permission, from M. Lehmann et al., 1993, "Influence of 6-week, 6 days per week, training on pituitary function in recreational athletes," *British Journal of Sports Medicine* 27: 186-192 (London: BMJ Publishing Group).

The adrenal cortex thus also shows reduced sensitivity to ACTH under overload training, like the adrenoreceptors and neuromuscular structures. The signal transmission from the hypophysis to the adrenals is impaired, which can be seen as protection for cells against overload-dependent damage. Biologically, we, like all animals, are constructed to rest when we are tired. Our physiology has many examples of negative feedback regulation. However, as social animals who have changed the role of play into the business of top sport, we frequently seem compelled to ignore our biological programming. Thus overtraining syndrome may be the ultimate negative feedback system in our physiological makeup.

The decreased adrenal cortisol response was correlated with slightly increased pituitary sensitivity to growth hormone releasing hormone (GHRH) and GH release (50); therefore, the increase in the ratio of GH release to cortisol release may reflect a counter-regulatory shift to a more anabolic endocrine responsibility (see figure 2.13) during overtraining.

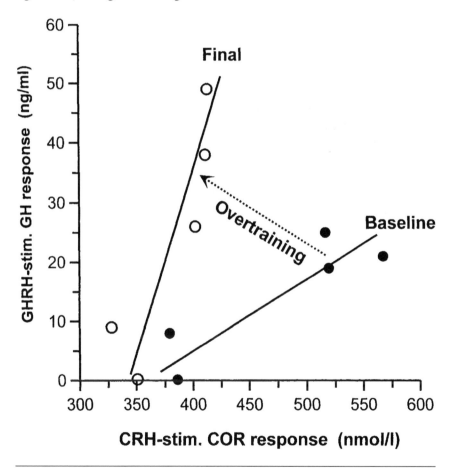

Figure 2.13 Since after 6 weeks of monotonous intensive ergometer training (IET), the ratio of a decreased CRH-stimulated adrenal cortisol response to an increased GHRH-stimulated pituitary response showed a significant increase, which may reflect an overreaching or overtraining response.

Molecular Biological Findings During Overreaching

Exercise causes heat, and metabolic and oxidative stress (shock) to muscle cells. The cells respond by synthesizing a set of protective polypeptides termed stress or shock proteins, particularly the so-called heat shock protein(s) (HSP) that increase stress tolerance in stressed cells or conduct cellular repair processes (69). The HSP belong to a group of polypeptide proteins, especially a group of 70 kDa proteins, which play a role in protein folding and assembly (4, 67). In rats, HSP 70 mRNA peaked 1 h after exercise and decreased towards baseline levels 6 h postexercise, but remained still slightly elevated (69).

Since muscular adaptation on a cellular and molecular biological level has not been examined sufficiently, particularly in human athletes, in addition to HSP 70, Opitz-Gress (64) studied changes in the expression on translation level of contractile proteins such as myosin heavy chain, slow and fast type (MHC slow, fast), in elite junior rowers preparing for the 1995 world championships. MHC is a major contractile protein that belongs to a multigene family (55, 57). Changes in MHC isoenzyme gene expression under different conditions like chronic electrical stimulation, stretch and force generation, hypertrophy, exercise, hormonal influences, or innervation patterns have been documented in animal models (6, 25, 26, 57), but seldom in humans and human athletes. In rabbits, chronic electrical stimulation of tibialis anterior and extensor digitorum longus muscle led to a rapid increase in slow MHC mRNA but only to a slow decrease after cessation of stimulation, followed by respective changes in the corresponding muscle proteins (6).

In nine elite junior rowers preparing for world championships, muscle biopsies were taken at baseline (before an intensified training period), at interim of the intensified or overreaching training period, and during the early tapering period. Blott analysis and chemiluminescence detection revealed significantly higher levels of MHC slow, fast, and of HSP 70 (inducible form) in the interim and final biopsies compared to baseline, and slightly lower final concentrations than at interim. There was no significant change in the interim and final constitutive form of HSP 70 (HSC 70). In particular, the increase of HSP 70 as observed on translation level during intensified training (overreaching period) may complement the complex adaptation and protective strategy of the organism against overload-dependent irreversible cellular damage.

Summary

Since athletes must push their limits in quest of the minimal differences in performance that separate the champion from the also-ran, overreaching (short-term overtraining) will remain a common part of training. If overreaching is too great, too monotonous, continued for too long a time, or coupled with too many competitions and nontraining stress factors, an overtraining syndrome may result and the athlete's performance may fail to return to expected levels, even if the regeneration is extended. On the basis of available studies, even in top athletes the maximal sustainable duration of overreaching in endurance sports can be assumed to be around three weeks of intensified or prolonged endurance training of more than 3 h/d. The risk of overtraining is increased by (1) one-sided, monotonous training without alternating hard and easy training days, (2) a lack of one complete rest day per week, (3) a high total and increasing training load combined with additional significant nontraining stress factors, and (4) too many competitions. So far as is known at present, physiological responses to overtraining in endurance athletes can be summarized as an inhibition of the transmission of stress (catabolic) signals to target organs such as (1) a decrease in neuromuscular excitability of overloaded muscles, (2) a decrease in adrenoreceptor density (decrease in sensitivity of the organism to catecholamines), (3) a decrease in adrenal sensitivity to ACTH during prolonged overtraining periods (decreased cortisol response) combined with an increased pituitary sensitivity to GHRH (increased GH response) reflecting a shift to a counterregulatory anabolic endocrine responsibility, (4) intracellular protection mechanisms such as an increased synthesis of heat-shock proteins to complement the complex strategy of the organism against overload-related cellular damage, and (5) an additional decrease in sympathetic intrinsic activity in advanced stages that may depend on an elevated concentration of inhibitory neurotransmitters in the brain (so-called central mechanism). The observed overtraining-related decrease in ß-adrenoreceptor density works like the nonselective beta-blockers, resulting in impaired glycogenolysis, glycolysis, lipolysis, heart rate response, and endurance performance ability.

Acknowledgment

Dedicated to Professor Dr. Josef Keul at the honor of his 65th birthday.

References

1. Adlercreutz, H., M. Härkönen, K. Kuoppasalmi, H. Näveri, I. Huhtamiemi, H. Tikkanen, K. Remes, A. Dessypris, J. Karvonen. 1986. Effect of training on plasma anabolic and catabolic steroid hormones and their response during physical exercise. *International Journal of Sports Medicine* 7: S27-28.
2. Alén, A., A. Pakarinen, K. Häkkinen, P.V. Komi. 1988. Responses of serum androgenic-anabolic and catabolic hormones to prolonged strength training. *International Journal of Sports Medicine* 9: 229-233.
3. Barron, J.L., T.D. Noakes, W. Levy, C. Smith, R.P. Millar. 1985. Hypothalamic dysfunction in overtrained athletes. *Journal of Clinical and Endocrinological Metabolism* 60: 803-806.
4. Beckmann, R.P., L.A. Mizzen, W.J. Welch. 1990. Interaction of HSP 70 with newly synthesized proteins: implications for protein folding and assembly. *Science* 248: 850-854.
5. Berg, A., D. Günther, J. Keul. 1986. Neuromuskuläre Erregbarkeit und körperliche Aktivität. I. Methodik, Reproduzierbarkeit, Tagesrhythmik. II. Riuhewerte bei Trainierten und Untrainierten. III. Abhängigkeit von beeinflussenden Faktoren. *Deutsche Zeitschrift Sportmedizin* 37: S4-22.
6. Brownson, C., P. Little, J.C. Jarvis, S. Salmons. 1992. Reciprocal changes in myosin isoform mRNAs of rabbit skeletal muscle in response to the initiation and cessation of chronic electrical stimulation. *Muscle and Nerve* 15: 694-700.

7. Bruin, G., H. Kuipers, H.A. Keizer, G.J. Vander Vusse. 1994. Adaptation and overtraining in horses subjected to increasing training loads. *Journal of Applied Physiology* 76: 1908-1913.

8. Budgett, R. 1990. Overtraining syndrome. *British Journal of Sports Medicine* 24: 231-236.

9. Costill, D.L., R. Bowers, G. Branam, K. Sparks. 1971. Muscle glycogen utilization during prolonged exercise on successive days. *Journal of Applied Physiology* 31: 834-838.

10. Costill, D.L., M.G. Flynn, J.P. Kirwan, J.A. Houmard, J.B. Mitchell, R. Thomas, S.H. Park. 1988. Effects of repeated days of intensified training on muscle glycogen and swimming performance. *Medicine and Science in Sports and Exercise* 20: 249-254.

11. Counsilman, J.E. 1955. Fatigue and staleness. *Athletic Journal* 15: 16-20.

12. Cumming, D.C., G.D. Wheeler, E.M. McColl. 1989. The effects of exercise on reproductive function in men. *Sports Medicine* 7: 1-17.

13. Daniels, J. 1989. Training distance runners: a primer. In *Sports science exchange*, 1: 11. Barrington, IL: Gatorade Sports Science Institute.

14. Dressendorfer, R.H., C.E. Wade, J. Claybaugh, S.A. Cucinell, G.C. Timmis. 1991. Effects of 7 successive days of unaccustomed prolonged exercise on aerobic performance and tissue damage in fitness joggers. *International Journal of Sports Medicine* 12: 55-61.

15. Foster, C., A.C. Snyder, N.N. Thompson, K. Kuettel. 1988. Normalization of the blood lactate profile in athletes. *International Journal of Sports Medicine* 9: 198-200.

16. Foster, C. 1994. Exercise session RPE reflects global exercise intensity. *Journal Cardiopulmonary Rehabilitation* 14: 332.

17. Foster, C., L.I. Hector, R. Welsh, M. Schrager, M.A. Green, A.C. Snyder. 1995. Effects of specific vs cross training on running performance. *European Journal of Applied Physiology* 70: 367-372.

18. Foster, C., M. Lehmann. 1997. Overtraining syndrome. In *Running injuries*, ed. G.N. Guten, 173-188. Philadelphia: Saunders.

19. Fry, R.W., A.R. Morton, D. Keast. 1991. Overtraining in athletes: an update. *Sports Medicine* 12: 32-65.

20. Fry, R.W., A.R. Morton, D. Keast. 1992. Periodisation and the prevention of overtraining. *Canadian Journal of Sports Science* 17: 241-248.

21. Fry, R.W., A.R. Morton, P. Garcia-Webb, G.P.W. Crawford, D. Keast. 1992. Biological responses to overload training in endurance sports. *European Journal of Applied Physiology* 64: 335-344.

22. Gastmann, U., M. Lehmann, J. Fleck, D. Jeschke, J. Keul. 1993. Der Einfluß eines sechswöchigen kontrollierten Trainings auf das Katecholaminverhalten und die Katecholamin-Sensitivität bei Freizeitsportlern. In *Sportmedizin: gestern-heute-morgen*, Hrsg. K. Tittel, K.H. Arndt, W. Hollmann, 191-193. Leipzig: Barth.

23. Gastmann, U., G. Schiestl, K. Schmidt, S. Bauer, J.M. Steinacker, M. Lehmann. 1995. Einfluß einer BCAA- und Saccharose-Substitution auf Leistung, Aminosäuren- und Hormonspiegel, Blutbild und blutchemische Parameter. In *Bewegung und Sport: eine Herausforderung für die Medizin*, Hrsg. W. Kindermann, W. Schwarz, 165. Basel: Ciba Geigy Verlag.

24. Gastmann, U., K. Schmidt, G. Schiestl, W. Lormes, J.M. Steinacker, M. Lehmann. 1995. Einfluß einer BCAA- und Saccharose-Substitution auf Leistung, neuromuskuläre Erregbarkeit, EMG und psychometrische Para-

meter. In *Bewegung und Sport: eine Herausforderung*, Hrsg. W. Kindermann, L. Schwarz, 175. Basel: Ciba Geigy Verlag.

25. Goldspink, G., A. Scutt, P.T. Loughna, D.J. Wells, T. Jaenicke, G.F. Gerlach. 1992. Gene expression in skeletal muscle in response to stretch and force generation. *American Journal of Physiology* 262: R356-363.

26. Gregory, P., R.B. Low, W.S. Striewalt. 1986. Changes in skeletal-muscle myosin isoenzymes with hypertrophy and exercise. *Biochemistry Journal* 238: 55-63.

27. Griffith, R., R.H. Dressendorfer, C.E. Wade. 1990. Testicular function during exhaustive exercise. *Physician and Sportmedicine* 18: 54-64.

28. Hackney, A.C., W.E Sinning, B.C. Bruot. 1990. Hypothalamic-pituitary-testicular axis function in endurance-trained males. *International Journal of Sports Medicine* 11: 298-303.

29. Hackney, A.C., S.N. Pearman III, J.M. Nowacki. 1990. Physiological profiles of overtrained and stale athletes: a review. *Applied Sport Psychology* 2: 21-33.

30. Hackney, A.C. 1991. Hormonal changes at rest in overtrained endurance athletes. *Biology of Sport* 8: 49-56.

31. Häkkinen, K., A. Pakarinen, M. Alén, H. Kauhanen, P.V. Komi. 1987. Relationships between training volume, physical performance capacity and serum hormone concentrations during prolonged weight training in elite weight lifters. *International Journal of Sports Medicine* 8: S61-65.

32. Häkkinen, K., A. Pakarinen, M. Alén, H. Kauhanen, P.V. Komi. 1988. Serum hormones during prolonged training of neuromuscular performance. *European Journal of Applied Physiology* 53: 287-293.

33. Häkkinen, K., A. Pakarinen, M. Alén, H. Kauhanen, P.V. Komi. 1988. Daily hormonal and neuromuscular responses to intensive strength training in 1 week. *International Journal of Sports Medicine* 9: 422-428.

34. Hooper, S.L., L.T. Mackinnon, R.D. Gordon, A.W. Bachmann. 1993. Hormonal responses of elite swimmers to overtraining. *Medicine and Science in Sports and Exercise* 25: 741-747.

35. Hooper, S.L., L.T. Mackinnon, A. Howard, R.D. Gordon, A.W. Bachmann. 1995. Markers for monitoring overtraining and recovery. *Medicine and Science in Sports and Exercise* 27: 106-112.

36. Israel, S. 1996. Zur Problematik des Übertrainings aus internistischer und leistungsphysiologischer Sicht. *Medizin und Sport* 16: 1-12.

37. Jeukendrup, A.E., M.K.C. Hesselink, A.C. Snyder, H. Kuipers, H.A. Keizer. 1992. Physiological changes in male competitive cyclists after two weeks of intensified training. *International Journal of Sports Medicine* 13: 534-541.

38. Jost, J., M. Weiss, H. Weicker. 1989. Unterschiedliche Regulation des adrenergen Rezeptorsystems in verschiedenen Trainingsphasen von Schwimmern und Langstreckenläufern. In *Sport, Rettung oder Risiko für die Gesundheit*, Hrsg. D. Böning, K.M. Braumann, M.W. Busse, N. Maassen, W. Schmidt, 141-145. Köln: Deutscher Ärzteverlag.

39. Kindermann, W. 1986. Das Übertraining: Ausdruck einer vegetativen Fehlsteuerung. *Deutsche Zeitschrift Sportmedizin* 37: 138-145.

40. Kirwan, J.P., D.L. Costill, J.A. Houmard, J.B. Mitchell, M.G. Flynn, W.J. Fink. 1990. Changes in selected blood measures during repeated days of intense training and carbohydrate control. *International Journal of Sports Medicine* 11: 362-366.

41. Kuipers, H., H.A. Keizer. 1988. Overtraining in elite athletes: review and directions for the future. *Sports Medicine* 6: 79-92.
42. Lehmann, M., J. Keul, K. Wybitul, H. Fischer. 1982. Effect of selective and non-selective adrenoceptor blockade during physical work on metabolism and sympatho-adrenergic system. *Drug Research* 32: 261-266.
43. Lehmann, M., H. Porzig, J. Keul. 1983. Determination of ß-receptors on live human polymorphonuclear leukocytes in autologous plasma. *Journal of Clinical Chemistry and Clinical Biochemistry* 21: 805-811.
44. Lehmann, M., H.H. Dickhuth, P. Schmid, H. Porzig, J. Keul. 1984. Plasma catecholamines, ß-adrenergic receptors, and isoproterenol sensitivity in endurance trained and non-endurance trained volunteers. *European Journal of Applied Physiology* 52: 362-369.
45. Lehmann, M., K.G. Petersen, A.N. Khalaf. 1991. Sympathetic autonomic dysfunction. Programmed subcutaneous noradrenaline administration via microdosing pump. *Klinische Wochenschrift* 69: 872-879.
46. Lehmann, M., W. Schnee, R. Scheu, W. Stockhausen, N. Bachl. 1992. Decreased nocturnal catecholamine excretion: parameter for an overtraining syndrome in athletes? *International Journal of Sports Medicine* 13: 236-242.
47. Lehmann, M., U. Gastmann, K.G. Petersen, N. Bachl, A. Seidel, A.N. Khalaf, S. Fischer, J. Keul. 1992. Training-overtraining: performance, and hormone levels, after a defined increase in training volume versus intensity in experienced middle- and long-distance runners. *British Journal of Sports Medicine* 26: 233-242.
48. Lehmann, M., P. Baumgartl, C. Wieseneck, A. Seidel, H. Baumann, S. Fischer, U. Spöri, G. Gendrisch, R. Kaminski, J. Keul. 1992. Training-overtraining: influence of a defined increase in training volume vs. training intensity on performance, catecholamines and some metabolic parameters in experienced middle- and long-distance runners. *European Journal of Applied Physiology* 64: 169-177.
49. Lehmann, M., C. Foster, J. Keul. 1993. Overtraining in endurance athletes: a brief review. *Medicine and Science in Sports and Exercise* 25: 854-862.
50. Lehmann, M., K. Knizia, U. Gastmann, K.G. Petersen, A.N. Khalaf, S. Bauer, L. Kerp, J. Keul. 1993. Influence of 6-week, 6 days per week, training on pituitary function in recreational athletes. *British Journal of Sports Medicine* 27: 186-192.
51. Lehmann, M., U. Gastmann, J.M. Steinacker, N. Heinz, F. Brouns. 1995. Overtraining in endurance sports: a short overview. *Medicina Sportiva Bohemica & Slovaca* 4: 1-6.
52. Lehmann, M., E. Jakob, U. Gastmann, J.M. Steinacker, J. Keul. 1995. Unaccustomed high mileage compared to high intensity training-related performance and neuromuscular responses in distance runners. *European Journal of Applied Physiology* 70: 457-461.
53. Lehmann, M., F. Brouns. 1995. Fatigue and amino acid imbalance hypothesis. In *9th International Triathlon Symposium Kiel 1994*, eds. M. Engelhardt, B. Franz, G. Neumann, A. Pfützner, 161-171. Hamburg: Czwalina.
54. Lehmann, M., C. Foster, N. Heinz, J. Keul. 1996. Overtraining in distance runners. In *Encyclopedia of sports medicine and exercise physiology*, ed. T.D. Fahey. New York: Garland. (Project was cancelled by Garland: manuscript can be requested from authors.)

55. Leinwand, L.A., R.E.K. Fournier, B. Nadal-Ginard, T.B. Shows. 1983. Multigene family for sarcomeric myosin heavy chain in mouse and human DNA: localization on a single chromosome. *Science* 221: 766-769.

56. Lormes, W., J.M. Steinacker, M. Lehmann. 1996. Short-term overtraining in elite rowers preparing for world championships. *International Journal of Sports Medicine*. In preparation.

57. Mahdavi, V., S. Izumo, B. Nadal-Ginard. 1987. Developmental and hormonal regulation of sarcomeric myosin heavy chain gene family. *Circulation Research* 60: 804-814.

58. Martin, W.H., E.F. Coyle, M. Joyner, D. Santeusanio, A.A. Ehsani, J.O. Holloszy. 1984. Effects of stopping exercise training on epinephrine-induced lipolysis in humans. *Journal of Applied Physiology* 56:845-848.

59. Miller, R.E., J.W. Mason. 1964. Changes in 17-hydroxy-corticosteroid excretion related to increased muscular work. *Walter Reed Institute Research* 137-151.

60. Morgan, W.P., D.R. Brown, J.S. Raglin, P.J. O'Connor, K.A. Ellickson. 1987. Psychological monitoring of overtraining and staleness. *British Journal of Sports Medicine* 21:107-114.

61. Naessens, G., J. Lefevre, M. Priessens. 1996. Practical and clinical relevance of urinary basal noradrenaline excretion in the follow-up of training processes in semiprofessional soccer players. *Clinical Journal of Sports Medicine* in press.

62. Newsholme, E.A., M. Parry-Billings, N. McAndrew, R. Budgett. 1991. A biochemical mechanism to explain some characteristics of overtraining. In Advances in nutrition and top sport, ed. F. Brouns. *Medicine Sport Science* 32: 79-83. Basel: Karger.

63. Noakes, T.D. 1991. *Love of running*. 263-361. Champaign, IL: Human Kinetics.

64. Opitz-Gress, A., C. Zeller, Y. Liu, M. Steinacker, W. Lormes, M. Lehmann. 1996. Muscular adaptation in elite rowers: gene expression of MHC and HSP 70 on translation level. In preparation.

65. Owen, I.R. 1964. Staleness. *Physical Education* 56: 35.

66. Parry-Billings, M., E. Blomstrand, N. McAndrew, N. Newsholme, E.A. Newsholme. 1980. A communicational link between skeletal muscle, brain, and cells of the immune system. *International Journal of Sports Medicine* 11: S122-128.

67. Parry-Billings, M., R. Budgett, Y. Koutedakis, E. Blomstrand, S. Brooks, C. Williams, P.C. Calder, S. Pilling, R. Baigrie, E.A. Newsholme. 1992. Plasma amino acid concentration in the overtraining syndrome: possible effects on the immune system. *Medicine and Science in Sports and Exercise* 24: 1353-1358.

68. Rowland, T.W. 1986. Exercise fatigue in adolescents: diagnosis of athlete burnout. *Physician and Sportsmedicine* 14: 69-77.

69. Salo, D.C., C.M. Donovan, K.J.A. Davies. 1991. HSP 70 and other possible heat shock or oxidative stress proteins are induced in skeletal muscle, heart, and liver during exercise. *Journal of Free Radicals in Biology and Medicine* 11: 239-246.

70. Schneider, F.J., K. Völker, H. Liesen. 1993. Zentral- versus peripher-nervale Belastungsreaktion der neuromuskulären Strukturen. In *Sportmedizin: gestern-heute-morgen*, Hrsg. K. Tittel, K.H. Arndt, W. Hollmann, 224-227. Leipzig: Barth.

71. Snyder, A.C., A.E. Jeukendrup, M.K.C. Hesselink, H. Kuipers, C. Foster. 1993. A physiological/psychological indicator of over-reaching during intensive training. *International Journal of Sports Medicine* 14: 29-32.
72. Snyder, A.C., H. Kuipers, B.O. Cheng, R. Servais, E. Fransen. 1995. Overtraining following intensified training with normal muscle glycogen. *Medicine and Science in Sports and Exercise* 27: 1063-1070.
73. Stone, M.H., R.E. Keith, J.T. Kearney, S.J. Fleck, G.D. Wilson, N.T. Triplett. 1991. Overtraining: a review of signs, symptoms and possible causes. *Journal of Applied Sport Science Research* 5: 35-50.
74. Tohmeh, J.F., P.E. Cryer. 1980. Biphasic adrenergic modulation of ß-adrenergic receptors in man. *Journal of Clinical Investigation* 65: 836-840.
75. Verma, S.K., S.R. Mahindroo, D.K. Kansal. 1978. Effect of four weeks of hard physical training on certain physiological and morphological parameters of basketball players. *Journal of Sports Medicine* 18: 379-384.
76. Vervoorn, C., A.M. Quist, L.J.M. Vermulst, W.B.M. Erich, W.R. de Vries, J.H.H. Thijssen. 1991. The behavior of the plasma free testosterone/cortisol ratio during a season of elite rowing training. *International Journal of Sports Medicine* 12: 257-263.
77. Wheeler, G.D., M. Singh, W.D. Pierce, S.F. Epling, D.C. Cumming. 1991. Endurance training decreases serum testosterone levels without change in LH pulsatile release. *Journal of Clinical Endocrinology and Metabolism* 72: 422-425.
78. Wilson, W., R.J. Maughan. 1993. Evidence for the role of 5-HT in fatigue during prolonged exercise. *International Journal of Sports Medicine* 14: 297-300.

CHAPTER 3

Monitoring and Preventing of Overreaching and Overtraining in Endurance Athletes

David G. Rowbottom, PhD, David Keast, PhD, and Alan R. Morton, PhD

Introduction

In the sporting arena of the 1990s the high incidence of overreaching and overtraining in athletes is well documented, and has become an increasing concern for athletes, coaches, and researchers (14, 37, 103). Equally noticeable is the remarkable lack of well-researched and effective treatment programs designed for the rehabilitation of overtrained athletes. Since the exact pathophysiological mechanisms leading to overtraining are still largely unknown, treatment programs have tended to lack a degree of scientific basis. The primary prescription offered to athletes would appear to be rest (60), perhaps for several weeks or even months, and often a prolonged break from training and competition. Furthermore, the difficulties in persuading affected athletes that they need to rest have been highlighted (13), even though this would seem to be recognized as the primary basis for treatment. While the success of rehabilitation programs remains at such a low level, it will also remain the consensus among researchers and coaches alike, that it is more critical to prevent overtraining than to treat it.

In terms of prevention, undertraining would seem to be the surest way of avoiding prolonged fatigue (90). However, undertraining will inevitably lead to underperformance. If an athlete is to reach full potential in the international sporting arena, performance, and consequently training stress, must be optimized (see figure 3.1). Optimal performance can only be achieved by the precise balance, on an individual basis, of training stress and adequate recovery. Unfortunately, it is

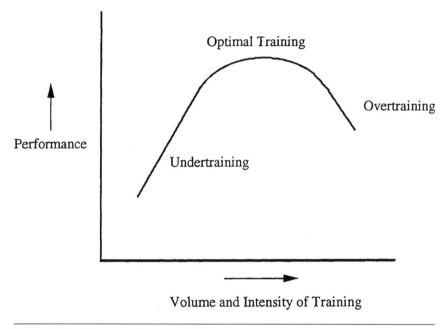

Figure 3.1   Schematic diagram of the relationship between training and performance.

clear that the boundary between optimal training and overtraining is not easily defined (52). With so many elite, often young, endurance athletes becoming prone to excessive training regimes in search of improved performance, the need has never been greater for the careful monitoring of their training and the prevention of overtraining.

General Considerations

Regular Monitoring of Endurance Athletes

A number of recent studies have attempted to purposefully overreach athletes by employing short periods of intensified training (35, 50, 54, 58, 65, 67). While these studies may, to some extent, be unrealistic in their use of extreme volumes and intensities of training, they serve to highlight important concerns. Many of these studies successfully used periods of intensified training to produce significant impairments of endurance performance (35, 50, 54) even if muscle glycogen status was maintained (101), with a number of days of recovery required before performance returned to baseline levels (35). More importantly, the majority of such studies were only of short duration, of five (58), ten (35), or fourteen (54) days. It is of concern that such short periods of intensified training can lead to profound performance impairments. Such observations clearly underline the need for endurance athletes to be assessed for early signs of overreaching and overtraining at regular and frequent intervals.

The Use of Performance Indicators

Performance decrements, along with chronic fatigue, are the most obvious indicators of both overreaching and overtraining, and have been used to diagnose the disorder in a number of studies (35, 36, 54, 67, 68, 100). Researchers have employed a number of different measures of endurance performance including maximal oxygen uptake (33, 67), running speed at 4 mmol/L lactate (67, 68, 112), a maximal effort time trial (24, 36, 49, 54), and a maximal run to exhaustion at a set treadmill speed (35, 36). It is essential that the performance indicator used be a true reflection of competition performance in the athlete's particular event. The variety of performance measures used by researchers is, in itself, a reflection of the lack of consensus on this issue.

A further complication appears to be the important issue of how much performance must deteriorate before overreaching or overtraining can be diagnosed (48). While prospective studies of overreaching have reported performance decrements of as much as 29% (35), it seems likely that the early signs of overreaching and overtraining would be far more subtle. A number of reports have even suggested that a stagnant performance may be sufficient evidence of overtraining (50, 62). Performance decrements of as little as 2.5% can make the difference between a world champion and a nonqualifier for the event (35, 69). Hence, an objective assessment of performance is critical, and it remains open to question whether a statistically significant performance decrement is required to diagnose overtraining (35). Therefore, difficulties in using performance markers to monitor athletes and prevent overtraining remain.

The Use of Biological Indicators

Once a performance decrement is clearly recognized, it may be too late to initiate training changes to avoid a period of enforced recovery. Thus, the usefulness of certain biological parameters as early markers of an overreached or overtrained state has attracted much attention among researchers. The ultimate aim of research in this field is to establish scientific procedures for monitoring an athlete's training regime and to identify objective methods for the early detection of overreaching and overtraining that can be used on a regular basis for routine assessment. In recent years, many researchers have attempted to identify reliable markers of overtraining, which would allow coaches and athletes to maximize training benefits yet avoid overtraining, thereby optimizing performance (48). No single, objective marker has been identified, however, and the multiplicity of recorded symptoms associated with overtraining syndrome (37) only serves to highlight the overwhelming complexity of the overtraining problem (see table 1.1). A wide range of hematological (29, 90), biochemical (36, 54, 56, 85, 86), hormonal (1, 6, 66, 67, 68, 107), and immunological (35, 72, 110) measures have been investigated as possible markers of overtraining.

Criteria for Markers of Overtraining

It is not the purpose of this chapter to present the current state of knowledge concerning specific biological markers of overtraining. Instead, if research in this field is to identify reliable indicators of both overreaching and overtraining, criteria for the selection of such markers must be established. This chapter will attempt to lay these essential foundations, in addition to outlining procedures for the effective use of these diagnostic tools.

Sensitivity of Overtraining Markers

Since performance stagnation or decrement is the most objective indicator of overreaching and overtraining, the most important criterion for any marker of overtraining has to be the demonstration of a clear-cut relationship to competition performance. On this basis, a number of parameters have been suggested since, theoretically at least, they have a close relationship to performance capability. For example, studies have reported a significant positive correlation between a hematological index such as red blood cell numbers, hematocrit, and hemoglobin concentration, and the duration of physical work to exhaustion (16, 117). Thus, these indexes have been proposed as markers of overtraining (90). While early studies reported a decrease in hemoglobin, hematocrit, and red blood cell numbers associated with both overreaching (29) and overtraining (90), a number of more recent studies have reported normal values for all hematological indexes as well as serum ferritin in overtrained athletes (48, 53, 88).

Similarly, muscle glycogen has long been appreciated as an important energy source for endurance performance (8); there is a close relationship between muscle glycogen depletion and fatigue. Athletes experience difficulty in maintaining muscle glycogen during consecutive days of intensive training or overreaching (22, 24, 57) even when carbohydrate consumption is matched to utilization (57), but especially when they fail to consume sufficient carbohydrate to meet the demands of training (24). Overtraining may be initiated by increased training combined with insufficient dietary carbohydrate, which results in muscle glycogen depletion (101). Reduced blood-lactate concentrations during both submaximal and maximal exercise in overreaching (24, 54, 65, 100) may be indicative of glycogen depletion. The degree of glycogen depletion required to impair performance is not clear, however, since it has been reported that following ten days of overreaching resulting in partial glycogen depletion, swimming time-trial performance was not impaired (24). Furthermore, it is still open to question whether overreaching or overtraining can occur in the absence of glycogen depletion as has recently been reported (101). While there is clearly a need to prevent both anemia (see chapter 7) and glycogen depletion (see chapter 14) among endurance athletes, the use of such indexes as markers of overreaching and overtraining appears to be limited.

Other studies have taken an observational approach to link decrements in performance following a period of intensified training or overreaching to changes in specific biological parameters. While direct physiological mechanisms linking biological changes to performance may be unclear in some cases (89), this by no means disqualifies them as potential indicators of overreaching and overtraining. Unfortunately, some markers that have been highlighted in this fashion may also be misleading. For example, elevated creatine kinase (CK) activity, as a marker of muscle damage (17), was at one time considered indicative of an overtrained state (90). Conversely, elevated CK activity has been reported in the absence of any performance impairment (30, 57) and in asymptomatic long-distance runners on routine blood tests (3). More recent studies have identified overtraining in athletes in the absence of elevated serum CK activity (36, 65, 88). In nonimpact sports such as swimming and cycling, serum CK may not rise in spite of severe fatigue (62).

Similar data exist for hormonal observations: elevated serum cortisol concentrations (6, 57), depressed serum testosterone concentrations (1, 33), and a depressed testosterone/cortisol ratio (1) have all been observed following overreaching or overtraining, and have been recommended as useful parameters for the early diagnosis of an overtrained state (1). Other studies have reported cortisol concentrations to decrease (35, 67) or show no change (49, 88) in overtrained athletes, while decreased testosterone/cortisol ratios have been observed in response to high-intensity training without performance decrements (109, 111).

These data serve to highlight the need for indicators that show consistent, predictable changes in response to overreaching and overtraining. To be useful as a means of preventing overtraining, any changes in such markers, at the very least, have to coincide with performance changes, and ultimately precede them. The presently accepted method of research, to deliberately intensify an athlete's training for a short period and then observe a series of biochemical changes, may not be sensitive, or realistic enough to identify the subtle changes that may be involved.

Markers of Optimal Training or Overtraining

If a biological marker is to be effective in diagnosing overtraining at an early stage, it must be possible to distinguish between changes associated with overtraining and changes associated with optimal training. This is a very fine distinction and there is clearly a difficulty in separating normal from abnormal responses. The main problem is that similar changes have been reported in both a training-related improvement in performance as well as an overtraining-induced deterioration in performance. For example, while a number of studies have reported low serum testosterone concentrations in overtrained athletes (1, 33, 67), similar responses have been observed when comparing trained male endurance runners to untrained controls (44).

Similarly, in studies comparing elite (mostly Olympic) endurance athletes to nonathletes, values for hematological indexes have been reported to be lower in the athletes (18, 102) suggesting that such markers may respond similarly to both optimal training and overtraining. While submaximal blood lactates may indicate overreaching or overtraining, it is equally clear that a normal training response would produce similar changes (96) and lead to a failure to diagnose overtraining (53). Some researchers have used the ratio of blood lactate to rating of perceived exertion (RPE) scores, arguing that RPE will decrease for a given workload with optimal training but not with overtraining (100), while others have stressed the need to measure maximal lactate to distinguish optimal training from overtraining (37, 53). More recently, researchers who have identified decreased plasma glutamine concentrations as an indicator of overtraining (56, 85, 88, 89) have observed increased plasma glutamine concentrations in well-trained athletes compared to untrained controls (88) and stale athletes (73).

Markers of Training Load or Overtraining

It is vitally important that data obtained during intensive phases of training are effectively separated from overtraining responses, where the condition is specifically diagnosed. Many biological indicators previously associated with overreaching and overtraining may be no more than physiological indicators of current training load, and may not remain outside the immediate recovery period. As such, it has been suggested that the behavior of cortisol (57), testosterone (107), and the testosterone/cortisol ratio (112) are only indicative of insufficient time to recover from the current training sessions. It has also been argued that muscle damage, and hence CK activity as a noninvasive marker (83), may only reflect the level of training load (37). Although a number of studies have reported elevated serum CK activity following periods of intensified training (4, 30), localized trauma may not have any systemic consequences. Even recent reports of elevated catecholamines following periods of intensified training (49, 50) may be more of a reaction to training stress than to overtraining per se.

If reliable indicators of overtraining are to be established in laboratory-based studies, there is a pressing need to re-evaluate the design of these programs. Many of the reports in the literature have assessed biological parameters at the conclusion of a period of intensified training, which does not effectively delineate training-load responses from overtraining responses per se. A more effective approach would require a period of enforced recovery over a number of days following the training period. This would allow changes associated with training stress to recover, and long-term changes associated with overtraining to be highlighted. Only if performance capability remains impaired at the end of this recovery period can overtraining, instead of overreaching, be diagnosed (see figure 3.2). Only a few studies to date have adopted this approach (35, 56), and it must be concluded that only overreaching was induced in these cases, since complete performance recovery had occurred within six days of training cessation (35). Alternatively, a number of researchers have specifically investigated athletes already suffering from the overtraining syndrome as a method of determining the long-term biological effects of overtraining. Results from our laboratories, and others, have highlighted depressed concentrations of plasma glutamine as the only biochemical marker to date that distinguishes these athletes from controls long after they have had to discontinue training (85, 88, 89).

Interfering Factors to Overtraining Markers

With any clinical assessment of a blood-borne index, a number of lifestyle and other factors must be taken into consideration when evaluating the result of a simple blood test. Such factors may all too easily confound the test-retest reproducibility of many indicators. An awareness of potential interfering factors, along with strategies to minimize possible effects, is essential if biochemical indicators are to be used on a routine basis. A number of major concerns specific to the overtraining situation have been addressed.

Acute Exercise

One of the major problems with the diagnosis of overtraining would appear to be separating the biological changes that result from a single, high intensity training session from the long-term changes associated with overtraining (35, 37). Exercise can be regarded as a disturbance of the body's homeostasis, and consequently a multitude of reported biochemical changes occur during, and following acute exercise sessions. If these same biological parameters are to be used as routine

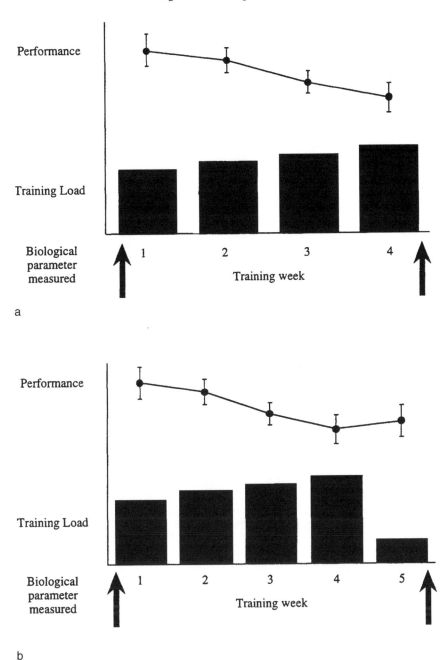

Figure 3.2   Two research study designs used to identify indicators of overtraining: (a) Assessment made at the conclusion of a period of intensified training—does not delineate training load (overreaching) from overtraining. (b) Period of enforced recovery following the training period—separates overtraining from overreaching.

markers of overtraining, it is clearly important that overtraining-induced changes be effectively separated from acute exercise-induced changes. An understanding of the time course of recovery of biological homeostasis following exercise is therefore essential.

Exercise-induced biochemical changes are only transient, returning to preexercise levels within a number of hours of exercise cessation. There is a general consensus that preexercise levels of plasma lactate (34, 47), ammonia (116), and catecholamines (5, 41) as well as a number of immune system parameters (38, 99) are restored within two hours. Serum concentrations of hormones such as cortisol (41, 75) and testosterone (64, 75) may take a number of hours for restoration of homeostasis. Depressed plasma concentrations of amino acids (87), particularly glutamine (56, 85), have also been reported following exercise, and may remain so for several hours during recovery (55, 87).

Conversely, a number of studies have reported that complete biochemical homeostasis was not fully restored even after 24 hours of recovery from a single exercise training session or competition. Plasma uric acid concentrations have consistently been found to remain elevated above resting levels 24 hours after exercise (2, 34, 42, 51). A number of other studies have also reported elevated epinephrine (74) and urea (108) concentrations, and depressed serum testosterone (108), cortisol (74), and amino acid (27) concentrations 24 hours postexercise. Elevated plasma creatine kinase activity has been widely reported postexercise (46, 77), in some cases for seven days or longer (31, 71), while the requirement of up to 72 hours to fully replenish muscle glycogen to resting values after prolonged exercise has long been recognized (8). Furthermore, muscle glycogen resynthesis may be further impaired by exercise-induced muscle damage (25, 84), with reports of incomplete restoration after six days recovery from a competitive marathon (98). Since the disturbances to biochemical homeostasis resulting from a single training session may remain for up to seven days postexercise, caution must be exercised in the interpretation of biological data collected during this time frame. Particular consideration should be given to reports that the degree of certain exercise-induced biochemical changes was dependent on the exercise mode (115), intensity (56, 92), and duration (43). In essence, the most appropriate schedule for the monitoring of overtraining should incorporate a period of enforced rest and recovery before each assessment, with testing being carried out before the start of the next period of training.

Biological Rhythms

Many biological functions are known to show characteristic circadian (24 hour), menstrual (four week) and circa-annual (one year) cycles (97). If potential markers of overtraining demonstrate such variations, this may have a major impact on the interpretation of data. Circadian rhythms have been identified for a range of biological parameters, including serum concentrations of cortisol (61, 114), free and total testosterone (21), uric acid (28), and glutamine (55), as well as muscle glycogen content (19), and a number of cellular immune function parameters (81). Studies that have reported coincidence between circadian rhythms of cortisol and both testosterone (21) and immune function (104) suggest a possible integration of rhythmical patterns. Given that there is considerable evidence for an intimate, molecular-based communication link between neuroendocrine and immune systems (9) these findings are not surprising. While it remains unclear to what extent daily cycles of physical activity and rest bear upon observations of circadian rhythm, a number of studies have also highlighted significant variations in physical performance of up to 3.5% (7) during a 24-hour cycle. It would therefore be recommended that any regular assessment for the early detection of overtraining, whether for performance or biological markers, should be standardized for the time of day. Since female athletes have been reported to have performance decrements (118), in addition to various hormonal changes, coincident with phases of the menstrual cycle, it would be further recommended that cycle phase be standardized for females.

These recommendations may present a somewhat simplistic view of a rather complex problem. Recent studies have demonstrated that alterations in early morning light exposure (110) and early awakening (106) can cause a one- to two-hour phase shift in some circadian rhythms, particularly for cortisol. Given that seasonal changes in photoperiod and light-dark cycles can be quite profound, particularly in the high latitudes, seasonal phase shifts in diurnal cycles should not be unexpected. Indeed, studies conducted in Alaska (70) and Norway (113) have observed significant seasonal variations in hormonal circadian rhythms (see figure 3.3). If biological parameters are to be regularly monitored, and seasonal phase shifts are not accounted for, there is obviously scope for misdiagnosis.

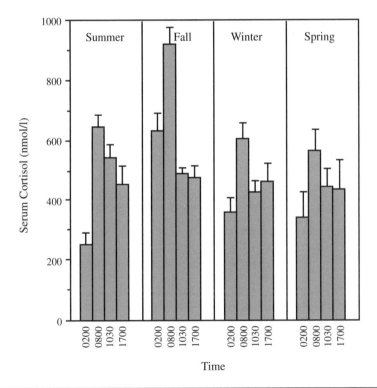

Figure 3.3   Circadian and seasonal variations in plasma cortisol concentration in a population of male soldiers in Alaska. There was a significant (p < .001) seasonal variation in the 0800 sample.
Reprinted from Levin, Milliron, and Duffy 1994.

Plasma Volume

A number of studies have highlighted considerable changes in the plasma volume during different environmental and physiological conditions. These changes are thought to be the result of transient fluid shifts into (hemodilution) and out of (hemoconcentration) the intravascular space (45). If the concentrations of plasma constituents are to be routinely assessed as early indicators of overtraining, then it is imperative that the dynamic nature of plasma volume be considered. Acute bouts of endurance exercise have been shown to produce a transient hemoconcentration, with decreased plasma volume reported in athletes immediately following long-distance running (26, 78), bicycle ergometry (23), and both maximal and submaximal swimming exercise (40). While these changes are transient, lasting only a few hours, other studies have reported a longer term hemodilution following acute exercise. Two days after a competitive marathon plasma volume has been shown to be elevated by as much as 16% from prerace values (74, 91). Irving et al. (51) reported that plasma volume remained elevated for at least six days following an ultramarathon run. It has been suggested for some time that these plasma volume changes may be a factor severely confounding the interpretation of biological data obtained in the recovery period from exercise (38, 40, 51). More importantly, endurance training programs are also generally thought to result in an overall expansion of plasma volume (20, 93). It would therefore seem essential that all plasma constituents regularly assessed as markers of overtraining should be corrected for plasma volume changes. Without this precaution, any interpretation of changes in plasma constituents based purely on their concentration in plasma may be invalidated. Recent data have also suggested that many biological markers, when corrected for plasma volume changes, may be unaffected by exercise stress, effectively reducing the pool of parameters requiring assessment (Kargotich et al., unpublished observations). To further accentuate this necessity, plasma volume changes have also been associated with heat acclimatization (95), hydration state (23), and postural changes (45), all of which may differ between routine assessments.

Age

The tendency of younger, often prepubescent athletes to strive towards optimal performance, and potentially to overtrain, poses additional problems for the monitoring of early markers of overtraining. Likewise, the increased participation in veterans' sporting competitions has meant that researchers must now consider the possible effects of age on the standard range of biochemical indicators. Age-dependent changes have been reported to affect a number of biological functions including hemoglobin, hematocrit, and serum ferritin levels (105), muscle glycogen content (80), a myriad of neuroendocrine alterations (82) particularly total and free serum testosterone in males (105), and a number of immune function parameters (79). Aging has also been shown to be associated with changes in circadian and seasonal rhythms, particularly of hormone secretion (11). Simply relating test results to normal population ranges would appear to be invalid, and may not be sensitive enough to identify the subtle changes that may be indicative of the onset of overtraining. To circumvent any potential age-related or other variability problems, it has been widely recommended that an athlete's own previous results must be used as his/her own standard (37, 90). Therefore, before data can be used to define overtraining, baseline data must be generated while the athlete is in a normal trained state (90).

Dietary Effects

The acute effects of food intake on plasma concentrations of catecholamines and cortisol (59), uric acid (12), amino acids (15, 32), as well as a variety of other biological parameters have been recorded, and it has long been recognized that blood samples for clinical assessment should be taken following a period of fasting. Similarly, some studies investigating changes in plasma constituents with overtraining have stipulated the use of fasting blood samples (33, 35, 109), although this is not universally stated. It is our recommendation that an early morning blood sample, following an overnight fast, presents the best opportunity to meet this criterion with the least possible disruption to an athlete's daily routine.

Prevention of Overtraining

Prevention of overtraining must begin with the need to structure individual athlete's training programs in such a way as to allow an adequate balance of training stress and recovery. Too often the recovery element is overlooked as an essential aspect of any training regime. Unfortunately, researchers are still some way from having a thorough insight into the quantitative relationship between training and performance (66). It remains a major problem for the coach and athlete to confidently prescribe training regimes to optimize performance, while avoiding overtraining. There is a great need for further, carefully controlled research into the systematic monitoring of training stress and performance. Furthermore, it has long been recognized that individual athletes can tolerate different levels of training and competition stress (14), and require differing lengths of recovery periods. Therefore, the need for individualized training programs remains paramount. Varied training is also essential to avoid the problem of monotonous overtraining, which has been identified particularly among swimmers (62).

One aspect of training that is still in need of widespread recognition is the cumulative nature of stress. During periods of high extraneous stress, whether environmental, occupational, educational, or social, training volumes and intensities may need to be modified and scaled down. Together, the combination of stresses may exceed the individual's capacity to adapt (94). It is entirely possible for overtraining to develop when a normally tolerable volume and intensity of training is undertaken, but in combination with other stresses from environment, profession, or private circumstances.

While coaches are still unable to confidently prescribe training for optimal performance, there is clearly a need for a scientific testing program to be regularly incorporated into an athlete's training regime. This program should consist of parameters suitable for the early detection of overtraining. Ultimately, due to the complexity of the interrelationships among variables associated with overreaching and overtraining, it is probable that there will never be a single definitive parameter that is diagnostic of an overtrained state. The challenge confronting researchers in this area is to develop an integrated set of tests based on several parameters known to be associated with a long-term imbalance of homeostasis. Unfortunately, laboratory testing of both athletic performance and blood-borne indicators can be not only time-consuming, but also expensive. Consequently, the selection of appropriate tests should be designed to produce the maximum amount of information necessary to diagnose overtraining at an early stage, and would normally include performance, biological, and also psychological parameters (see chapter 16).

Such testing procedures must, however, be incorporated into a training program in such a way that changes in homeostasis associated with training are not confused with changes associated with overtraining. The key to this delineation of responses may lie in the structuring of the training program (37). Within a framework of periodization of training, appropriate time points during the training regime can be established when full regeneration should be complete, and testing procedures can be carried out.

The work of Russian sport scientists has laid many of the foundations for the concept of training periodization, with the suggestion that the training year should be broken down into four major periods (63, 76) termed macrocycles. The concept is based on the assumption that the development of certain athletic abilities is considered prerequisite to other abilities (10). More importantly, in terms of the prevention of overtraining, each macrocycle is divided into a number of mezocycles of three to six weeks duration, that in turn are divided into microcycles (one week) consisting of the individual training units (see figure 3.4). At each level, the format incorporates a structured, progressive increase in training stress and appropriate, timely, quantified periods of reduced training or recovery to allow regeneration.

Within each mezocycle, normally of four weeks duration (63), there are three weeks of increasing volume and intensity of training, followed by a fourth week of rehabilitation and recovery. This structure may not necessarily allow complete recovery between each training session or even between microcycles and may lead to a period of overreaching. It has been suggested that this cumulative fatigue may provide a more powerful stimulus for adaptation and supercompensation during regeneration periods (63). However, the mezocycle controls the cumulative nature of training stress by enforcing regeneration periods on the athlete. The fourth week is characterized by low volumes and intensities of training with an increased number of rest days (63, 76), and even complete abstinence from training (62), hence reducing the physiological and psychological demands on the athlete. This period provides the prime opportunity for the incorporation of regular testing procedures, during every four week mezocycle within the training program. The athlete should be completely regenerated at the end of this recovery week; the test results should be unaffected by training stress, and will provide the most representative information about an athlete's progress. If homeostasis remains imbalanced at this time, overtraining can be diagnosed at an early stage, extended recovery can be recommended before the start of the next mezocycle, and training program adjustments can be made accordingly.

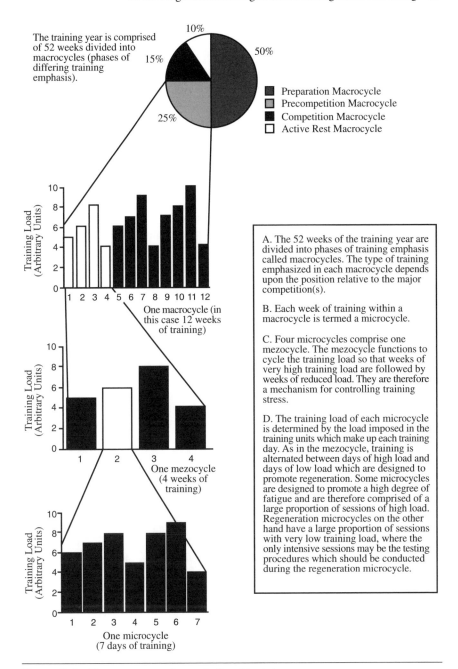

The training year is comprised of 52 weeks divided into macrocycles (phases of differing training emphasis).

10%
15%
50%
25%

■ Preparation Macrocycle
▨ Precompetition Macrocycle
■ Competition Macrocycle
□ Active Rest Macrocycle

One macrocycle (in this case 12 weeks of training)

One mezocycle (4 weeks of training)

One microcycle (7 days of training)

A. The 52 weeks of the training year are divided into phases of training emphasis called macrocycles. The type of training emphasized in each macrocycle depends upon the position relative to the major competition(s).

B. Each week of training within a macrocycle is termed a microcycle.

C. Four microcycles comprise one mezocycle. The mezocycle functions to cycle the training load so that weeks of very high training load are followed by weeks of reduced load. They are therefore a mechanism for controlling training stress.

D. The training load of each microcycle is determined by the load imposed in the training units which make up each training day. As in the mezocycle, training is alternated between days of high load and days of low load which are designed to promote regeneration. Some microcycles are designed to promote a high degree of fatigue and are therefore comprised of a large proportion of sessions of high load. Regeneration microcycles on the other hand have a large proportion of sessions with very low training load, where the only intensive sessions may be the testing procedures which should be conducted during the regeneration microcycle.

Figure 3.4   The structure of a periodized training program.

Reprinted from Fry, Morton, and Keast 1991.

Summary

The procedure of periodization of an athlete's training provides a structured approach to training, with regularly scheduled periods of recovery. It also enables overtraining to be monitored regularly and effectively within a framework that eliminates many of the potential problems of such assessment. Unfortunately, much of the published work in this area is no more than experiential evidence and conjecture (39). There is still a real need for widely held opinions to be tested with well-designed, systematic research studies. The periodization approach also represents a more effective structure to overtraining research design, whereby a period of recovery is enforced prior to performance, biological, and psychological assessment. In this way, those parameters most closely associated with overtraining can be more readily identified and incorporated into an athlete's regular battery of tests, while potential interfering factors can be minimized.

References

1. Adlercreutz, H., M. Härkönen, K. Kuoppasalmi, H. Näveri, I. Huhtaniemi, H. Tikkanen, K. Remes, A. Dessypris, J. Karvonen. 1986. Effect of training on plasma anabolic and catabolic steroid hormones and their response during physical exercise. *International Journal of Sports Medicine* 7: S27-28.
2. Allen, G.D., D. Keenan. 1988. Uric acid production and excretion with exercise. *Australian Journal of Science and Medicine in Sport* 20: 3-6.
3. Apple, F.S. 1981. Presence of creatine kinase MB isoenzyme during marathon training. *New England Journal of Medicine* 305: 764-765.
4. Apple, F.S. 1992. The creatine kinase system in the serum of runners following a doubling of training mileage. *Clinical Physiology* 12: 419-424.
5. Bahr, R., A.T. Hostmark, E.A. Newsholme, O. Gronnerod, O.M. Sejersted. 1991. Effect of exercise on recovery changes in plasma levels of FFA, glycerol, glucose and catecholamines. *Acta Physiologica Scandinavica* 143: 105-115.
6. Barron, J.L., T.D. Noakes, W. Levy, C. Smith, R.P. Millar. 1985. Hypothalamic dysfunction in overtrained athletes. *Journal of Clinical Endocrinology and Metabolism* 60: 803-806.
7. Baxter, C., T. Reilly. 1983. Influence of time of day on all-out swimming. *British Journal of Sports Medicine* 17: 122-127.
8. Bergstrom, J., L. Hermansen, E. Hultman, B. Saltin. 1967. Diet, muscle glycogen and physical performance. *Acta Physiologica Scandinavica* 71: 140-150.
9. Blalock, J.E. 1989. A molecular basis for bidirectional communication between the immune and neuroendocrine systems. *Physiological Reviews* 69: 1-32.
10. Bompa, T.O. 1983. *Theory and methodology of training*. Dubuque, IA: Kendall/Hunt.
11. Brock, M.A. 1991. Chronobiology and aging. *Journal of the American Geriatrics Society* 39: 74-91.
12. Brule, D., G. Sarwar, L. Savoie. 1992. Changes in serum and urinary uric acid levels in normal human subjects fed purine-rich foods containing different amounts of adenine and hypoxanthine. *Journal of the American College of Nutrition* 11: 353-358.

13. Budgett, R. 1990. Overtraining syndrome. *British Journal of Sports Medicine* 24: 231-236.
14. Budgett, R. 1994. The overtraining syndrome. *British Medical Journal* 309: 465-468.
15. Castell, L.M., C.T. Liu, E.A. Newsholme. 1995. Diurnal variation of plasma glutamine and arginine in normal and fasting subjects. *Proceedings of the Nutrition Society* 54: 118A
16. Celsing, F., J. Svedenhag, P. Pihlstedt, B. Ekblom. 1987. Effects of anaemia and stepwise-induced polycythaemia on maximal aerobic power in individuals with high and low haemoglobin concentrations. *Acta Physiologica Scandinavica* 129: 47-54.
17. Clarkson, P.M., C. Ebbling. 1988. Investigation of serum creatine kinase variability after muscle damaging exercise. *Clinical Science* 75: 257-261.
18. Clement, D.B., D.R. Lloyd-Smith, J.G. Macintyre, G.O. Matheson, R. Brock, M. Dupont. 1987. Iron status in Winter Olympic sports. *Journal of Sports Sciences* 5: 261-271.
19. Conlee, R.K., M.J. Rennie, W.W. Winder. 1976. Skeletal muscle glycogen: diurnal variation and effects of fasting. *American Journal of Physiology* 231: 614-618.
20. Convertino, V.A. 1991. Blood volume: its adaptation to endurance training. *Medicine and Science in Sport and Exercise* 23: 1338-1348.
21. Cooke, R.R., J.E.A. McIntosh, R.P. McIntosh. 1993. Circadian variation in serum free and non-SHBG-bound testosterone in normal men: measurements, and simulation using a mass action model. *Clinical Endocrinology* 39: 163-171.
22. Costill, D.L., R. Bowers, G. Branam, K. Sparks. 1971. Muscle glycogen utilisation during prolonged exercise on successive days. *Journal of Applied Physiology* 31: 834-838.
23. Costill, D.L., W.J. Fink. 1974. Plasma volume changes following exercise and thermal dehydration. *Journal of Applied Physiology* 37: 521-525.
24. Costill, D.L., M.G. Flynn, J.P. Kirwan, J.A. Houmard, J.B. Mitchell, R. Thomas, S.H. Park. 1988. Effects of repeated days of intensified training on muscle glycogen and swimming performance. *Medicine and Science in Sports and Exercise* 20: 249-254.
25. Costill, D.L., D.D. Pascoe, W.J. Fink, R.A. Robergs, S.I. Barr, D. Pearson. 1990. Impaired muscle glycogen resynthesis after eccentric exercise. *Journal of Applied Physiology* 69: 46-50.
26. Davidson, R.J.L., J.D. Robertson, G. Galea, R.J. Maughan. 1987. Hematological changes associated with marathon running. *International Journal of Sports Medicine* 8: 19-25.
27. Décombaz, J., P. Reinhardt, K. Anantharaman, G. von Glutz, J.R. Poortmans. 1979. Biochemical changes in a 100 km run: free amino acids, urea, and creatinine. *European Journal of Applied Physiology* 41: 61-72.
28. Devgun, M.S., H.S. Dhillon. 1992. Importance of diurnal variations on clinical value and interpretation of serum urate measurements. *Journal of Clinical Pathology* 45: 110-113.
29. Dressendorfer, E.H., C.E. Wade, E.A. Amsterdam. 1981. Development of pseudoanemia in marathon runners during a 20-day road race. *Journal of the American Medical Association* 246: 1215-1218.
30. Dressendorfer, R.H., C.E. Wade. 1983. The muscular overuse syndrome in long-distance runners. *Physician and Sportsmedicine* 11: 116-130.

31. Evans, W.J., C.N. Meredith, J.G. Cannon, D.A. Dinarello, W.R. Frontera, V.A. Hughes, B.H. Jones, H.G. Knuttgen. 1986. Metabolic changes following eccentric exercise in trained and untrained men. *Journal of Applied Physiology* 61: 1864-1868.

32. Feigin, R.D., A.S. Klainer, W.R. Beisel. 1968. Factors affecting circadian periodicity of blood amino acids in man. *Metabolism* 17: 764-775.

33. Flynn, M.G., F.X. Pizza, J.B. Boone, F.F. Andres, T.A. Michaud, J.R. Rodriguez-Zayas. 1994. Indices of training stress during competitive running and swimming seasons. *International Journal of Sports Medicine* 15: 21-26.

34. Fry, R.W., A.R. Morton, P. Garcia-Webb, G.P.M. Crawford, D. Keast. 1991. Monitoring exercise stress by changes in metabolic and hormonal responses over a 24-h period. *European Journal of Applied Physiology* 63: 228-234.

35. Fry, R.W., A.R. Morton, P. Garcia-Webb, G.P.M. Crawford, D. Keast. 1992. Biological responses to overload training in endurance sports. *European Journal of Applied Physiology* 64: 335-344.

36. Fry, R.W., S.R. Lawrence, A.R. Morton, A.B. Schreiner, T.D. Polglaze, D. Keast. 1993. Monitoring training stress in endurance sports using biological parameters. *Clinical Journal of Sport Medicine* 3: 6-13.

37. Fry, R.W., A.R. Morton, D. Keast. 1991. Overtraining in athletes: an update. *Sports Medicine* 12: 32-65.

38. Fry, R.W., A.R. Morton, D. Keast. 1992. Acute intensive interval exercise and T-lymphocyte function. *Medicine and Science in Sport and Exercise* 24: 339-345.

39. Fry, R.W., A.R. Morton, D. Keast. 1992. Periodisation of training stress: a review. *Canadian Journal of Sport Sciences* 17: 234-240.

40. Goodman, C., G.G. Rogers, H. Vermaak, M.R. Goodman. 1985. Biochemical responses during recovery from maximal and submaximal swimming exercise. *European Journal of Applied Physiology* 54: 436-441.

41. Gray, A.B., R.D. Telford, M. Collins, M.J. Weidemann. 1993. The response of leukocyte subsets and plasma hormones to interval exercise. *Medicine and Science in Sports and Exercise* 25: 1252-1258.

42. Green, H.J., I.G. Fraser. 1988. Differential effects of exercise intensity on serum uric acid concentration. *Medicine and Science in Sports Exercise* 20: 55-59.

43. Guglielmini, C., A.R. Paolini, F. Conconi. 1984. Variations in serum testosterone concentrations after physical exercises of different duration. *International Journal of Sports Medicine* 5: 246-249.

44. Hackney, A.C., W.E. Sinning, B.C. Bruot. 1988. Reproductive hormonal profiles of endurance-trained and untrained males. *Medicine and Science in Sports and Exercise* 20: 60-65.

45. Harrison, M.H. 1985. Effects of thermal stress and exercise on blood volumes in humans. *Physiological Reviews* 65: 149-209.

46. Hellsten-Westing, Y., A. Sollevi, B. Sjodin. 1991. Plasma accumulation of hypoxanthine, uric acid, and creatine kinase following exhausting runs of differing durations in man. *European Journal of Applied Physiology* 62: 380-384.

47. Hermansen, L., I. Stensvold. 1972. Production and removal of lactate during exercise in man. *Acta Physiologica Scandinavica* 86: 191-201.

48. Hooper, S.L., L.T. Mackinnon. 1995. Monitoring overtraining in athletes. *Sports Medicine* 20: 321-327.

49. Hooper, S.L., L.T. Mackinnon, R.D. Gordon, A.W. Bachmann. 1993. Hormonal responses of elite swimmers to overtraining. *Medicine and Science in Sports and Exercise* 25: 741-747.

50. Hooper, S.L., L.T. Mackinnon, A. Howard, R.D. Gordon, A.W. Bachmann. 1995. Markers for monitoring overtraining and recovery. *Medicine and Science in Sports and Exercise* 27: 106-112.

51. Irving, R.A., T.D. Noakes, S.C. Burger, K.H. Myburgh, D. Querido, R. Van Zyl Smit. 1990. Plasma volume and renal function during and after ultramarathon running. *Medicine and Science in Sports and Exercise* 22: 581-587.

52. Jakeman, P.M., E.M. Winter, J. Doust. 1994. A review of research in sports physiology. *Journal of Sports Science* 12: 33-60.

53. Jeukendrup, A.E., M.K.C. Hesselink. 1994. Overtraining: what do lactate curves tell us? *British Journal of Sports Medicine* 28: 239-240.

54. Jeukendrup, A.E., M.K.C. Hesselink, A.C. Snyder, H. Kuipers, H.A. Keizer. 1992. Physiological changes in male competitive cyclists after two weeks of intensified training. *International Journal of Sports Medicine* 13: 534-541.

55. Kargotich, S., D.G. Rowbottom, D. Keast, C. Goodman, A.R. Morton. 1996. Plasma glutamine changes after high intensity exercise in elite male swimmers. *Medicine and Science in Sports and Exercise* 28: S133.

56. Keast, D., D. Arstein, W. Harper, R.W. Fry, A.R. Morton. 1995. Depression of plasma glutamine following exercise stress and its possible influence on the immune system. *Medical Journal of Australia* 162: 15-18.

57. Kirwan, J.P., D.L. Costill, M.G. Flynn, J.B. Mitchell, W.J. Fink, P.D. Neufer, J.A. Houmard. 1988. Physiological responses to successive days of intense training in competitive swimmers. *Medicine and Science in Sports and Exercise* 20: 255-259.

58. Kirwan, J.P., D.L. Costill, J.A. Houmard, J.B. Mitchell, M.G. Flynn, W.J. Fink. 1990. Changes in selected blood measures during repeated days of intense training and carbohydrate control. *International Journal of Sports Medicine* 11: 362-366.

59. Knoll, E., F.W. Muller, D. Ratge, W. Bauersfeld, H. Wisser. 1984. Influence of food intake on concentrations of plasma catecholamines and cortisol. *Journal of Clinical Chemistry and Clinical Biochemistry* 22: 597-602.

60. Koutedakis, Y., R. Budgett, L. Faulmann. 1990. Rest in underperforming elite competitors. *British Journal of Sports Medicine* 24: 248-252.

61. Krieger, D.T., W. Allen, F. Rizzo, H.P. Krieger. 1971. Characterisation of the normal temporal pattern of plasma corticosteroid levels. *Journal of Endocrinology* 32: 266-284.

62. Kuipers, H., H.A. Keizer. 1988. Overtraining in elite athletes: review and directions for the future. *Sports Medicine* 6: 79-92.

63. Kukushkin, G. 1983. *The system of physical education in the U.S.S.R.* Moscow: Radugi.

64. Kuoppasalmi, K., H. Näveri, M. Härkönen, H. Adlercreutz. 1980. Plasma cortisol, androstenedione, testosterone, luteinising hormone in running exercise of different intensities. *Scandinavian Journal of Clinical Laboratory Investigation* 40: 403-409.

65. Lehmann, M., H.H. Dickhuth, G. Gendrisch, W. Jazar, M. Thum, R. Kaminski, J.F. Aramendi, E. Peterke, W. Wieland, J. Keul. 1991. Training-overtraining: a prospective, experimental study with experienced middle- and long-distance runners. *International Journal of Sports Medicine* 12: 444-452.

66. Lehmann, M., C. Foster, J. Keul. 1993. Overtraining in endurance athletes: a brief review. *Medicine and Science in Sports and Exercise* 25: 854-862.

67. Lehmann, M., U. Gastmann, K.G. Petersen, N. Bachl, A. Seidel, A.N. Khalaf, S. Fischer, J. Keul. 1992. Training-overtraining: performance, and hormone levels, after a defined increase in training volume versus intensity in experienced middle- and long-distance runners. *British Journal of Sports Medicine* 26: 233-242.

68. Lehmann, M., W. Schnee, R. Scheu, W. Stockhausen, N. Bachl. 1992. Decreased nocturnal catecholamine excretion: parameter for an overtraining in athletes? *International Journal of Sports Medicine* 13: 236-242.

69. Levin, S. 1991. Overtraining causes Olympic-sized problems. *Physician and Sportsmedicine* 19: 112-118.

70. Levine, M.E., A.N. Milliron, L.K. Duffy. 1994. Diurnal and seasonal rhythms of melatonin, cortisol and testosterone in interior Alaska. *Arctic Medical Research* 53: 25-34.

71. Lijnen, P., P. Hespel, R. Fagard, R. Lysens, E. Vanden-Eynde, M. Goris, W. Goossens, W. Lissens, A. Amery. 1988. Indicators of cell breakdown in plasma of men during and after a marathon race. *International Journal of Sports Medicine* 9: 108-113.

72. Mackinnon, L.T., S. Hooper. 1994. Mucosal (secretory) immune system responses to exercise of varying intensity and during overtraining. *International Journal of Sports Medicine* 15: S179-S183.

73. Mackinnon, L.T., S.L. Hooper. 1996. Plasma glutamine and upper respiratory tract infection during intensified training in swimmers. *Medicine and Science in Sports and Exercise* 28: 285-290.

74. Maron, M.B., S.M. Horvath, J.E. Wilkerson. 1977. Blood biochemical alterations during recovery from competitive marathon running. *European Journal of Applied Physiology* 36: 231-238.

75. Mathur, R.S., M.R. Neff, S.C. Landgrebe, L.O. Moody, R.F. Kirk, R.H. Gadsden, P.F. Rust. 1986. Time-related changes in the plasma concentrations of prolactin, gonadotrophins, sex hormone-binding globulin, and certain steroid hormones in female runners after a long-distance race. *Fertility and Sterility* 46: 1067-1070.

76. Matveyev, L. 1981. *Fundamentals of sports training.* (Translated from the Russian). Moscow: Progress.

77. Maughan, R.J., A.E. Donnelly, M. Gleeson, P.H. Whiting, K.A. Walker, P.J. Clough. 1989. Delayed-onset muscle damage and lipid peroxidation in man after a downhill run. *Muscle and Nerve* 12: 332-336.

78. Maughan, R.J., P.H. Whiting, R.J.L. Davidson. 1985. Estimation of plasma volume changes during marathon running. *British Journal of Sports Medicine* 19: 138-141.

79. Mazzeo, R.S. 1994. The influence of exercise and aging on immune function. *Medicine and Science in Sports and Exercise* 26: 586-592.

80. Meredith, C.N., W.R. Frontera, E.C. Fisher, V.A. Hughes, J.C. Herland, J. Edwards, W.J. Evans. 1989. Peripheral effects of endurance training in young and old subjects. *Journal of Applied Physiology* 66: 2844-2849.

81. Moldofsky, H. 1994. Central nervous system and peripheral immune functions and the sleep-wake system. *Journal of Psychiatry and Neuroscience* 19: 368-374.

82. Mooradian, A.D. 1993. Mechanisms of age-related endocrine alterations. Part I. *Drugs and Aging* 3: 81-97.

83. Noakes, T.D. 1987. Effect of exercise on serum enzyme activities in humans. *Sports Medicine* 4: 245-267.

84. O'Reilly, K.P., M.J. Warhol, R.A. Fielding, W.R. Frontera, C.N. Meredith, W.J. Evans. 1987. Eccentric exercise-induced muscle damage impairs muscle glycogen repletion. *Journal of Applied Physiology* 63: 252-256.

85. Parry-Billings, M., R. Budgett, Y. Koutedakis, E. Blomstrand, S. Brooks, C. Williams, P. Calder, S. Pilling, R. Baigrie, E. Newsholme. 1992. Plasma amino acid concentrations in the overtraining syndrome: possible effects on the immune system. *Medicine and Science in Sports and Exercise* 24: 1353-1358.

86. Pyne, D.B. 1993. Uric acid as an indicator of training stress. *Sport Health* 11: 26-27.

87. Rennie M.J., R.H.T. Edwards, S. Krywawych, C.T.M. Davies, D. Halliday, J.C. Waterlow, D.J. Millward. 1981. Effect of exercise on protein turnover in man. *Clinical Science* 61: 627-639.

88. Rowbottom, D.G., D. Keast, C. Goodman, A.R. Morton. 1995. The haematological, biochemical and immunological profile of athletes suffering from the overtraining syndrome. *European Journal of Applied Physiology* 70: 502-509.

89. Rowbottom, D.G., D. Keast, A.R. Morton. 1996. The emerging role of glutamine as an indicator of exercise stress and overtraining. *Sports Medicine* 21: 80-97.

90. Ryan, A.J., R.L. Brown, E.C. Frederick, H.L. Falseti, E.R. Burke. 1983. Overtraining of athletes. *Physician and Sportsmedicine* 11: 93-110.

91. Schmidt, W., N. Maassen, U. Tegtbur, K.M. Braumann. 1989. Changes in plasma volume and red cell formation after a marathon competition. *European Journal of Applied Physiology* 58: 453-458.

92. Schwarz, L., W. Kindermann. 1990. Beta-endorphin, adrenocorticotropic hormone, cortisol and catecholamines during aerobic and anaerobic exercise. *European Journal of Applied Physiology* 61: 165-171.

93. Selby, G.B., E.R. Eichner. 1994. Hematocrit and performance: the effect of endurance training on blood volume. *Seminars in Hematology* 31: 122-127.

94. Selye, H. 1957. *The stress of life*. London: Longmans Green.

95. Senay, L.C., D. Mitchell, C.H. Wyndham. 1976. Acclimatization in a hot, humid environment: body fluid adjustments. *Journal of Physiology* 40: 786-796.

96. Sharp, R.L., C.A. Vitelli, D.L. Costill, R. Thomas. 1984. Comparison between blood lactate and heart rate profiles during a season of competitive swim training. *Journal of Swimming Research* 1: 17-20.

97. Shephard, R.J. 1984. Sleep, biorhythms and human performance. *Sports Medicine* 1: 11-37.

98. Sherman, W., D. Costill, W. Fink, F. Hagerman, L. Armstrong, T. Murray. 1983. Effect of 42.2-km foot race and subsequent rest or exercise on muscle glycogen and enzymes. *Journal of Applied Physiology* 55: 1219-1224.

99. Shinkai, S., S. Shore, P.N. Shek, R.J. Shephard. 1992. Acute exercise and immune function: relationship between lymphocyte activity and changes in subset counts. *International Journal of Sports Medicine* 13: 452-461.

100. Snyder, A.C., A.E. Jeukendrup, M.K.C. Hesselink, H. Kuipers, C. Foster. 1993. A physiological/psychological indicator of over-reaching during intensive training. *International Journal of Sports Medicine* 14: 29-32.

101. Snyder, A.C., H. Kuipers, B. Cheng, R. Servais, E. Fransen. 1995. Overtraining following intensified training with normal muscle glycogen. *Medicine and Science in Sports and Exercise* 27: 1063-1070.

102. Stewart, G.A., J.E. Steel, A.H. Toyne, M.J. Stewart. 1972. Observations on the haematology and the iron and protein intake of Australian Olympic athletes. *Medical Journal of Australia* 2: 1339-1343.
103. Stone, M.H., R.E. Keith, J.T. Kearney, S.J. Fleck, G.D. Wilson, N.T. Triplett. 1991. Overtraining: a review of the signs, symptoms and possible causes. *Journal of Applied Sport Science Research* 5: 35-50.
104. Tavadia, H.B., K.A. Fleming, P.D. Hume, H.W. Simpson. 1975. Circadian rhythmicity of human plasma cortisol and PHA-induced lymphocyte transformation. *Clinical and Experimental Immunology* 22: 190-193.
105. Tietz, N.W., D.F. Shuey, D.R. Wekstein. 1992. Laboratory values in fit aging individuals—sexagenarians through centenarians. *Clinical Chemistry* 38: 1167-1185.
106. Touitou, Y., O. Benoit, J. Foret, A. Aguirre, A. Bogdan, M. Clodore, C. Touitou. 1992. Effects of a two-hour early awakening and of bright light exposure on plasma patterns of cortisol, melatonin, prolactin and testosterone in man. *Acta Endocrinologica* 126: 201-205.
107. Urhausen, A., H. Gabriel, W. Kindermann. 1995. Blood hormones as markers of training stress and overtraining. *Sports Medicine* 20: 251-276.
108. Urhausen, A., W. Kinderman. 1987. Behaviour of testosterone, sex hormone binding globulin (SHBG) and cortisol before and after a triathlon competition. *International Journal of Sports Medicine* 8: 305-308.
109. Urhausen, A., T. Kullmer, W. Kindermann. 1987. A 7-week follow-up study of the behaviour of testosterone and cortisol during the competition period in rowers. *European Journal of Applied Physiology* 56: 528-533.
110. Van Cauter, E., J. Sturis, M.M. Byrne, J.D. Blackman, N.H. Scherberg, R. Leproult, S. Refetoff, O. Van Reeth. 1993. Preliminary studies on the immediate phase-shifting effects of light and exercise on the human circadian clock. *Journal of Biological Rhythms* 8: S99-108.
111. Verde, T., S. Thomas, R.J. Shephard. 1992. Potential markers of heavy training in highly trained distance runners. *British Journal of Sports Medicine* 26: 167-175.
112. Vervoorn, C., A.M. Quist, L.J.M. Vermulst, W.B.M. Erich, W.R. deVries, J.H.H. Thijssen. 1991. The behaviour of the plasma free testosterone/cortisol ratio during a season of elite rowing training. *International Journal of Sports Medicine* 12: 257-263.
113. Weitzman, E.D. 1975. Seasonal patterns of sleep stages and secretion of cortisol and growth hormone during 24 hour periods in Northern Norway. *Acta Endocrinologica* 78: 65-76.
114. Weitzman, E.D., D. Fukushima, C. Nogeire, H. Roffwarg, T.F. Gallagher, L. Hellman. 1971. Twenty-four hour pattern of the episodic secretion of cortisol in normal subjects. *Journal of Clinical Endocrinology* 33: 14-22.
115. Weltman, A., C.M. Wood, C.J. Womack, S.E. Davis, J.L. Blumer, J. Alvarez, K. Sauer, G.A. Gaesser. 1994. Catecholamine and blood lactate responses to incremental rowing and running exercise. *Journal of Applied Physiology* 76: 1144-1149.
116. Wilkerson, J.E., D.L. Batterton, S.M. Horvath. 1977. Exercise-induced changes in blood ammonia levels in humans. *European Journal of Applied Physiology* 37: 255-263.
117. Woodson, R.D. 1984. Hemoglobin concentration and exercise capacity. *American Review of Respiratory Disease* 129: S72-75.
118. Zaharieva, E. 1965. Survey of sportswomen at the Tokyo Olympics. *Journal of Sports Medicine and Physical Fitness* 5: 215-219.

Physiology of Overtraining in Strength/Power Athletes

CHAPTER 4

Factors Involved With Overtraining for Strength and Power

William J. Kraemer, PhD, and Bradley C. Nindl, MS

Introduction

Overtraining in strength and power training is a problem that arises when mistakes are made in the training program. Mistakes are related to the acute program variables that define the exercise stimuli for the workout. When inappropriate workouts are repeated over a chronic period of time, the potential for developing an overtraining syndrome exists. Overtraining can be defined as a decrease in performance due to a maladaptation to the exercise stimuli. Furthermore, it is possible for performance to plateau well under the theoretical genetic potential for performance of the athlete. A decrease in performance might also be intentionally created by the performance of a difficult training program with the intention to correct the mistake and allow a rebound in performance to take place. This concept can be called overreaching or supercompensation (12). Figure 4.1 shows responses to training in relationship to the adaptive potential. While training does not induce a linear type response, the optimal training program approaches over time the adaptive potential and remains close for a set period of time (e.g., peaking phase in periodized training) for optimal performance. The second curve shows the overreaching phenomenon where a planned suppression in performance due to greater workout demands is followed by a rebound increase even closer to the optimal genetic potential. Finally, if training mistakes are made that are either unknown or are due to the athlete's or coach's miscalculations, a depression in performance can continue for weeks or months and the athlete is then considered to be in an overtraining syndrome. The use of an overreaching technique might be considered a higher risk training approach and is dependent on the understanding of the training

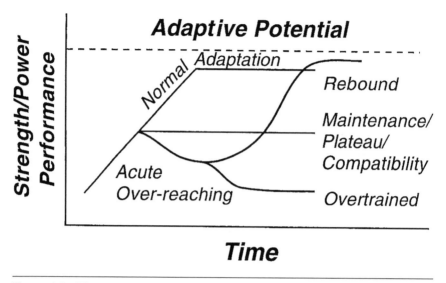

Figure 4.1   Theoretical responses to training in relation to the adaptive potential.

stimuli and the timing of the rebound is vital to its success. Yet in elite competitions a small performance increase may be the difference between the gold medal and 100th place. Thus, the high stakes game of training for elite performance requires even more careful planning and understanding of the training stimuli than does training for fitness and health.

The most common mistake made in training is most likely related to the rate of progression. If mechanical and chemical loads are created which damage the fundamental morphological structures involved with the adaptational changes (e.g., increased muscle size) required for improved performance, overtraining can occur rapidly. For example, a shot putter performs a supplemental speed workout by sprinting both the straight-aways and turns on a small indoor cement track (e.g., 20 laps/mi). Such a workout can create high levels of shearing forces resulting in tremendous stress and damage to muscles and connective tissue of the lower legs. The morphological damage of vital structures results in a decrease in shot putting performance and consequently acute overtraining is observed. One might argue that it is a training mistake and therefore it could contribute to the development of an overtraining problem. Thus, overtraining could arise from mistakes in supplemental training (i.e., sprint training). This indicates that sources of overtraining must be examined from both within the specific sport-related training modalities (e.g., resistance training programs for strength/power athletes) and outside of the specific training modality for the sport (e.g., run training for weight events).

Overtraining and Overreaching Studies in Strength/Power Athletes

It must be realized that only a few studies have been performed that have been able to examine various types of overtraining models in the laboratory. Furthermore, the intentional improper programming needed to create a chronic depression in performance has not been easily accomplished in the laboratory environment and may need the use of tracking studies of athletes in training to document overtraining mistakes and effects. However, in order to systematically study the overtraining and overreaching phenomena both approaches will be needed. While some of these studies will be examined in much greater detail later in this book, a few comments are instructive in better understanding some of the fundamental factors involved with overtraining and overreaching.

In a study by Callister et al. (3) it was first suggested that anaerobic training may present a different set of symptoms than endurance overtraining studies. The investigators in this study increased the volume of resistance training, run training, and practice in U.S. national-level judo athletes over a six-week program and the expected increases in resting heart rate or blood pressure (halter-monitored) did not occur. What these investigators observed were significantly decreased performances in a set of 300-m sprint performance times and decreases in isokinetic force production. Furthermore, the resistance training performed was not effective in showing any signs of enhancing maximal strength (i.e., 1 repetition maximum (1 RM)) performances.

In two studies by A.C. Fry et al. (11, 12), it also became obvious that in order to create a successful intensity-specific overtraining protocol for the laboratory, certain factors were needed in order to create a reduction in performance. In the first attempt, the impact and importance of rest on the development of an intensity-specific overtraining model was discovered.

Recreationally strength trained athletes performed after warm-up eight sets of machine-squats at 95% of 1 RM. In this protocol Sunday was utilized for rest after 6 days of overtraining. No reduction in the 1 RM performance was observed. What did result were nonspecific performance decrements in isolated knee joint isokinetic torque production at 60 degrees per second, and increases in sprint times and agility speed to the nondominant side. Therefore, it was found that the body protects the high force production of the 1 RM and the use of one day of rest appeared to enhance toleration to the protocol. Thus the importance of rest and recovery as programmed into the theoretical development of periodization protocols was shown in this study to be a vital principle to effective training. Yet the decrements in peripheral performances loom as a caution that something was still being negatively affected by the overtraining protocol even in this study. Thus, it is important for athletes, coaches, and sport scientists to examine the global performance abilities of the athlete and not to assume that if nothing negative is happening to one parameter, overtraining may not occur. While the decrease in performance in sprints and isokinetic torque production may not have been important, for example, in a power lifter, if this occurred in an American football player, concern would be obvious.

In the subsequent study by A.C. Fry et al. (12) the one day of rest was eliminated from the overtraining protocol and each subject was expected after warm-up sets to lift 10 sets of their 1 RM with rest between sets. This resulted in a significant decrease (greater than 4.5 kg) in the 1 RM performance in 73% of the subjects who participated in the overtraining study. What is interesting is that 23% of the subjects decreased less than 4.5 kg and were in some cases still making progress and not yet in an overtraining state. This demonstrated that the time-course of the physiological development of an overtraining condition is highly dependent upon individual responses and genetic endowment. While follow-up was not done to determine if this model produced overtraining or overreaching, it was apparent that intensity overtraining would be difficult to sustain for extended periods of time due to the localized stress on the thigh musculature and knee joint. Nevertheless, this study showed that a relatively low-volume exercise stress, but one with a high (100%+) relative intensity load without any day of rest can result in dramatic performance decrements in both nonspecific test variables (e.g., isokinetic knee extension peak torque, sprint times) and training mode-specific decreases.

The development in my laboratory (WJK) of this intensity-overtraining model under Dr. A.C. Fry's direction provided the first laboratory model to study overtraining under controlled conditions. This allowed us to evaluate associated changes in the neuroendocrinological and neuromuscular environments in order to search for potential markers and mechanisms of action. Surprisingly, the neuroendocrine pattern did not follow what might have been expected from prior knowledge of the endurance-overtraining studies, e.g., no decreases in resting testosterone or increases in resting cortisol were commonly observed. In fact it might be theorized that the neuroendocrine responses were first-line responses to fight the overtraining stress created by the exercise. Therefore, while this data will be looked at in greater detail in chapter 6, it might be that in the initial stages of overtraining the body engages all of the potential repair processes to offset the extreme physical demands. If the body is unable to cope with the alarm responses of the overtraining stressor and consistent with Selye's model of distress (40), a reduction in the anabolic hormone and an increase in catabolic hormone response will result. Figure 4.2 illustrates this theoretical hormonal response to a strength overtraining model. Therefore, in strength and power intensity-related overtraining or overreaching, a

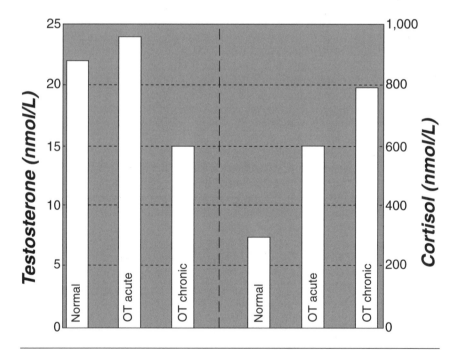

Figure 4.2   Theoretical hormonal response of testosterone and cortisol to a strength overtraining model. OT acute = acute overreaching; OT chronic = chronic overtraining.

suppressed testosterone and elevated cortisol endocrine profile may actually follow the negative response in performance due to the body's inability to obviate the physiological maladaptation associated with inappropriate exercise stimuli (35). Thus, the use of such markers to predict this type of overtraining may be questioned due to their time-course of appearance in relationship to the physical performance decrement. Questions still remain that need to be documented as to the nature of the performance suppression. Would the eventual outcome become overreaching or truly overtraining in nature? Such studies, with overt overtraining as the endpoint (i.e., suppressed performance for weeks) will be ethically difficult and physically demanding even in athletic populations where performance is a valued commodity.

Mistakes in Training

As previously mentioned, the advent of overtraining is really a function of mistakes in the resistance training program or in the supplemental training for the sport. What are the factors that make up the exercise stimuli? What possible mistakes might be made that would lead to an overtraining state? Answers to such questions remain somewhat speculative but further analysis of the resistance exercise stimuli might provide some insights into the problem of creating an overtraining stimulus. Furthermore, an ineffective program that plateaus or maintains a performance parameter well under its optimal performance potential may also contribute to a pseudo-type of overtraining state. Thus, mistakes related to too much or too little could both have similar performance results but the overtraining state will be associated with greater damage or negative physiological alterations in the neuromuscular system.

The acute program variables (9, 10) are: (1) choice of exercise, (2) order of exercise, (3) intensity of exercise, (4) number of sets, and (5) amount of rest allowed between sets and exercises. Mistakes in any of these variables in the progression of a program could theoretically result in an overtraining syndrome, but the mechanisms that mediate the performance decrements might be quite different.

In classic terms, overtraining has been thought to be a function of mistakes in intensity (an acute program variable) and volume of exercise (total work which is a function of the number of sets and the intensity). The use of periodization approaches to training have attempted to utilize these two variables in attempts to carefully plan a progression that would not lead to overtraining. One of the biggest factors in the success of the periodization approach is the use of planned rest in each training cycle. As we have seen, even one day of rest can have beneficial effects on force production.

Choice of Exercise

The choice of exercise involves a host of decisions from the type of muscle action to the type of equipment used to perform the exercise. Muscle actions can be performed using isometric, dynamic concentric or dynamic eccentric actions (25). The common types of exercise equipment are so-called isotonic, variable resistance, isokinetic, and hydraulic or pneumatic. In overtraining, performing an exercise with inappropriate equipment could be the cause of a decrease in performance.

In a study by Newton et al. (34), in a bench press exercise, inappropriately hanging on to the bar and trying to perform speed repetitions resulted in a deceleration of movement and a reduction in power. This was opposite of what was intended from using the concept of speed reps and was a mistake in the choice of equipment, i.e., selecting free weights or machines that create a momentum wave due to the accelerating mass that must be decelerated if you hang onto the bar to the end of the joints range of motion. Therefore, the use of equipment that allows the weight to be released at the end of the range of motion (e.g., medicine balls), or that does not have a mass that needs to be slowed down at the end of the range of motion (isokinetic, hydraulic, or pneumatic devices) may have been a better choice, because such equipment would most likely not create a deceleration response and reduction in power when the intention was to train the body how to accelerate a mass and develop greater power. Figure 4.3 shows the response of a subject to two types of exercise conditions; in the first condition the subject holds onto the bar, and in the second condition the subject is able to release the mass at the end of the range of motion. Notice the dramatic effects on acceleration and power through the range of motion.

One may classify equipment as being fixed form or free form in nature. Fixed form equipment maintains the user's pattern of movement over a range of motion. Proper position and fit are vital for proper use and effective training or overuse problems can arise. Free form exercise equipment includes free weights and machines that allow movement through multiple planes and that require balance. When a machines fixes the pattern of movement in an exercise it fixes the tissue that will be recruited. Fixed form exercise equipment typically isolates muscles at a specific joint and is very effective in training the isolated muscle or muscles. In fact, if one is not careful in the choice of exercise around a joint (e.g., both quad and hamstring exercises in a program), strength imbalances can occur faster than is observed with free weight training. A lack of variation in the pattern of muscle fiber recruitment, a lack of a requirement for balance in multiple planes of movement, and less synergistic muscle use during the activity can diminish the specificity of the carryover to actual sport or daily life activity. In addition, with fixed form machines, one must make sure the individual fits the equipment so that if the resistance is varied over the range of motion (variable resistance), inappropriate forces will not contribute to overuse syndromes due to a dramatic difference between the strength curve of the individual and that of the machine.

Thus, each piece of equipment must be analyzed for the strengths and weaknesses it brings to the exercise program and to determine if it can contribute to any negative effects that could enhance the potential for overtraining due to injury or ineffective physiological stimuli.

Order of Exercise

Putting a squat exercise either at the beginning of a workout or at the end of a workout will affect the amount of resistance that can be used. Therefore, one must prioritize the exercises in the workout so that the most important ones related to the training goal are placed up front in a workout. A plateau in the training response could be a result of exercises being performed in a fatigued state so that

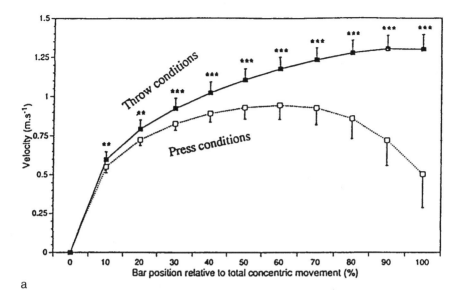

a

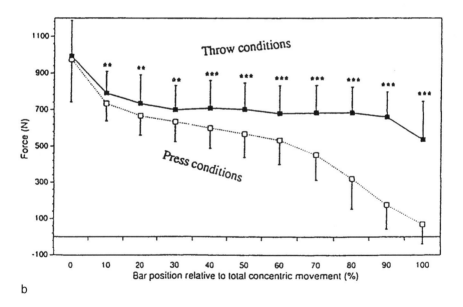

b

Figure 4.3 (a) Mean (±SD) bar velocity in relation to total concentric bar movement for press (open box) and throw (solid box) conditions. (b) Mean (±SD) vertical force in relation to total concentric bar movement for press (open boxes) and throw (solid boxes) conditions. **p < .01; ***p < .001.

Adapted from Newton et al. 1996.

optimal recruitment is not achieved and maximal loads are not utilized in training. While preexhaustion techniques have used this order effect, the fact is that if maximal strength and power performance are desired, maximal loads must be used in training to adapt the nervous system optimally.

Order effects, even in circuit training going from arm to arm and leg to leg rather than going from arm to leg allowing more recovery, can create a problem in the rate of progression for stimulation of the targeted musculature. If an individual is not progressively adapted to the consecutive use of the same limbs, the force capability of the subsequent exercise will be reduced and thus the intensity affected. Furthermore, more complex exercises and large muscle group exercises must be typically placed in the beginning of the workout to optimize their performance and intensity. So the order of exercise is a part of the workout design that must be given careful consideration as to its physiological effects. Thus, the impact of order on a resistance-exercise workout is related to rate of fatigue and the quality of the intensity utilized.

Intensity of Exercise

As previously discussed, the intensity of a resistance exercise interacts with the volume of exercise (i.e., sets × reps or sets × reps × intensity) to provide a potent variable, which may contribute to an overtraining phenomenon if mistakes are made. The intensity of the exercise is typically mode specific. For example, an external mass is lifted in free weights and machines, submaximal or maximal effort is exerted in isometrics, a velocity of movement with maximal exertion is used in isokinetic resistance, and specific force settings (i.e., hydraulics or pneumatics) or band thickness are used in a variety of other resistance modalities for resistance settings.

Typically a repetition maximum is utilized (e.g., 10 RM) or a percentage of a repetition maximum (e.g., % of 1 RM or 10 RM) is utilized to quantify the demands of the exercise. Too rapid a rate of progression in the resistance used can lead to less than optimal training adaptations, or as we have seen in what might be considered a workout that is not normal (i.e., 10 sets of 1 RM every day for 2 weeks), results in overtraining. Mistakes in the real world are many times due to highly motivated athletes using a high volume of heavy loads and taking too little rest to recover from such workouts. As one ages the recovery process also slows down, and overtraining may be even more frequent in master athletes who have not modified their progression formats from their younger years. This could be due in part to an altered hormonal environment that does not support the remodeling of muscle and connective tissue as quickly as the hormonal environment of younger athletes. This is especially true as one goes from the fourth to the sixth decade of life.

Number of Sets

The number of sets is an important factor in the calculation of total work (J) or volume of exercise in a workout. The volume of exercise has been found to be important for continued gains in programs that are classified as building programs.

Building programs are attempting to optimally improve strength, power, and muscle size. Conversely, an inappropriate volume of exercise can create a program stimulus that overshoots the physiological ability of the individual to recover from the stress. In other words the total work is too great and an alarm response occurs resulting in excessive soreness or residual fatigue that does not disappear within 48 hours. Inadequate adaptational responses result, therefore, if the mistake is made over a long-term training program and overtraining results. Mistakes related to the number of sets typically impact the amount of muscle tissue recruited; increased total work requires a greater involvement of more motor units as fatigue accumulates thus utilizing greater tissue mass. The volume of exercise also interacts with the intensity variable to create a volume-intensity interaction, which is why many people estimate total work as a function of sets × reps × intensity. If too much work is done too quickly, overtraining can occur due to a lack of tissue recovery, too much total body physiological stress, psychological stress, or energy substrate depletion.

Amount of Rest Allowed Between Sets and Exercises

One of the more overlooked variables in the design of a workout program is the amount of rest between sets and exercises. This variable can have a dramatic effect on the acid-base status of the muscle and blood and energy substrate utilization and depletion patterns. In addition, Tharion et al. (44) have demonstrated that state anxiety can be increased prior to short-rest workouts, and that anger with the associated pain can also occur. Thus, the physiological strain created by the short-rest protocols with higher intensities of resistance (i.e., 10 RM) results in a psychological component that must be dealt with or overtraining may well occur due to mental fatigue for workout toleration. Workouts using multiple exercises and multiple sets along with one-minute rest periods can result in blood lactate concentrations that are greater than 10 mmol/L and in elite lifters short-rest (i.e., < 1 min rest between sets and exercises) protocols have been shown to create lactate concentrations in the blood of over 20 mmol/L, which places a tremendous stress on the body's ability to buffer the acidity in both the muscle and blood (26-28). Since direct links to muscle fatigue have now been attributed to both hydrogen ions and lactate itself, adequate time (about 8 weeks) is needed to develop the body's acid-base buffering systems (e.g., sodium bicarbonate system, phosphate mechanisms in the cell) (14).

Again, the rate of progression with short-rest protocols will produce symptoms (nausea, dizziness, vomiting, fainting) that are counterproductive to optimal intensity and healthy lifting. Overtraining can occur due to maladaptation from too great an acid load too early in a training program. Furthermore, it is possible that free radical formation can further contribute to tissue breakdown, and serum cortisol elevation with exercise may not return to resting levels following the workout thereby enhancing the catabolic environment. The quality of the exercise session is not related to the symptomology (e.g., nausea, vomiting). In other words, the so-called adverse symptoms are a direct sign of too much too soon and indicate an inappropriate stress for a particular individual. Mistakes in rate of progression for the decrease in rest period lengths can augment an overtraining syndrome that may be starting from another source (e.g., too much exercise volume).

Multiple Interactive Mistakes in Training

It has become obvious that overtraining can be due to multiple causes resulting in an additive phenomenon. While no research has looked at multiple-factor overtraining models, it is easy to see from the many variables involved with the development of a single workout that a tremendous potential exists for multiple mistakes in this process. Thus, multiple mistakes of program ineffectiveness or of physiological overload may produce an overtraining state. How small errors in different variables may add up to cause an overtraining problem remains unclear. It is easy to speculate that psychological stress can augment already existing physiological stress in systems such as adrenal activity. In our example the training stimulates higher levels of cortisol production, and extreme psychological stress can stimulate even more cortisol production. By itself each stress may not significantly elevate the cortisol concentration above normal ranges but together the responses may put cortisol values well out of the normal range. In turn, this elevated cortisol then is at significant molar concentrations to negatively interact with protein metabolism and the immune system (e.g., reduce T cell function). A multiple component model of overtraining in strength/power training deserves further research, especially within the context of sport, due to the multiple stressors that can be present.

Compatibility of Training: A Model of Supplemental Training Interference

The simultaneous training for both maximal strength and aerobic endurance might compromise the development of strength and power. In 1980, Hickson (16) demonstrated that when an endurance program was added to a strength training program, 1 RM strength improvement was suppressed after about two months of training. This study demonstrated that the addition of an aerobic endurance training program to a strength training program contributed to at least a plateauing effect for strength. Since that time the physiological compatibility of simultaneous strength and endurance training has been a subject of great interest (4, 6). A variety of studies have demonstrated that strength can be either compromised (6, 15, 16, 19, 31, 32) or increased (2, 18, 39), while no decreases in aerobic endurance capabilities have been observed. A couple of studies have demonstrated that both strength and endurance capabilities can be attenuated, especially over longer periods of simultaneous training or in highly trained athletes (15, 30, 33, 45).

The physiological mechanisms that may mediate such adaptational responses to simultaneous training remain unclear but the stimulus to the muscle fiber is related to alterations in neural recruitment patterns and attenuation of muscle hypertrophy (4, 7, 8). Such physiological attenuation may, in fact, result in overtraining (i.e., a decrease in performance) (15, 33). It is also possible that if the simultaneous exercise training programs are properly designed, it may just require a longer period of time for the strength and power to express itself due to the simultaneous development of several ultrastructural and enzymatic adaptations for multiple performance demands. Furthermore, the magnitude of the adaptation (e.g., the amount of protein accretion in certain muscle fibers) may not be the same in a system (e.g., muscle) that is being trained by two high-intensity training programs for both strength and endurance.

Few cellular data are available to elucidate changes at the muscle fiber level with concurrent strength and endurance training (38, 41, 42). A recent study by Kraemer et al. (21) provides insights into the responses of the muscle cell in men to the use of simultaneous strength and endurance training. Anabolic and catabolic hormones (e.g., testosterone and cortisol, respectively) may play a vital role in mediating any differential responses to simultaneous strength and endurance training. Kraemer et al. (24) had previously demonstrated that simultaneous sprint and endurance training produced differential cortisol responses when compared to sprint or endurance training only. And a study by Callister et al. (3) also showed that simultaneous sprint and endurance training affected sprint speed and jump power development. Thus, the interference to anaerobic training when high levels of aerobic training are undertaken appears consistent.

High-intensity strength training produces a potent stimulus for muscle cell hypertrophy, that appears to be mediated via increased protein synthesis and accretion of contractile proteins (8). Conversely, an oxidative endurance training stress causes muscle to respond in an opposite fashion by ultimately degrading and sloughing myofibril protein to optimize the kinetics of oxygen uptake (20, 43, 46). Anabolic and catabolic hormones play a key role in such metabolic phenomena (13).

The majority of studies in the literature have utilized relatively untrained subjects to examine the physiological effects of simultaneous strength and endurance training (4, 7). Few data are available regarding the effects of simultaneous strength and endurance training utilizing previously active or fit individuals who are able to tolerate much higher intensity exercise training programs (16). In a study by Kraemer et al. (21), three months of training were performed in order to examine the physiological adaptations to simultaneous high-intensity strength and endurance training in physically active men. Five groups of subjects were utilized. A combination group (C), which performed resistance training and endurance training, an upper body-only resistance training and endurance training group (UC), a resistance training only group (S), an endurance training only group (E), that performed both interval and continuous run training, and a control group. The results demonstrated that in the groups performing the resistance training, musculature increased 1 RM strength in those lifts. One might argue that the rate of strength development was quicker in the S group because the percentage increase in the leg press 1 RM was greater than in the C group. Interestingly, only the S group saw increases in Wingate peak power outputs in both upper and lower body tests. None of the other groups observed any significant peak power changes. It was clear from this study, as Dudley and Djamil (6) had previously observed for high speed isokinetic torque production, that power capabilities may be more susceptible to overtraining because of the addition of an intense supplemental endurance training program. This may be due to a wide variety of factors differentially related to neuromuscular function (4, 5, 7, 36-39). Thus, it may be that power development is much more susceptible to the antagonistic effects of combined strength and endurance training programs than slow velocity strength (7, 12, 19). The fact that cycle endurance training was used in Dudley and Djamil (6) study, indicates that it may be the extreme oxidative stress on the tissue even more than the modality that creates the alterations in the neuromuscular system that are counterproductive to power and high speed development.

In the Kraemer et al. (21) study, and consistent with almost all of the other studies in the literature, maximal oxygen consumption improvements were not affected by the simultaneous training. This was substantiated by almost identical improvements in two-mile run times. This suggests that even the inclusion of a strength training program well beyond the type that might be used for runners did not adversely affect the oxidative or running capabilities. Thus, no overtraining state was apparent for the endurance aspect of the individual and further supports the use by runners of a resistance training program (more sport-specific) to offset the tremendous pounding on the lower body musculature.

One of the more fascinating findings in the study by Kraemer et al. (21) was the changes that were observed in the muscle fiber characteristics of the thigh musculature. As shown in figure 4.4, a transformation of type II muscle sub-types occurred in the training groups. The transformation of type IIB to type IIA has been observed in prior studies (1, 41, 42). The transformation of type IIB to type IIA fibers was almost complete in the S and C groups. The UC and E groups, which only performed run training (including interval training) with the lower body musculature, showed that about 9% of the type IIB fiber sub-types remained after the three months of training. This indicates that heavy resistance training recruits more of the type IIB fiber populations than endurance training that even includes interval training (200-800 m distances). In addition, a small number (<3%) of type IIA fibers were converted to type IIC fibers in the groups that performed the endurance training. Of interest to an overtraining concept, the C group saw muscle fiber size increases only in the type IIA fibers while the S group demonstrated increases in type I, IIC, and IIA fiber types. The E groups saw exercise-induced decreases in the type I and IIC muscle fiber types, most likely due to the higher cortisol levels observed in the study for the E group and the physiological need at shorter distances to enhance oxygen kinetics in the capillary-to-cell movement of gases. It is also interesting that the UC group saw no changes in muscle fiber sizes. The subtle influence of only isometric force development for stabilizing the body for upper body exercises underscores the sensitivity of muscle fibers to resistance stimuli.

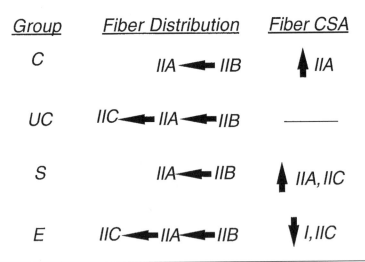

Figure 4.4   Transformation of muscle fiber sub-types for the training groups C, UC, S, and E after training. C = combination of total body weight training and running; UC = combination of upper body weight training and running; S = total body weight training only; E = endurance training only; CSA = cross-sectional area.

The data demonstrate that endurance training places significant regulatory controls on size changes in the type I muscle fibers. Since type I muscle fibers typically hypertrophy via a reduction in protein degradation, no reduction appears to have occurred over the three months.

Simultaneous training has been shown to result either in no changes or in increases in type I and type II muscle fiber areas (38, 42). Our data support the concept that muscle fiber-type area adaptations to simultaneous training differs from the single training mode adaptations. The fact that all muscle fibers hypertrophied in the S group (except for type II B fibers which were all transformed to type IIA) demonstrates that recruitment patterns followed typical size principle order with type I fibers being recruited with this type of program that included both hypertrophy and strength/power components (21). The lack of power development in the C group appears to be a function of neural mechanisms or unknown changes in the type IIA fibers because power typically is related to the fast-twitch fiber population. Thus, changes in fiber populations are different with simultaneous training of both strength and endurance. Furthermore, due to overuse injury some resistance training might be warranted for runners because the reduction in type I and type IIC fiber sizes was not indicative of any advantage in maximal oxygen consumption or 2-mile run times. These findings of size antagonism on the cellular level are unique. It appears the type I and type II muscle fibers were differentially responsible for the endurance and strength training adaptations in the C group. Type I muscle fibers in the C group did not hypertrophy in response to the strength training program, nor did they decrease in response to the endurance training program as was observed respectively in the S and E groups. Such an intermediate response of the type I muscle fibers and the inability of the type II muscle fibers to apparently compensate for the needed magnitude of hypertrophy required for some 1 RM strength and power performances indicates support for the hypothesis that strength, power, and endurance performance decrements may be influenced to some extent over 12 weeks of training due to differential muscle fiber adaptations.

Overtraining may start by the differential training effect at the level of the muscle fiber. The mechanisms that mediate such differential adaptations in muscle fiber areas remain speculative. In a recent study, Deschenes et al. (5) demonstrated soleus muscle fiber atrophy in endurance-trained rats. In addition, they observed differential alterations in the morphology of the neuromuscular junction (e.g., in the high-intensity trained group, more dispersed synapses, greater total length of branching) with different intensities of endurance training. Previous studies in humans have also shown decreases in muscle fiber size with endurance training (20, 43). Decreases in muscle fiber size and increased nerve cell branching and morphology may contribute to more optimal kinetics for oxygen utilization and innervation patterns promoting endurance capabilities (46). Conversely, such changes would be hypothesized to compromise muscle size and strength adaptations (4, 7). The lack of change in type I muscle fiber areas and the increase in type IIA muscle fiber areas solely in group C appears to be a cellular adaptation representative of the antagonism of simultaneous strength and endurance stimuli, since strength training alone produced increases in both type I and type II muscle fiber areas.

Testosterone and cortisol are representative of anabolic and catabolic hormones in the body and have been used to reflect training adaptations of the endocrine system (13, 22, 35). The training programs produced a different hormonal environment for muscle and nerve cells over the course of the training programs. Such differences in the hormonal environment can influence the cellular changes related to protein synthesis, neurotransmitter synthesis, and subsequent muscle fiber adaptations as well as substrate utilization and endurance capabilities (13, 17, 22-24, 29). In the study by Kraemer et al. (21), the C group demonstrated changes in both testosterone and cortisol over the training period. In the E group testosterone stayed constant but undulations occurred in the cortisol response over the training period. Conversely, the S group kept testosterone constant while producing a decrease in cortisol response. Such changes indicate that the overall anabolic and catabolic environment are driven by heavy resistance exercise and endurance exercise, respectively. Thus, incompatibility of training may be attributed to a large extent to the extreme stress of adrenal activation due to the total amount of high-intensity exercise, whether endurance or strength training. Whether or not successful adaptations can occur remains dependent upon the ability of various anabolic compensatory mechanisms (e.g., testosterone, insulin-like growth factors, growth hormone) to eventually override a catabolic environment (10, 22). This ability to overcome the catabolic environment was in part demonstrated by the UC group, which performed the upper body strength training program along with the endurance training program. By week 12, the UC group demonstrated a total cortisol exposure response (i.e., AUC) that was no different from the pretraining level. Not performing the lower body strength training program resulted in a reduction in the total work that was associated with the program. Similar to the S and E groups, in the UC group no changes occurred in the concomitant testosterone response over the 12 weeks of training. While no decrease or increase in the testosterone to cortisol ratio was observed, the training did not enhance the catabolic environment and may again have influenced the lack of changes in type I and type IIC muscle fiber areas.

Is incompatibility an overtraining phenomenon? It appears that the inappropriate use of supplemental training modes such as endurance exercise may have the potential to interfere with the optimal adaptational response of a given system (e.g., neuromuscular system) for a specific performance variable. If performance becomes depressed or plateaus well below expectations, it may be due to a training error in the volume of supplemental exercise performed. At present, only endurance and strength training compatibility modes of exercise have been combined to examine simultaneous training.

Sports Competition

Other factors such as extreme volumes of physical competition may also contribute to an overtraining state in a matter of days. In our laboratory we have observed elite collegiate wrestlers tolerate a loss of 6% body mass and maintain performance. These same wrestlers, however, dramatically lose isometric force capabilities of the grip strength and upper body strength over two days of tournament wrestling (unpublished data). Concomitantly, whole body power as measured via power production on a force plate demonstrates no significant changes. Such observations indicate that athletes may make special adaptations to the rigors of their sport, and then only certain performance characteristics may be susceptible to the further physical and emotional stress of competition.

Summary

It is apparent that for strength and power athletes overtraining is a complex multivariate environment where training mistakes can be made in sport-specific, supplemental training, in resistance training, and in training combinations. By being aware of the highly interactive nature of training and competition, a solution as simple as a few days of rest may prevent an overtraining problem (9). The periodization of training and sport competition starts to take on dramatic importance when one considers that intensity and volume of exercise and competition are two of the primary contributors to overtraining. Any program variable has the potential to contribute to an overtraining state but the variable can also be periodized if one is aware that it may be a problem (e.g., rest period lengths between sets and exercises and acid-base toleration). We are far from understanding all of the biological bases of overtraining but clues to the factors and variables in the configuration of a workout should help in the very careful detective work needed to intentionally manipulate (overreaching) or prevent it.

References

1. Adams, G.R., B.M. Hather, K.M. Baldwin, G.A. Dudley. 1993. Skeletal muscle myosin heavy chain composition and resistance training. *Journal of Applied Physiology* 74: 911-915.
2. Bell, G.J., S.R. Petersen, J. Wessel, K. Bagnall, H.A. Quinney. 1991. Physiological adaptations to concurrent endurance training and low velocity resistance training. *International Journal of Sports Medicine* 12: 384-390.
3. Callister, R., R.J. Callister, S.J. Fleck, G.A. Dudley. 1990. Physiological and performance responses to overtraining in elite judo athletes. *Medicine and Science in Sports and Exercise* 22: 816-824.
4. Chromiak, J.A. and D.R. Mulvaney. 1990. A review: the effects of combined strength and endurance training on strength development. *Journal of Applied Sport Science Research* 4: 55-60.
5. Deschenes, M. R., C.M. Maresh, J.F. Crivello, L.E. Armstrong, W.J. Kraemer, J. Covault. 1993. The effects of exercise training of different intensities on neuromuscular junction morphology. *Journal of Neurocytology* 22: 603-615.
6. Dudley, G.A. and R. Djamil. 1985. Incompatibility of endurance and strength training modes of exercise. *Journal of Applied Physiology* 59: 1446-1451.
7. Dudley, G.A. and S.J. Fleck. 1987. Strength and endurance training: are they mutually exclusive? *Sports Medicine* 4: 79-85.
8. Dudley, G.A., P.A. Tesch, B.J. Miller, P. Buchanan. 1991. Importance of eccentric actions in performance adaptations to resistance training. *Aviation, Space and Environmental Medicine* 62: 543-550.
9. Fleck, S.J., W.J. Kraemer. 1996. *Periodization breakthrough!* Ronkonkona: Advanced Reach Press.
10. Fleck, S.J. and W.J. Kraemer. 1997. *Designing resistance training programs.* Champaign, IL: Human Kinetics.

11. Fry, A.C., W.J. Kraemer, J. M. Lynch, N.T. Triplett, L.P. Koziris. 1994. Does short-term near-maximal intensity machine resistance training induce overtraining? *Journal of Strength and Conditioning Research* 8: 188-191.

12. Fry, A.C., W.J. Kraemer, F. van Borselen, J.M. Lynch, J.L. Marsit, E. Pierre Roy, N.T. Triplett, H.G. Knuttgen. 1994. Performance decrements with high-intensity resistance exercise overtraining. *Medicine and Science in Sports and Exercise* 26: 1165-1173.

13. Galbo, H. 1983. *Hormonal and metabolic adaptation to exercise.* New York: Thieme-Stratton.

14. Gordon, S.E., W.J. Kraemer, N.H. Vos, J.M. Lynch, H.G. Knuttgen. 1994. Effect of acid-base balance on the growth hormone response to acute high-intensity cycle exercise. *Journal of Applied Physiology* 76:821-829.

15. Hennessy, L.C., A.W.S. Watson. 1994. The interference effects of training for strength and endurance simultaneously. *Journal of Strength and Conditioning Research* 8: 12-19.

16. Hickson, R.C. 1980. Interference of strength development by simultaneously training for strength and endurance. *European Journal of Applied Physiology* 45: 255-269.

17. Hickson, R.C., K. Hidaka, C. Foster, M.T. Falduto, T.T. Chatterton Jr. 1994. Successive time courses of strength development and steroid hormone responses to heavy-resistance training. *Journal of Applied Physiology* 76: 663-670.

18. Hortobágyi, T., F.I. Katch, P.F. LaChance. 1991. Effects of simultaneous training for strength and endurance on upper and lower body strength and running performance. *Journal of Sports Medicine and Physical Fitness* 31: 20-30.

19. Hunter, G., R. Demment, D. Miller. 1987. Development of strength and maximum oxygen uptake during simultaneous training for strength and endurance. *Journal of Sports Medicine and Physical Fitness* 27: 269-275.

20. Klausen, K., L.B. Anderson, I. Pelle. 1981. Adaptive changes in work capacity, skeletal muscle capillarization and enzyme levels during training and detraining. *Acta Physiologica Scandinavica* 113: 9-16.

21. Kraemer, W.K., J.F. Patton, S.E. Gordon, E.A. Harman, M.R. Deschenes, K. Reynolds, R.U. Newton, N.T. Triplett, J.E. Dziados. 1995. Compatibility of high-intensity strength and endurance training on hormonal and skeletal muscle adaptations. *Journal of Applied Physiology* 78: 976-989.

22. Kraemer, W.J. 1992. Endocrine responses and adaptations to strength training. In *The encyclopedia of sports medicine: strength and power*, ed. P.V. Komi, 291-304. Oxford: Blackwell Scientific.

23. Kraemer, W.J., J.E. Dziados, L.J. Marchitelli, S.E. Gordon, E.A. Harman, R. Mello, S.J. Fleck, P.N. Frykman, N.T. Triplett. 1993. Effects of different heavy-resistance exercise protocols on plasma ß-endorphin concentrations. *Journal of Applied Physiology* 74: 450-459.

24. Kraemer, W.J., S.J. Fleck, R. Callister, M. Shealy, G. Dudley, C.M. Maresh, L. Marchitelli, C. Cruthirds, T. Murray, J.E. Falkel. 1989. Training responses of plasma beta-endorphin, adrenocorticotropin and cortisol. *Medicine and Science in Sports and Exercise* 21: 146-153.

25. Kraemer, W.J., A.C. Fry. 1994. Strength testing: development and evaluation of methodology. In *Physiological assessment of human fitness*, eds. P. Maud, C. Foster. Champaign, IL: Human Kinetics.

26. Kraemer, W.J., S.E. Gordon, S.J. Fleck, L.J. Marchitelli, R. Mello, J.E. Dziados, K. Friedl, E. Harman, C. Maresh, A.C. Fry. 1991. Endogenous

anabolic hormonal and growth factor responses to heavy resistance exercise in males and females. *International Journal of Sports Medicine* 12: 228-235.

27. Kraemer, W.J., B.J. Noble, B.W. Culver, M.J. Clark. 1987. Physiologic responses to heavy-resistance exercise with very short rest periods. *International Journal of Sports Medicine* 8: 247-252.

28. Kraemer, W.J., L. Marchitelli, D. McCurry, R. Mello, J.E. Dziados, E. Harman, P. Frykman, S.E. Gordon, S.J. Fleck. 1990. Hormonal and growth factor responses to heavy resistance exercise. *Journal of Applied Physiology* 69: 1442-1450.

29. Kraemer, W.J., J.F. Patton, H.G. Knuttgen, L.J. Marchitelli, C. Cruthirds. 1989. Pituitary-adrenal responses to short-term high intensity cycle exercise. *Journal of Applied Physiology* 66: 161-166.

30. McCarthy, J.P., J.C. Agre, B.K. Graf, M.A. Pozniak, A.C. Vailas. 1990. Compatibility of adaptive responses with combining strength and endurance training. *Medicine and Science in Sports and Exercise* 27: 429-436.

31. Mitchell, J.H., B.J. Sprule, C.B. Chapman. 1958. The physiological meaning of the maximal oxygen intake test. *Journal of Clinical Investigations* 37: 538-547.

32. Murphy, M.M., J.F. Patton, F.A. Frederick. 1986. Comparative anaerobic power of men and women. *Aviation, Space and Environmental Medicine* 57: 636-641.

33. Nelson, G.A., D.A. Arnall, S.F. Loy, L.J. Silvester, R.K. Conlee. 1990. Consequences of combining strength and endurance training regimens. *Physical Therapy* 70: 287-294.

34. Newton, R.U., W.J. Kraemer, K. Häkkinen, B.J. Humphries, A.J. Murphy. 1996. Kinematics, kinetics, muscle activation during explosive upper body movements. *Journal of Applied Biomechanics* 12: 31-43.

35. Nindl, B.C., K.E. Friedl, P.N. Frykman, L.J. Marchitelli, R.L. Shippee, J.F. Patton. n.d. Physical performance and metabolic recovery among lean, healthy males following a prolonged energy deficit. *International Journal of Sports Medicine* in press.

36. Patton, J.F., W.J. Kraemer, H.G. Knuttgen, E.A. Harman. Factors in maximal power production and in exercise endurance relative to maximal power. 1990. *European Journal of Applied Physiology* 60: 222-227.

37. Perrine, J.J., V.R. Edgerton. 1978. Muscle force-velocity and power-velocity relationships under isokinetic loading. *Medicine and Science in Sports* 10:159-166.

38. Sale, D.G., I. Jacobs, J.D. MacDougall, S. Garner. 1990. Comparison of two regimens of concurrent strength and endurance training. *Medicine and Science in Sports and Exercise* 22: 348-356.

39. Sale, D.G., J.D. MacDougall, I. Jacobs, S. Garner. 1990. Interaction between concurrent strength and endurance training. *Journal of Applied Physiology* 68: 260-270.

40. Selye, H. 1976. *The stress of life.* New York: McGraw-Hill.

41. Staron, R.S., R.S. Hikida, F.C. Hagerman. 1983. Myofibrillar ATPase activity in human muscle fast-twitch subtypes. *Histochemistry* 78: 405-408.

42. Staron, R.S., D.L. Karapondo, W.J. Kraemer, A.C. Fry, S.E. Gordon, J.E. Falkel, F.C. Hagerman, R.S. Hikida. 1994. Skeletal muscle adaptations during the early phase of heavy-resistance training in men and women. *Journal of Applied Physiology* 76: 1247-1255.

43. Terados, N., J. Melichna, C. Sylven, E. Jansson. 1986. Decrease in skeletal muscle myoglobin with intensive training in man. *Acta Physiologica Scandinavica* 128: 651-652.

44. Tharion, W.J., E.A. Harman, W.J. Kraemer, T.M. Rauch. 1991. Effect of different resistance exercise protocols on mood states. *Journal of Strength and Conditioning Research* 5:60-65.

45. Volpe, S.L., J. Walberg-Rankin, K.W. Rodman, D.R. Sebolt. 1993. The effect of endurance running on training adaptations in women participating in a weightlifting program. *Journal of Strength and Conditioning Research* 7: 101-107.

46. Weibel, E.R., C.R. Taylor, P. Gehr, H. Hoppeler, O. Mathieu, G.M.O. Maloi. 1991. Design of the mammalian respiratory system. IX. Functional and structural limits for oxygen flow. *Respiratory Physiology* 44: 151-164.

CHAPTER 5

Increased Training Volume in Strength/Power Athletes

Michael H. Stone, PhD, and Andrew C. Fry, PhD

Introduction

The volume of training is important for various reasons including a relationship to improved performance, changes in body composition, and health parameters (45). Additionally, the volume of training can be associated with both overreaching and overtraining (13, 46). Fry et al. (13) and Stone et al. (46) have suggested that overtraining signs and symptoms may represent a continuum that is related to changes in training volume (see figure 5.1). However, much of the data concerning overtraining and overreaching is based on observations or studies involving aerobically trained athletes. Few studies have investigated the effect of increased training volume on weightlifting performance.

It has been suggested that performance may be enhanced by short periods of overtraining or overreaching (13, 28, 46). For strength and power athletes overreaching typically consists of a few weeks (1-3) of increased training volume or intensity resulting in a greater training volume load (repetitions × mass lifted). Although overreaching may produce some symptoms of chronic overwork including decreased performance, it is not so severe that recovery cannot be accomplished within a few days after returning to normal training (13, 28, 46). The increased short-term training volume load causes physiological adaptations leading eventually to increased performance after returning to normal training volume loads. Theoretically, 2-5 weeks after returning to normal training, a delayed increase in performance can occur (13, 28, 46). This increased performance may be enhanced by using a training taper (46).

Performance enhancement is the primary goal of sport training. The purpose of this chapter is to present experimental evidence that weightlifting performance may be enhanced by manipulating training volume consistent with the concept of

Progression and symptom severity

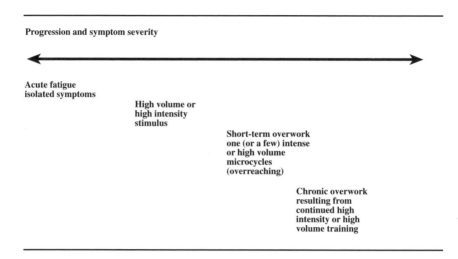

Figure 5.1   Overtraining symptom continuum.

overreaching. Additionally, data will be presented pertaining to the physiological adaptations associated with volume manipulation.

Overview and Methodology

An observation of elite junior weightlifters was undertaken in order to study the physiological and performance effects of training volume manipulation. Twenty-four elite junior weightlifters (ages 14-20 years) participated in this observation. Descriptive data over the course of the study are shown in table 5.1. At the junior nationals, these 24 athletes finished among the top three in their respective weight classes and age groups. All 24 weightlifters were national-class and several were world-class level athletes. The athletes participated in a month-long training camp at the USOC training center in Colorado Springs. Their training and test regimen is shown in figure 5.2. The protocol for a test day is shown in figure 5.3.

The testing exercise protocol (see figure 5.3) was used because previous observation had suggested that high intensity exercise with an endurance component (anaerobic challenge) may more adequately reflect overtraining symptoms (5, 55). Additionally, this protocol was chosen because of its observed practicality and efficacy in a pilot study (55).

Performance variables were measured using standard techniques. Measurements of blood-borne variables were performed using accepted analytical procedures. Lactate and ammonia were measured at the USOC sport science laboratory. Glucose was measured in the department of nutrition and foods at Auburn University under the supervision of Dr. Robert Keith. Hormone analysis was performed using RIA methods at the Center for Sports Medicine, Pennsylvania State University, under the supervision of Dr. William Kraemer. The coefficient of variation was within acceptable limits for all blood-borne variables. Dietary records were collected three times during the experiment. The

Table 5.1 Physical Changes Over 1 Month of Training (N=24)

	Age (yr)	Height (cm)	Body mass (kg)	Body fat (%)	Lean body mass (kg)
T1	17.7 ± 1.6	171.0 ± 10.0	74.3 ± 15.9	7.9 ± 4.9	67.6 ± 11.9
T2	17.7 ± 1.6	171.0 ± 10.0	73.7 ± 16.2	7.9 ± 5.0	67.4 ± 12.1
T3	17.9 ± 1.6	171.0 ± 10.0	74.0 ± 16.1	8.0 ± 5.0	67.5 ± 11.8

Data are means ± standard deviations.
No significant differences were observed.

Training/Testing Protocol

Day	Protocol description
1	Subject meeting—explanation of study, training, obtain informed consents
2-4	Normal training
5	**T1 Testing**
6-11	Increased volume and intensity of training. The volume of training was approximately doubled over typical training for this group.
12	**T2 Testing**
13-26	Normal training
27	**T3 Testing**

Figure 5.2   Training and testing protocol.

24-h dietary records were collected on Wednesday and Saturday of weeks two, three, and four. Proper collection of dietary records, including estimation of food portion sizes, was discussed with the subjects the day before the first collection day. An experimenter was present at all meals to aid the subjects in collecting their dietary records. Diet was analyzed for total kcals and the percentage of protein, carbohydrate, and fat in the diet.

Training for these athletes consisted primarily of large muscle mass exercises made up of various types of squats, overhead lifts, the snatch, clean and jerk, and their variations. During normal training consisting of 0-2 workouts per day, the volume of training as measured by the volume load (repetitions × mass lifted) ranged from 5,000 to 12,000 kg per training day (days 2-3 and days 13-26) with a relative intensity (% 1 RM) of approximately 79% for the target sets. During the high-volume phase, 3-4 workouts per day were used and the volume of training ranged from 15,000 to 35,000 kg per training day (days 6-11) with a relative intensity of approximately 76% for the target sets.

Test Day Protocol (Days 5 (T1), 12 (T2), and 27 (T3))

Rest - 7:00 a.m. (after 12 hour fast)

1. Measure seated heart rate (HR) and blood pressure (BP) after a 5-minute rest.
2. Baseline venous blood collection (30 ml)
3. Body composition (skinfolds)

Lunch

Pretesting - 2:30 p.m.

1. Measure standing HR and BP after a 5-minute rest.
2. Preexercise venous blood collection (30 ml)
3. Vertical jump test (standard measuring techniques using a Vertec)
   a. Warm-up using stretching and 2-3 practice jumps
   b. 15 maximal effort vertical jumps—determine the height of the 1st, 3rd, and 15th jump with one jump performed every 3 seconds.
4. Snatch test
   a. Warm-up (1 × 10 repetitions, 40% of 1 repetition maximum (1 RM) snatch)
   b. Begin with 50% of 1 RM snatch, perform 1 snatch every 15 seconds until 2 consecutive misses, increase 5 kg between repetitions.
5. 3 × 10 repetitions, snatch pulls with 65% of 1 RM snatch, perform 1 snatch every 4 seconds
6. **Posttesting**—Immediate postexercise HR
7. **5-Posttesting**—Postexercise (5 minutes) venous blood collection (30 ml)
8. **15-Posttesting**—Repeat #6-7 at 15 minute postexercise

Figure 5.3   Test day protocol for day 5 (T1), day 12 (T2), and day 27 (T3).

Training Adaptations

It should be pointed out that no control group was available for the observations concerning the training program. Thus, the subjects served as their own controls. These were elite athletes, however, well trained and familiar with the training exercises and the tests, which increases the validity of the observations and conclusions. Dietary records showed no significant changes in dietary factors and indicated that diet was not a factor responsible for changes observed in performance or physiological variables.

The purpose of the training program design was to induce overreaching. The training program included a one-week high-volume phase (see figure 5.2). Performance may have been expected to remain unchanged or decrease from T1, day 5, until T2, day 12, due to the increased demands of the program increasing fatigue levels (13, 18, 46); however, this did not occur. Performance as measured by the snatch tests and vertical jump (VJ) tests improved steadily over the month's training period (see tables 5.2 , 5.3, and 5.4). Previous experimentation did not produce performance improvements after a week of high-volume training in a similar group of subjects (55). The reasons for the difference in outcome of the two studies are unknown but may include differences in the trained state of the subjects prior to beginning the high-volume training phase.

Table 5.2 Performance Variables: Snatch Test Results in kg (N=24)

	T1	T2	T3
Snatch	89.8 ± 15.5	91.7 ± 13.0	93.2 ± 16.2
Reps/fail	8.9 ± 1.3	9.2 ± 1.5	9.6 ± 1.4

Data are means ± standard deviations.
A significant time effect (p ≤ .05) for snatch and reps/fail was observed; snatch: T1 < T3; reps: T1 < T3.

Table 5.3 Performance Variables: Average Vertical Jump (VJ) Values for VJ 1, 3, and 15 Expressed in cm (N=24)

	VJ1	VJ3	VJ15
T1	56.9 ± 7.6	56.9 ± 6.7	53.2 ± 6.2
T2	58.7 ± 8.3	58.7 ± 7.1	54.9 ± 7.1
T3	56.6 ± 8.2	58.7 ± 7.7	54.8 ± 7.6

Data are means ± standard deviations.
A significant time effect (p ≤ .05) was observed: T1 < T2 and T3 for each trial.

Table 5.4 Performance Variables: Average Vertical Jump Power Index (VJP) Values for Vertical Jump 1, 3, and 15 Expressed in kgm/sec (N=24)

	VJP1	VJP3	VJP15
T1	123.4 ± 27.3	123.3 ± 26.3	119.0 ± 24.2
T2	123.0 ± 26.3	124.5 ± 27.5	119.7 ± 24.2
T3	124.1 ± 27.7	124.7 ± 27.0	120.3 ± 25.2

Data are means ± standard deviations.
A significant time effect (p ≤ .05) was observed: T1 < T2; T1 < T3 for each trial.

The increase in performance variables was accomplished without change in body mass or body composition (see table 5.1). This would suggest that the increased performance was the result of neural factors. A learning effect as the sole cause of increased performance is unlikely because the subjects were very familiar with the tests performed, and in fact regularly performed the snatch and VJ or variations of these exercises in training. However, it is possible that much of the improvement noted was due to increased neural activation of motor units as a result of the experimental training program (21, 39).

Heart Rate and Blood Pressure

Although blood pressure did not change significantly, heart rate (HR) showed steady decreases at rest and postexercise across the month's training program (see table 5.5). This agrees with other observations of weight training-induced decreases in HR (19, 36, 47). Decreases in HR may be due to training-induced changes in sympathetic and parasympathetic input and probably do not reflect changes in ventricular volume (10, 45). The decreased HR postexercise suggests a beneficial adaptation to the exercise performed. No significant differences were observed in systolic or diastolic blood pressure responses (see tables 5.6 and 5.7).

Blood-Borne Variables

Resting and preexercise (pre) blood lactate concentrations (see table 5.8) showed a slight increase at T2. The reason for an elevated resting lactate is unknown. Postexercise lactic acid concentrations showed the expected decrease over time due to training (36, 55). The total decrease (T1, day 5, to T3, day 27) for postexercise values was 29% for the five-minute postexercise (5-post) value, and 40% for the 15-minute postexercise (15-post) values. By T2, 65% of the total 5-post and 90% of the total 15-post decrease had occurred. It is possible that a lower postexercise lactate on T2 may have represented a symptom of overwork as a result of chronic

Table 5.5 Heart Rate Responses Expressed in Beats/Min (N=24)

	Rest	Pre	5-Post	15-Post
T1	74.6 ± 7.3	86.9 ± 9.5	121.7 ± 10.2	112.0 ± 9.5
T2	71.2 ± 11.7	78.6 ± 10.6	112.3 ± 11.3	97.7 ± 13.3
T3	62.3 ± 8.3	77.5 ± 12.1	108.3 ± 12.1	97.8 ± 11.7

Data are means $\pm$ standard deviations.
Exercise Effects: Significant time effect (p ≤ .05): 5-post and 15-post > rest and pre. Significant test day × time effect (p ≤ .05): 5-post > rest and pre on T1, T2, and T3; 15-post > rest on T1, T2, and T3; 15-post > pre on T1 and T3.
Training Effects: Significant time effect (p ≤ .05): T1 > T2 > T3. Significant test day × time effect (p ≤ .05), all times: T1 > T2 > T3.

Table 5.6 Systolic Blood Pressure Responses (mmHg) to Exercise (N=24)

	Rest	15-Post
T1	113.4 ± 11.0	119.9 ± 12.0
T2	112.5 ± 9.6	122.7 ± 10.0
T3	111.6 ± 10.0	118.5 ± 11.7

Data are means ± standard deviations.
No significant differences.
T2 > T3 (p = .07).

Table 5.7 Diastolic Blood Pressure Responses to Exercise (N=24)

	Rest	15-Post
T1	76.2 ± 8.5	80.7 ± 8.6
T2	75.0 ± 8.1	81.5 ± 5.8
T3	73.5 ± 8.7	78.9 ± 6.9

Data are means ± standard deviations.
No significant differences.
T2 > T3 (p = .08).

glycogen depletion (41, 46); however, this is not probable because (1) the subjects were eating sufficient carbohydrates, (2) performance did not decrease at T2, and (3) no aberrant measurements suggesting overwork were noted for the other variables. Thus, the postexercise reductions in lactate represent beneficial adaptations to the exercise performed.

Ammonia can exert a toxic effect on the bioenergetic and nervous systems and may be associated with fatigue (46). Thus, changes in ammonia concentration (see table 5.9) as a result of training could alter the physiological responses to exercise and performance capabilities of the athlete. Ammonia concentrations tended to be lower on T2 at all measurement times. In a previous observation with a similar group of subjects ammonia tended to rise at rest as a result of a week of high-volume training (55). However, in both the present study and in the study of Warren et al. (55), all fasting values were within the normal range (4) and likely did not contribute to excessive fatigue associated with chronic overwork (46). Postexercise concentrations (5-post and 15-post) were markedly lower by T2 and this trend continued through T3. Lower ammonia concentrations as a result of training may result from increased urea production, increased buffering capability, decreased sympathetic activity, or a decreased ammonia production (8, 20, 54). The lower postexercise concentrations suggest a beneficial training adaptation (4).

Table 5.8 Lactate Responses Expressed in mmol/L (N=24)

	Rest	Pre	5-Post	15-Post
T1	0.8 ± 0.2	1.0 ± 0.3	8.0 ± 1.9	6.2 ± 2.2
T2	1.4 ± 0.1	1.3 ± 0.6	6.6 ± 1.4	3.9 ± 1.2
T3	0.4 ± 0.2	0.7 ± 0.2	5.7 ± 1.3	3.6 ± 1.3

Data are means ± standard deviations.
Exercise Effects: Significant time effect (p ≤ .05): 5-post and 15-post > rest and pre. Significant test day × time effect (p ≤ .05): 5-post and 15-post > rest and pre on T1, T2, and T3.
Training Effects: Significant time effect (p ≤ .05): T1 > T2 > T3; significant test day × time effect (p ≤ .05): T2 > T3; 5-post and 15-post: T1 >T2 and T3.

Table 5.9 Ammonia Responses Expressed in μmol/L (N=24)

	Rest	Pre	5-Post	15-Post
T1	67.6 ± 16.4	45.5 ± 21.6	130.9 ± 46.2	80.5 ± 34.0
T2	51.6 ± 12.1	36.0 ± 14.7	81.5 ± 19.8	55.5 ± 17.3
T3	75.8 ± 26.5	93.3 ± 24.2	111.2 ± 26.2	47.6 ± 16.0

Data are means ± standard deviations.
Exercise Effects: Significant time effect (p ≤ .05): 5-post > pre. Significant test day × time effect (p ≤ .05): 5-post > rest and pre on T1, T2, and T3; 15-post < rest on T3.
Training Effects: Significant time effect (p ≤ .05): T1 and T3 > T2; significant test day × time effect (p ≤ .05): pre: T3 > T1 and T2; 5-post: T1 > T2; 15-post: T1 > T2.

Blood glucose concentrations can remain unchanged (14, 27) or increase as a result of weight training exercise (33). Differences in blood glucose response to resistance training exercise may be related to the exercise protocol and the size of the muscle mass engaged. Both Kraemer et al. (27) and Guezennec et al. (14) used smaller muscle mass exercises compared to those used in the McMillan et al. (33) study or in the present study. Although the exercise protocol in the present study was similar to that used by McMillan et al. (33), blood glucose responses were variable (see table 5.10). Blood glucose did not show the expected postexercise increase on T2. Blood glucose concentrations respond to changes in cortisol and catecholamine concentrations (6, 34).

Postexercise cortisol concentrations were markedly reduced at T2 and showed little further change by T3. Although catecholamines were not measured in this study they are known to rise in response to weight training exercise (33). Plasma epinephrine at rest and after submaximum (6 × 8 repetitions at 70% 1 RM) and maximum (1 set to exhaustion at 70% 1 RM) bench pressing has been shown to decrease after two months of training (14). It is possible that the high-volume week of training in the present study decreased the exercise catecholamine response and therefore contributed to decreased postexercise blood glucose concentrations on T2.

In the present study resting free fatty acid (FFA) concentrations (see table 5.11) were higher than those observed in older (age 26) weightlifters (33). Fasting FFA concentrations have been observed to be higher in weightlifters than sedentary controls (20, 33), a result that is likely caused by chronic weight training rather than by diet (33). McMillan et al. (33) noted that resting FFA concentrations can be significantly elevated after a single training session as long as 20 h later. A prolonged exercise-induced elevation of FFA likely contributes to the elevated fasting FFA values. Weight training exercise-induced increases in FFA are consistent with insulin antagonism and increased concentrations of catecholamines, growth hormone (hGH), glucagon, and cortisol as noted by McMillan et al. (33), and the rise in hGH and cortisol concentrations observed postexercise in the present study.

Table 5.10 Blood Glucose Responses Expressed in mg/dl (N=24)

	Rest	Pre	5-Post	15-Post
T1	79.6 ± 12.7	78.4 ± 9.0	90.3 ± 17.7	89.4 ± 14.9
T2	92.5 ± 4.6	89.2 ± 9.2	85.7 ± 22.2	76.5 ± 13.6
T3	61.4 ± 5.0	68.0 ± 10.4	67.7 ± 13.4	75.2 ± 12.2

Data are means ± standard deviations.
Exercise Effects: No significant time effect. Significant test day × time effect (p ≤ .05): 15-post > pre on T1 and T3.
Training Effects: Significant time effect (p ≤ .05): T1 and T2 > T3. Significant test day × time effect (p ≤ .05); rest: T2 > T3.

Table 5.11 Free Fatty Acid Responses Expressed in mg/L (N=24)

	Rest	Pre	5-Post	15-Post
T1	153 ± 79	111 ± 51	150 ± 51	111 ± 41
T2	201 ± 76	105 ± 45	151 ± 50	109 ± 30
T3	96 ± 46	90 ± 52	127 ± 47	128 ± 44

Data are means ± standard deviations.
Exercise Effects: Significant time effect (p ≤ .05): 5-post > pre. Significant test day × time effect (p ≤ .05): 5-post > pre on T1, T2, and T3; 15-post > pre on T3.
Training Effects: Significant time effect (p ≤ .05): T2 > T3. Significant test day × time effect (p ≤ .05); rest: T2 > T3.

A single session of weight training can result in considerable muscle glycogen depletion, especially in type II muscle fibers (30, 37). Many sports use multiple high-intensity training sessions per day, similar to that used in the present study, that could result in a marked glycogen depletion. A postexercise elevation of FFA (resulting in a chronic resting elevation) and a resultant use of FFA as an energy substrate would be important during recovery from weight training exercise (33) or other high-intensity forms of training. A sufficient use of FFA for energy during recovery (33) may enhance the loss of body fat and may at least partially explain the low body fats observed among many weight trainers including the subjects in the present study. It has been speculated that the degree of increased fasting FFA concentration may be related to the volume of training (33, 48). Fasting (resting) FFA concentrations were somewhat higher on T2 compared to T3, suggesting that high-volume weight training increased fasting FFA concentrations compared to a normal volume of training. It is possible that higher volumes of training (if training is prolonged) such as that used in the present study, could markedly contribute to body fat losses.

Growth hormone affects lipid and carbohydrate metabolism, stimulates the production of adipolytic lipase, and generally promotes a positive nitrogen balance and protein synthesis (11, 15, 23, 24, 38). Many of the protein synthesizing properties of hGH occur through hGH-induced increases in insulin-like growth factor (IGF$_1$) (11, 23). Exercise can result in hGH concentrations increasing 20 to 40 times depending upon the trained state of the subject, type of exercise (large versus small muscle mass), and the duration and intensity of the exercise (24, 31, 42, 51).

Large muscle mass exercises and higher repetitions per set, as used in the testing protocol of this study, produce higher hGH postexercise increases than smaller muscle mass exercises (22, 24). In the present study, weight training exercise increased hGH (see table 5.12) postexercise; however, the postexercise response diminished over time (T1 to T3). Of the total postexercise decrease in hGH, 68% of the 5-post and 76% of the 15-post occurred by T2, suggesting that the volume of weight training may be an important factor in long-term hGH adaptations. A decreased hGH response to weight training exercise is consistent with the observation of McMillan et al. (33) showing a lower hGH response in weightlifters compared to untrained controls. This diminished hGH response may reflect a greater hormone sensitivity (33). While hGH increases with exercise, its exact role in promoting increased protein synthesis is unclear (24). The increased hGH likely does play a significant role in mobilizing FFA (33).

Although fasting concentrations of hGH were unchanged across time, resting concentrations of IGF$_1$ (see table 5.13) were highest at T2, which were significantly higher than at T3. It is possible that the high-volume training stimulated increased IGF$_1$ resting concentrations in some manner. The mechanisms controlling IGF$_1$ concentrations (production and clearance) are unclear. A direct stimulation of IGF$_1$ by hGH may have occurred through increases in hGH peaks, which were undetected by the protocol used in this study. Additionally, some evidence suggests that IGF$_1$ concentrations could respond to synergistic effects of non-growth hormones or exercise-induced disruption of cells such as adipose or muscle cells (26). It is possible that the higher volume of weight training from T1 until T2 may have increased cellular disruption and may have affected increases in IGF$_1$ fasting concentration; the subsequent return to low-volume training (T2 until T3) caused less cellular disruption allowing IGF$_1$ concentrations to decrease. The increased IGF$_1$ could affect increased protein synthesis. If the elevated IGF$_1$

Table 5.12 Growth Hormone (hGH) Response Expressed in μg/L (N=24)

	Rest	Pre	5-Post	15-Post
T1	0.2 ± 0.2	2.0 ± 3.5	16.1 ± 11.9	16.2 ± 13.3
T2	0.2 ± 0.1	1.7 ± 2.5	11.2 ± 10.4	9.3 ± 7.8
T3	0.1 ± 0.2	1.3 ± 4.8	8.9 ± 8.0	7.1 ± 5.5

Data are means ± standard deviations.
Exercise Effects: Significant time effect: pre, 5-post, and 15-post > rest; 5-post and 15-post > pre. Significant test day × time effect: pre, 5-post, and 15-post > rest on T1, T2, and T3; 5-post and 15-post > pre on T1, T2, and T3.
Training Effects: Significant time effect: T1 > T2 and T3. Significant test day × time effect: 5-1post: T1 > T3; 15-post: T1 > T2 and T3.

Table 5.13 Insulin-Like Growth Factor (IGF$_1$) Responses Expressed in nmol/L (N=24)

	Rest
T1	33.3 ± 6.1
T2	34.6 ± 6.5
T3	30.9 ± 7.7

Data are means ± standard deviations.
Significant time effect: T2 > T3.

persisted as a result of longer periods of high-volume training, then gains in lean body mass may occur.

Testosterone plays an important role in stimulating protein synthesis, may produce anticatabolic effects by interacting with cortisol receptors (32), influences neural factors, possibly promotes a greater glycogenolytic profile in type II muscle fibers (24), and promotes glycogen resynthesis during recovery by stimulating production of glycogen synthetase (1, 3, 29). Thus, exercise responses and training adaptations in testosterone concentrations could alter a variety of factors related to increased performance.

The exercise-induced increases in testosterone (see table 5.14) are in agreement with the increases noted resulting from other forms of high-intensity exercise (24). As with hGH, large muscle mass exercises as used in this study produce larger postexercise increases than small muscle mass exercises (24, 26). However, weight training does not always elicit an increase in testosterone in this age group.

Fahey et al. (9) found that testosterone concentrations were not altered in high school students (16 ± 0.8 y) as a result of performing five sets of five repetitions of

Table 5.14 Testosterone (T) Response Expressed in nmol/L (N=24)

	Rest	Pre	5-Post	15-Post
T1	24.8 ± 9.3	12.7 ± 4.5	18.0 ± 7.0	16.2 ± 7.4
T2	26.9 ± 9.7	19.1 ± 6.5	22.1 ± 10.0	20.8 ± 10.1
T3	27.9 ± 6.9	20.6 ± 7.6	25.4 ± 7.9	21.9 ± 6.9

Data are means $\pm$ standard deviations.
Exercise Effects: Significant time effect: pre < 5-post and rest. Significant test day $\times$ time effect: pre < 5-post and rest on T1 and T3.
Training Effects: Significant time effect: T1 < T2 and T3; pre: T1 < T3; 5-post: T1 < T3; 15-post: T1 < T3.

the dead lift. Differences between that study (9) and the present study may be related to weight training experience or the trained state (12, 26).

Data from the present study suggest that both resting and postexercise values of testosterone can be increased as a result of weight training. Of the total increase in testosterone (T1-T3), most of the increase occurred by T2. For example, 80% of the pre, 56% of the 5-post, and 81% of the 15-post total increase occurred by T2 suggesting that the high-volume training has a greater effect on increasing resting and postexercise concentrations of testosterone than low-volume training. This chronic increase in testosterone concentration would likely increase the exposure of receptors to higher hormonal concentrations and suggests that protein anabolism may be stimulated. It should be noted that prolonged high-volume training may reduce resting testosterone concentrations and reduce protein synthesis stimulation (17).

Cortisol suppresses the immune system, effects carbohydrate, fat, and protein metabolism, stimulates gluconeogenesis (34, 42), antagonizes testosterone production (7, 56), and generally has catabolic effects (24). High-volume resistance exercise can cause marked increases in cortisol (23-25), especially when large muscle mass exercises are used (33, 36). Resistance or other exercise-induced elevated cortisol concentrations may remain above baseline for longer than an hour (33, 53). The exercise-induced increases in cortisol concentrations noted in this study (see table 5.15) diminished over time (T1-T3). Diminished postexercise cortisol concentration agrees with the observation of lower concentrations postsubmaximal exercise in both aerobically-trained subjects (51) and in weight trained subjects (33, 36) compared to untrained subjects. It is unclear what effect increased cortisol concentrations may have during weight training exercise; however, postexercise increases may be involved in the recovery mechanism (33), causing increased FFA mobilization or inhibition of the immune or inflammatory response system (34). A diminished exercise response would then reflect relatively small changes in homeostasis as a result of the exercise session.

A state of hypercortisolism has been noted among well-trained distance runners, especially during precompetition (53). Additionally, resting cortisol concentrations have been shown to increase as a result of increased weight training volume (17). In the present study resting concentrations increased significantly by T2

Table 5.15 Cortisol Response Expressed in nmol/L (N=24)

	Rest	Pre	5-Post	15-Post
T1	503 ± 109	280 ± 78	546 ± 153	610 ± 163
T2	585 ± 149	336 ± 102	476 ± 176	570 ± 183
T3	573 ± 120	326 ± 98	472 ± 148	521 ± 111

Data are means ± standard deviations.
Exercise Effects: Significant time effect: pre < 5-post, 15-post, and rest. Significant test day × time effect: pre < 5-post, 15-post, and rest on T1, T2, and T3.
Training Effects: Significant time effect: T1 > T2 and T3. Significant test day × time effect: rest: T1 < T2 and T3; pre: T1 < T2; 15-post: T3 < T1.

and remained elevated until T3 and preexercise concentrations increased from T1 to T2 but returned toward initial values at T3. It is possible that the elevated resting and preexercise values are an adaptation to added physiological stress (i.e., chronic disturbance of homeostasis) associated with large muscle mass high-volume resistance training. The increase in cortisol as a result of high-intensity or high-volume training has been speculated to be due to increased corticotropin-releasing hormone secretion (52). Relatively large increases in resting cortisol, especially if prolonged over several weeks, may result in a diminished anabolic state leading to an increase in protein catabolism (16), which can be associated with the overtrained state (46). However, hypercortisolism may also produce beneficial effects in terms of exercise response. Some increase in resting (or preexercise) cortisol concentrations may represent a partial adaptation to upcoming physical stress such as competition or a training session (53).

The testosterone/cortisol (T/C) ratio has been correlated to lean body mass (LBM) and to measures of maximum weightlifting performance (16, 17). These relationships may partially explain and characterize the importance of the balance between anabolic activity (testosterone) and catabolic activity (cortisol) during prolonged weight training in weightlifters or other strength and power athletes (2, 17, 24). In the present study (see table 5.16) the T/C ratio decreased as a result of exercise, largely due to marked increases in cortisol. However, as a result of training the T/C ratio increased at all measurement points over time (T1-T2), with the greatest increase occurring by T2. Of the total increase, 68% of the rest, 121% of the pre, 64% of the 5-post, and 86% of the 15-post occurred by T2. This suggests that high-volume training may increase the T/C ratio to a greater extent than low-volume training. Increases in the T/C ratio may indicate a more anabolic state.

While exact functions are unclear, beta endorphins have been associated with a variety of related functions including suppression of pain, glucose regulation, and immune system suppression (40). Resistance exercise-induced increases in beta endorphins are intensity and duration dependent (25). Elevations in beta endorphins as a result of exercise may last up to an hour (40, 53).

Although resting beta endorphins were unaffected, 15-post concentrations decreased over time. Of the total decrease (T1-T3), 80% occurred by T2, again suggesting that high-volume training had a greater effect on altering hormonal concentrations than low-volume training. Adrenocorticotropic hormone and beta endorphin share the same precursor molecule (proopiomelanocortin); the reductions in postexercise cortisol and beta endorphin suggest that a change in hypothalamic-pituitary function is at least partially responsible for the decreases noted (see tables 5.15 and 5.17). Decreases in postexercise beta endorphin values suggest that the exercise is less physiologically stressful. Infections are often noted shortly after emotional or physical stress including physical exercise (35, 40). Part of the reason for the increased incidence of infection may be a result of increased postexercise concentrations of beta endorphin and cortisol which suppress the immune system (40). The reductions in postexercise concentrations of cortisol and beta endorphin noted in this study may reduce immunosuppression and reduce the incidence of infection.

Table 5.16 Testosterone:Cortisol (T:C) Response × 100 (N=24)

	Rest	Pre	5-Post	15-Post
T1	4.9 ± 1.4	5.0 ± 2.5	3.7 ± 2.1	2.9 ± 1.4
T2	4.7 ± 1.4	6.1 ± 2.6	5.1 ± 3.7	4.1 ± 2.4
T3	5.0 ± 2.4	7.0 ± 3.3	6.3 ± 1.4	4.3 ± 1.3

Data are means ± standard deviations.
Exercise Effects: Significant time effect: rest > 15-post; pre > 5-post and 15-post. Significant test day × time effect: rest > 5-post and 15-post on T1; pre > 15-post on T1.
Training Effects: Significant time effect: T1 < T2 < T3. Significant test day × time effect: rest, pre, 5-post, and 15-post: T < T2 < T3.

Table 5.17 Beta Endorphin Responses to Exercise Expressed in pmol/L (N=24)

	Pre	15-Post
T1	6.0 ± 2.3	33.0 ± 18.2
T2	6.1 ± 1.9	20.9 ± 15.4
T3	6.1 ± 2.9	17.8 ± 11.3

Data are means ± standard deviations.
Exercise Effects: Significant time effect: 15-post > pre. Significant test day × time effect: 15-post > pre on T1, T2, and T3.
Training Effects: Significant time effect: T1 > T2 and T3. Significant test day × time effect: 15-post: T1 > T2 and T3.

Conclusions

The training program did positively alter performance and beneficially alter physiology in a manner suggesting improved exercise tolerance. In general, the beneficial alterations were affected to a greater degree as a result of the high-volume training compared to the low-volume training.

Practical Applications

Longitudinal (36, 44, 47, 50) and cross-sectional (33, 49) studies have noted beneficial adaptations in various physiological parameters as a result of weight training suggesting an improved tolerance to exercise. High-volume training using large muscle mass exercises appears to stimulate these beneficial adaptations to a greater extent than low-volume training or using small muscle mass exercises (45). Periodized training programs begin with a short (weeks) high-volume phase (43, 46, 48). The purpose of this high-volume phase is to prepare the athletes physiologically so that they can better tolerate high-intensity exercise (which comes later in a training program). However, the in-depth study of the physiological effects of high-volume weight training has received little attention, especially as it concerns elite strength athletes (33, 46). The present study offers evidence that high-volume weight training of short duration can markedly enhance physiology, which reduces the physiological stress associated with weight training exercise. In this study, several factors, including changes in postexercise heart rate, lactic acid, growth hormone, cortisol, testosterone, the testosterone/cortisol ratio, and the increase in fasting IGF_1 concentrations, suggest that physiological adaptations were most apparent as a result of the high-volume training at T2. Furthermore, many of these adaptations persisted or increased from T2 to T3. In this context we would suggest that higher volumes of training can enhance several factors better than low-volume training. These factors include (1) increased protein synthesis, which may result in increased LBM and increased rates of tissue repair, (2) FFA mobilization and use during recovery, which may result in lower body fat content, and (3) reduced exercise stress (i.e., a smaller disturbance in homeostasis) and improved exercise tolerance.

Furthermore, this study offers evidence that the concept of overreaching is valid. First, a basic tenet of the overreaching concept is that the high-volume phase will beneficially alter physiology, leading to less physiological stress as a result of weight training exercise; evidence from the present study suggests that this is possible. Second, the overreaching period can cause no change or a decrease in performance; after a return to low-volume training a delayed increase in performance can be expected (approximately 2-5 weeks after the high-volume phase). Unlike the previous one-week overreaching observation from our laboratories with a similar group of athletes in which performance did not improve (55), in the present study, performance variables (snatch test and VJ) tended to increase as a result of high-volume weight training. This suggests that an overreaching protocol as used in this study does not necessarily adversely effect performance.

Conformation of the practical application and effectiveness of an overreaching protocol would gain support if performance increases could be demonstrated during typical competition conditions. The present experiment was held in conjunction with a weightlifting camp for 44 junior elite weightlifters. Ten lifters did not participate in the experiment and did not engage in the high-volume portion of training. Twenty-one of the 24 lifters serving as subjects and the 10 lifters not taking part in the experiment participated in a weightlifting meet within 2-4 days after the conclusion of the experiment. Of the participants at the meet, 14 of 21 (67%) made personal records; of the nonparticipants, 3 of 10 (30%) made personal records. Similar competition results were noted during weightlifting meets two weeks following the study of Warren et al. (unpublished data). These observations suggest that the delayed performance increase did occur, or that the increased volume did contribute to increased performance several weeks after returning to lower volume training.

Acknowledgments

The authors of this chapter are indebted to several colleages who were instrumental in the collection, analysis of data, and the preparation of this chapter: S.J. Fleck and J.T. Kearney, USOC, Colorado Springs; W.J. Kraemer and S. Gordon, Center for Sports Medicine, Pennsylvania State University; R.E. Keith, Department of Nutrition and Foods, Auburn University; R.L. Johnson, Exercise Science, Appalachian State University; and a special thanks to John Thrush, USWF Junior National Coach, for his insights into designing the training protocol and for his commitment to USWF Sports Science.

References

1. Adolphson, S. 1973. Effects of insulin and testosterone on glycogen synthesis and glycogen synthetase activity. *Acta Physiologica Scandinavica* 88: 234-247.
2. Alén, M., K. Häkkinen. 1987. Androgenic steroid effects on several hormones and on maximal force development in strength athletes. *Journal of Sports Medicine and Physical Fitness* 27: 38-46.
3. Allenberg, K., S.G. Holmquist, P. Johnsen, J. Bennett, J. Niehlsen, H. Galbo, N.H. Secher. 1982. Effects of exercise and testosterone on the active form of glycogen synthetase in human skeletal muscle. In *Biochemistry of exercise,* vol. 13, eds. H.G Knuttgen, J.A. Vogel, 625-630. Champaign, IL: Human Kinetics.
4. Bannister, E., W. Rajendra, J. Mutch. 1985. Ammonia as an indicator of exercise stress: implications of recent findings to sports medicine. *Sports Medicine* 2: 34-46.
5. Callister, R., R.J. Callister, S. Fleck, G. Dudley. 1990. Physiological and performance responses to overtraining in elite judo athletes. *Medicine and Science in Sports and Exercise* 22: 816-824.
6. Christensen, N., H. Galbo. 1983. Sympathetic nervous activity during exercise. *Annual Review of Physiology* 45: 139-153.

7. Doerr, P., K.M. Pirke. 1976. Cortisol-induced suppression of plasma test-osterone in normal adult males. *Journal of Clinical Endocrinology and Metabolism* 43: 622-629.

8. Dudley, G., R. Staron, T. Murray, F. Hagerman, A. Luginbuhl. 1983. Muscle fiber composition and blood ammonia levels after intense exercise in humans. *Journal of Applied Physiology* 54: 582-586.

9. Fahey, T.D., R. Rolph, P. Moungmee, J. Nagel, S. Mortara. 1976. Serum testosterone, body composition and strength of young adults. *Medicine and Science in Sports and Exercise* 8: 31-34.

10. Fleck, S.J. 1993. Cardiovascular responses to strength training. In *Strength and power in sport*, 305-315. Oxford, UK: Blackwell Scientific.

11. Florini, J.R. 1987. Hormonal control of muscle growth. *Muscle and Nerve* 10: 577-598.

12. Fry, A.C., W.J. Kraemer, M.H. Stone, S.J. Fleck, B. Warren, B.P. Conroy, C.A. Weseman, S.E. Gordon. 1990. Acute endocrine response in elite junior weightlifters. *Medicine and Science in Sports and Exercise* 22: S54.

13. Fry, R.W., A.R. Morton, D. Keast. 1991. Overtraining in athletes. *Sports Medicine* 12: 32-65.

14. Guezennec, Y., L. Leger, F. Lhoste, M. Aymood, P.C. Pesquies. 1986. Hormone response to weight-lifting training sessions. *International Journal of Sports Medicine* 7: 100-105.

15. Haynes, S.P. 1986. Review: growth hormone. *Australian Journal of Science and Medicine in Sport* 20: 3-15.

16. Häkkinen, K., A. Pakarinen, M. Alén, P.V. Komi. 1985. Serum hormones during prolonged training of neuromuscular performance. *European Journal of Applied Physiology* 53: 287-293.

17. Häkkinen, K., A. Pakarinen, M. Alén, H. Kauhanen, P. Komi. 1987. Relationships between training volume, physical performance capacity, and serum hormone concentrations during prolonged training in elite weight lifters. *International Journal of Sports Medicine* 8: 61-65.

18. Häkkinen, K., A. Pakarinen, M. Alén, H. Kauhanen, P. Komi. 1988. Daily hormonal and neuromuscular response to intensive strength training in one week. *International Journal of Sports Medicine* 9: 422-428.

19. Kanakis, C., R.C. Hickson. 1980. Left ventricular responses to a program of lower limb strength training. *Chest* 78: 618-621.

20. Keul J., G. Haralambie, M. Bruder, H.J. Gottstein. 1978. The effect of weight lifting experience on heart rate and metabolism in experienced lifters. *Medicine and Science in Sports and Exercise* 10: 13-15.

21. Komi, P.V., K. Häkkinen. 1989. Strength and power. *Proceedings of the Weightlifting Symposium*, 159-175. Siofok, Budapest, Hungary: International Weightlifting Federation.

22. Kraemer, R., J.L. Kilgore, G.R. Kraemer, D. Castracane. 1992. Growth hormone, IGF-1 and testosterone responses to resistive exercise. *Medicine and Science in Sports and Exercise* 24: 1346-1352.

23. Kraemer, W.J. 1993a. Hormonal mechanisms related to the expression of muscular strength and power. In *Strength and power in sport*, ed. P.V. Komi, 64-76. Oxford,UK: Blackwell Scientific.

24. Kraemer, W.J. 1993b. Endocrine response and adaptations to strength training. In *Strength and power in sport*, ed. P.V. Komi, 291-304. Oxford, UK: Blackwell Scientific.

25. Kraemer, W.J., J.E. Dziados, L.J. Marchitelli, S.E. Gordon, E.A. Harmon, R. Mello, S.J. Fleck, P.N. Frykman, N.T. Triplett. 1993. Effects of different heavy-resistance exercise protocols on plasma β-endorphin concentrations. *Journal of Applied Physiology* 74: 450-459.
26. Kraemer, W.J., A.C. Fry, B.J. Warren, M.H. Stone, S.J. Fleck, J.T. Kearney, B.P. Conroy, C.M. Maresh, C.A. Weseman, N.T. Triplett, S.E. Gordon. 1992. Acute hormonal responses in elite junior weightlifters. *International Journal of Sports Medicine* 13: 103-109.
27. Kraemer, W.J., L. Marchitelli, S.E. Gordon, E. Harman, J.E. Dziados, R. Mello, P. Frykman, D. McMurry, S.J. Fleck. 1990. Hormonal and growth factor responses to heavy exercise protocols. *Journal of Applied Physiology* 69: 1442-1450.
28. Kuipers, H., H.A. Keizer. 1988. Overtraining in elite athletes. *Sports Medicine* 6: 79-92.
29. Lambert, C.P., M.G. Flynn, J.B. Boone, T.J. Michaud, J. Rodriguez-Zayas. 1991. Effects of carbohydrate feeding on multiple-bout resistance exercise. *Journal of Applied Sports Science Research* 5: 192-197.
30. Lambert, M.I., S. Jaabar, T.D. Noakes. 1990. The effect of anabolic steroid administration on running performance in sprint trained rats. *Medicine and Science in Sports and Exercise* 22: S64.
31. Lukaszewska, J., B. Biczowa, D. Bobilewixz, M. Wilk, B. Bouchowixz-Fidelus. 1976. Effect of physical exercise on plasma cortisol and growth hormone levels in weight lifters. *Endokrynologia Polska* 2: 140-158.
32. Mayer, M., F. Rosen. 1975. Interaction of anabolic steroids with glucosteroid steroid receptor sites in rat muscle control. *American Journal of Physiology* 229:1381-1386.
33. McMillan, J., M.H. Stone, J. Sartain, D. Marple, R. Keith, D. Lewis, C. Brown. 1993. The 20-hr hormonal response to a single session of weight-training. *Journal of Strength and Conditioning Research* 7: 9-21.
34. Munck, A., P.M. Guyre, N.J. Holbrook. 1974. Physiological functions of glucocorticoids in stress and their relation to pharmacological actions. *Endocrine Reviews* 5: 25-44.
35. Nieman, D., S.L. Nehlsen-Cannarella. 1992. Exercise and infection. In *Exercise and disease*, eds. R.R. Watson, M. Eisinger, 121-148. Boca Raton, FL: CCR Publishers.
36. Pierce, K., R. Rozenek, M.H. Stone, D. Blessing. 1987. The effects of weight-training on plasma cortisol, lactate and heart rate. *Journal of Applied Sports Science Research* 5: 58.
37. Robergs, R.A., D.R. Pearson, D.L. Costill, W.J. Fink, D.D. Pascoe, M.A. Benedict, C.P. Lambert, J.J. Zacheija. 1991. Muscle glycogenolysis during differing intensities of weight- resistance exercise. *Journal of Applied Physiology* 70: 1700-1706.
38. Rogol, A.D. 1989. Growth hormone physiology, therapeutic use and potential for abuse. In *Exercise and sports sciences reviews*, ed. K.B. Pandolf, 353-377. Baltimore: Williams & Wilkins.
39. Sale, D.G. 1993. Neural adaptation to strength training. In *Strength and power in sport*, ed. P.V. Komi, 246-265. Oxford, UK: Blackwell Scientific.
40. Sforzo, G.A. 1988. Opiods and exercise. *Sports Medicine* 7: 109-124.
41. Sherman, W., A. Doyle, D. Lamb. 1991. Training, dietary carbohydrate and muscle glycogen. *Medicine and Science in Sports and Exercise* 23: S101.

42. Shephard, R.J., K.H. Sidney. 1975. Effects of physical exercise on plasma growth hormone and cortisol levels in human subjects. In *Exercise and sports sciences reviews*, vol. 3, eds. J. Wilmore, J. Keough, 1-30. New York: Academic Press.

43. Stone, M.H. 1990. Muscle conditioning and muscle injuries. *Medicine and Science in Sports and Exercise* 22: 457-462.

44. Stone, M.H., D.L. Blessing, R. Byrd, D. Boatwright, J. Tew, L. Johnson, A. Lopez-S. 1982. Physiological effects of a short term resistive training program on middle-aged sedentary men. *National Strength and Conditioning Association Journal* 4: 16-20.

45. Stone, M.H., S.J. Fleck, N.T. Triplett, W.J. Kraemer. 1991. Health- and performance-related potential of resistance training. *Sports Medicine* 11: 210-231.

46. Stone, M.H., R. Keith, J.T. Kearney, G.D. Wilson, S.J. Fleck. 1991. Overtraining: a review of the signs and symptoms of overtraining. *Journal of Applied Sports Science Research* 5: 35-50.

47. Stone, M.H., J.R. Nelson, S. Nader, D. Carter. 1983. Short term weight training effects on resting and recovery heart rates. *Athletic Training* 18: 69-71.

48. Stone M.H., H.S. O'Bryant. 1987. *Weight training: a scientific approach.* Minneapolis: Burgess International.

49. Stone, M.H., K. Pierce, R. Godsen, G.D. Wilson, D. Blessing, R. Rozenek. 1987. Heart rate and lactate levels during weight training exercise in trained and untrained males. *Physician and Sports Medicine* 15: 97-106.

50. Stone, M.H., G.D. Wilson, D. Blessing, R. Rozenek. 1983. Cardiovascular responses to short-term Olympic style weight training in young men. *Canadian Journal of Applied Sports Science* 8: 134-139.

51. Sutton, J.R. 1978. Hormonal and metabolic responses to exercise in subjects of high and low work capacities. *Medicine and Science in Sports and Exercise* 10: 1-6.

52. Sutton, J.R., J.H. Casey. 1975. The adrenocortical response to competitive athletics in veteran athletes. *Journal of Clinical Endocrinology and Metabolism* 40: 135-138.

53. Sutton, J.R., P.A. Farrel, V.J. Haber. 1990. Hormonal adaptation to physical activity. In *Exercise, fitness and health*, eds. C. Bouchard, R.J. Shephard, T. Stephans, J.R. Sutton, B. McPherson, 217-257. Champaign, IL: Human Kinetics.

54. Triplett, N.T., M.H. Stone, C. Adams, K.D. Allran, T.W. Smith. 1990. Effects of aspartic acid salts on fatigue parameters during weight training exercise and recovery. *Journal of Applied Sports Science Research* 4: 141-147.

55. Warren, B.J., M.H. Stone, J.T. Kearney, S.J. Fleck, R.L. Johnson, G.D. Wilson, W.J. Kraemer. 1992. Performance measures, blood lactate and plasma ammonia as indicators of overwork in elite junior weightlifters. *International Journal of Sports Medicine* 13: 372-376.

56. Wilkerson, J.G., L. Swain, J.C. Howard. 1988. Endurance training, steroid interactions and skeletal interactions. *Medicine and Science in Sports and Exercise* 20: S59.

CHAPTER 6

The Role of Training Intensity in Resistance Exercise Overtraining and Overreaching

Andrew C. Fry, PhD

Introduction

It is a well accepted concept that development of optimal muscular strength or power requires a certain amount of resistance exercise using maximal, or near-maximal, loads. Athletes who require high levels of strength and power when performing their events routinely incorporate training loads that approach 100% of their one repetition maximum (1 RM). Coaches and athletes have recognized that such training methods can be very effective in eliciting relatively fast increases in strength and power, but can an individual rely solely on such a high-intensity training protocol for prolonged improvements in performance? Those of you who have spent any time in the strength training environment have most likely observed athletes who spend an inordinate amount of their strength training time using extremely heavy resistance approaching 1 RM loads. While the short-term effect can be quite gratifying, continued use of such training methods will often prove unsatisfactory for continued gains. The question now becomes, how much is too much when it comes to training with maximal loads? From a scientific perspective, the question becomes, what are the physiological maladaptations that contribute to attenuated performances due to excessive use of maximal training loads?

Scientific research interest in the phenomenon of overtraining has grown considerably over the last several decades. Up to this point in time, however, the majority of overtraining study has focused on endurance, or aerobically-based, activities (3, 23, 45, 47). In general, this appears to reflect the past trends in the exercise sciences that have only recently begun to emphasize study of resistance exercise activities. When looking at the entire body of scientific literature on exercise, it becomes clear that the study of this type of exercise has grown along with its increased popularity with the general public and the athletic community. Although numerous myths and fallacies have accompanied this increased popularity of resistance exercise, the continued study of this training modality will extend our understanding of the physiological benefits and adaptations of such training.

It is quite evident that the adaptations to aerobic, or endurance, types of exercise are quite different from the adaptations to anaerobic exercise, in this case resistance exercise. Each form of exercise results in a differing adaptational profile, which is why each has an important place in the conditioning of an athlete. What has not been as apparent in the overtraining literature, is that overtraining with endurance exercise is also quite different from overtraining with resistance exercise (12, 19, 45, 72). While proper exercise prescriptions for either endurance or resistance exercise produce different physiological adaptations, likewise the overtraining responses for each modality are also different (12, 19, 72). As a result, one must be wary of using the endurance-overtraining literature to infer what happens during overtraining with resistance exercise.

Although there is little data to support the contention, it has been suggested that anaerobic activities may be more susceptible to overtraining than aerobic activities (23). If this is indeed the case, it becomes imperative that scientific investigations follow the development of overtraining as it progresses from its initial stages to the point of undeniably impaired performances. This is, of course, much easier to say than to do. The study of overtrained individuals is sometimes limited to monitoring a person after overtraining has already been diagnosed (2). However, the development of one particular type of endurance overtraining has been nicely studied by Lehmann and colleagues as they monitored distance runners progressing through increasing volumes of training (46, 48, 49). In this manner, it was possible to monitor the various stages of the development of overtraining and the accompanying impaired performances. Attempts to longitudinally study stressful phases of resistance exercise have also been made using a variety of training modalities and protocols, although not all of these protocols have definitively resulted in overtraining (14-16, 19, 20, 26, 29, 74, 75).

Operational Definitions

A critical issue at this point is to operationally define several relevant terms related to overtraining. Such an understanding and agreement is critical for the organized study of this phenomenon, and is repeatedly addressed by the other authors in this text. For the purposes of the present chapter, overtraining will be defined as any increase in volume and/or intensity of exercise training resulting in long-term (e.g., several weeks or months, or longer) performance decrements (12, 19, 45, 68). Such a scenario can be described by the well-known general adaptation syndrome described by Hans Selye (64), where the individual is unable to adequately recover from a stressor, and impaired function or even death results. Numerous other terms have been used or suggested for overtraining (3, 12, 23, 26, 45, 47, 53, 55, 67, 68, 71, 72) and the reader must be careful not to get mired in confusion over the differing terminologies. A milder form of overtraining can also occur on a short-term basis that is easily recovered from over several days. This is known as overreaching, and is often a planned part of many training programs (12, 45, 68, 72).

The term resistance exercise refers to exercise that uses heavy resistance. Such exercise can utilize barbells, dumbbells, and other types of free weights, but it can also use various types of weight machines or even body weight exercises such as calisthenics. Regardless of the training modality, all resistance exercise protocols can be described by the five acute training variables described by Fleck and Kraemer (10): choice of exercise, order of exercise, the load or resistance used, the volume of exercise, and the rest between sets. Each of these variables can have a profound impact on the physiological effect of the training program, and can be manipulated to attain the desired results. Such purposeful variation of the training program is called periodization, and has been shown to be very effective for the attainment of maximal performances (10, 24, 68, 69). A more detailed accounting of the characteristics of a resistance exercise program is presented in chapter 4 of this text.

Of particular interest to the present chapter is the definition of intensity. In a very simplistic approach, intensity is sometimes confused with difficulty. For example, a coach may state that his/her athletes are using a high-intensity training program, when they actually mean it is a very stressful or difficult phase of training. Training intensity can be correctly defined in several ways, but the two of interest here are absolute intensity and relative intensity. Relative intensity is based on an individual's 1 RM, where training is performed at a set or known percentage of their 1 RM (10, 12, 68). Although two different athletes may possess different maximal strength levels (e.g., 1 RM = 100 kg and 150 kg, respectively), they may be exercising at the same relative intensity (e.g., 80% 1 RM, or 80 kg and 120 kg, respectively). This is quite similar to aerobic exercise where an individual may be training at a set percentage of their $\dot{V}O_2$max or their HRmax (37, 46, 48). Absolute intensity refers to the actual load on the barbell or the machine. Thus with our previous example of two athletes, the one training with 120 kg on the barbell is exercising at a greater absolute intensity than the athlete using 80 kg (10, 68). Since absolute intensity can be defined as the amount of work (force $\times$ distance; Joules) performed in a set time period (work/time), it is actually a measure of power, and is an important training variable (10, 68). Likewise, absolute intensity is also related to the volume of work performed, and when studying any training program, volume and intensity can not be completely separated from each other. From a practical standpoint, most resistance exercise training programs prescribe training loads using a measure of relative intensity, and it is relative intensity that has been manipulated in many of the studies cited in this chapter.

One final point of concern for the study of resistance exercise overtraining is whether overtraining actually occurred; in other words, were performance decrements evident? Although important information relevant to overtraining can be obtained from studying various types of stressful training, it is important to differentiate between those scenarios where performance is actually impaired (12, 14, 18-22, 39, 57, 58, 72), and where the athletes/subjects have successfully adapted to the stressful training without impaired performances (15-17, 27, 29). Studies of both types, however, are critical for the complete understanding of what the actual training tolerance is for any particular type of exercise.

Volume Versus Intensity Overtraining for Resistance Exercise

As might be expected, resistance exercise overtraining is a multivariate phenomenon, and as such, it can affect many physiological systems. In this chapter, we will take a look at how resistance exercise overtraining using excessively high relative training intensities may influence muscular strength and power performances (14, 19, 21, 75), motor control and performance (14, 19, 75), endocrine profiles (12, 18, 57, 58), catecholamine responses (20, 57), psychological profiles (14, 22), and orthopedic injuries (21). Where appropriate, comparisons will be made with resistance exercise overtraining resulting from other etiologies (e.g., high training volumes, combined aerobic and anaerobic training). For more in-depth discussions on these types of resistance exercise overtraining, see chapters 4 and 5.

As previously mentioned, much of the resistance exercise overtraining-related literature has been concerned with increasing or decreasing volumes of training (4, 11, 15-17, 27, 29, 39, 43, 74). Increasing the training volume is not unusual among athletes and those who are seriously training. A common misconception is that more is better, and that by increasing the number of exercises, the number of sets or repetitions, or the number of training sessions, the result will be better performances. While short-term improvements in performance may occur, such training can easily result in performance decrements and overtraining (see chapter 5). Overtraining research with endurance athletes has addressed the issue of excessive training intensity in only a few studies (46, 48), suggesting that training volume may be a more critical issue with those athletes. Many sports require a combination of both aerobic and anaerobic capabilities, and thus necessitate training for both qualities. The role of such cross-training, and the capacity to train in this manner, is also now being studied (43). Such complex training, however, is difficult for the sport scientist to study at this time, since there are so many training variables that are difficult to control or to monitor.

Empirical observations by coaches and athletes suggest that it is not unusual for an individual to rely on extremely high relative training intensities for extended periods of time. Examples of this might be the individual who performs sets at 100% of 1 RM to 100% of 3 RM loads every training session. Another example might be the competitive lifter who has experienced satisfactory gains by using heavy sets of 1-2 repetitions when peaking for a competition. This short-term gain convinces the lifter to use this type of training scheme consistently, resulting in too great a relative training intensity for extended periods and either a plateau or a decrease in strength and power performance. Use of consistently high relative intensity training has also been reported for several very successful eastern European weightlifting squads (snatch, and clean and jerk). Such success again begs the question, how much high relative intensity weight training is too much? Recent research addressing this question will be presented in this chapter (14, 18-22, 57, 58, 75). The possible role of pharmaceutical interventions on high relative intensity resistance exercise training will not be directly addressed in this chapter, although it is likely that toleration of such training by some individuals may be due to drug use. Figure 6.1 illustrates the relationship between resistance exercise training volume and relative training intensity as it concerns the problem of overtraining.

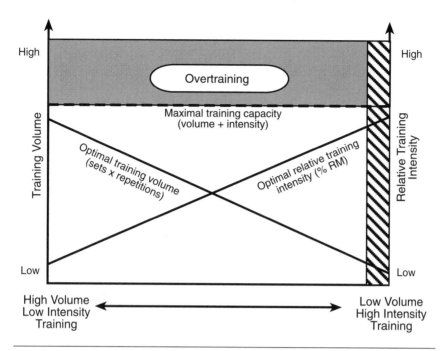

Figure 6.1 Relationship between relative training intensity (% repetition maximum (RM)) and training volume (sets × repetitions) as they relate to resistance exercise overtraining. Maximal resistance exercise training capacity is a combination of the training volume and the relative training intensity. Training loads greater than an individual's capacity may result in overtraining. The shaded area on the right side indicates the training stresses discussed in this chapter.
Based on data from references 17-21, 56, and 57.

Protocols for Studying the Development of Overtraining

Recent reports have addressed the development of intensity-dependent resistance exercise overtraining as it progresses from its initial stages until performance decrements are evident. A critical issue has been the determination of an operationally defined state of overtraining. For research purposes, it was determined that overtraining occurs when performance on a training-specific task exhibits a statistically significant decrease. This, of course, ignores the problem of plateaus, or stagnation, in performance results, which has been suggested as part of the definition of overtraining (23, 68). It should be noted here that plateaus in performance improvements can often occur in even the best training programs. Thus, to avoid any gray area between whether overtraining actually occurred or not, statistically significant decreases in muscular strength were required to determine that overtraining actually occurred (14).

The initial problem was the development of a high relative intensity resistance training protocol that would produce a significant decrease in muscular strength. Such a task would appear to be simple, given the frequently mistimed peaking and unsuccessful results of competitive lifters. However, pilot work was necessary to determine if muscular strength would actually decrease due to inappropriately prescribed relative training intensity. Another important point was the avoidance of muscle damage. It has been well documented that muscle damage can result from excessive eccentric muscular work, and that this can cause dramatic decreases in strength (8). Muscle damage, however, does not appear to be the primary problem with highly trained individuals who have developed the tolerance for the workloads regularly used. It was therefore critical to determine a high relative intensity training program that would cause muscular strength decreases without causing muscular damage. A final concern was the training status of the subjects. Overtraining is primarily a problem for more advanced athletes and individuals. Although untrained and novice individuals can certainly have problems with inappropriately prescribed training programs, much of their difficulty is due to the initial adaptation to any form of exercise. As a result, subjects for the study of overtraining must be previously trained. On the other hand, it would not be appropriate to recruit currently competing athletes for studies such as these. Such training would certainly make it difficult for the athlete to perform well in their sport, and would create a hostile relationship between the researcher and the coach and athlete. The best solution was to recruit individuals who were currently training, had been training for several years, and could demonstrate a minimal level of strength (1.5 × body weight) on a criteria lift, the parallel barbell back squat.

An initial study was carried out to determine an appropriate overtraining protocol (14). A weight training machine that somewhat mimicked the squat motion was used for training, and allowed control for gross motor patterns (13). In this manner, changes in the kinetics of the lifting movement would not be a factor during the course of the investigation. All subjects performed two weeks of familiarization training (3 d/wk; 1 × 5-32 kg, 1 × 5-50% BW, 3 × 5-70% BW), and were then randomly divided into a control group (2 d/wk; same protocol as during familiarization), or an experimental group (5 d/wk; 1 × 5-32 kg, 1 × 5-40% 1 RM, 1 × 3-60% 1 RM, 1 × 2-80% 1 RM, 8 × 1-95% 1 RM; 2 min rest intervals). The experimental protocol was modified from a barbell squat program used by elite weightlifters that had proven to be overly stressful, but used slightly greater relative intensity and a greater number of repetitions for each training session. Three weeks of training were performed using these protocols, with test batteries administered at four times during this period.

The critical result of this pilot study was that 1 RM strength on the squat machine did not decrease, but actually increased for the experimental group by the end of the three weeks (M ± SE; test 1 = 109.8 ± 9.8 kg; test 4 = 117.0 ± 10.1 kg) (14). Strictly speaking, overtraining did not occur, and at first glance it might be stated that the training protocol was successful for improved physical performance. It is important to note, however, that other types of performance were adversely affected. As a result, it could be argued that a state of overreaching may have developed, although short-term strength decrements did not occur. First, decreased isokinetic knee extension strength at 1.05 rad/sec occurred, the velocity closest to the actual training velocity. Second, sprint speed decreased as indicated by losses of 0.09 sec over 9.1 m, and 0.17 sec over 36.6 m. Based on self-reports, muscle soreness was not experienced by any subject at any time during the three-week period. In addition, the commonly reported overtraining symptoms of disrupted sleep patterns and increased resting heart rate were not observed.

So what are the implications of this pilot work? First of all, it was surprising that the subjects could successfully perform as many repetitions as they did at 95% 1 RM loads. Although the differences between machine and free weight modalities are readily acknowledged (10), it appears that the use of the squat machine permitted a much greater training load than was originally expected based on the previous empirical observations with the elite weightlifters who used free weights. It is speculated that the overall stresses of free weight training may be greater than for machines. The second finding of this pilot work was that the use of the squat machine modality for such overtraining research would permit either a greater training volume (i.e., more sets), or a greater relative training intensity (i.e., 100% 1 RM). The final finding was that maximal strength appears to be preserved or enhanced longer than other performance variables (e.g., velocity-controlled strength and sprint speed). The practical implications of these findings are critical for the practitioner. These results demonstrate that it is not appropriate to base the effectiveness of a strength and conditioning program, such as used by many sports, solely on 1 RM performance in the weight room. Although maximal strength may not be compromised, improper resistance exercise prescription in the weight room may attenuate performance on other important physical tasks.

Although this pilot study was relatively simplistic in nature, it points out several directions for follow-up study. A similar investigation has been recently completed using a free weight training modality (75). Resistance exercise was performed using primarily the parallel barbell squat, supplemented with leg curls. Since it was speculated that the training capacity would be less when using a free weight modality, the lifting protocol was modified accordingly. Again, previously weight trained individuals who were capable of at least a 1.5 × BW barbell squat served as subjects. A four-week training period using a normal training protocol was followed by three weeks of a program designed to induce overtraining and the concomitant decrements in 1 RM strength. The normal training was performed 2 d/wk, with 3 × 10 RM loads for the squat performed on Mondays, and 3 × 5 RM loads performed on Thursdays. The overtraining period included training 3 d/wk, with 2 × 1-95% 1 RM and 3 × 1-90% 1 RM performed during each training session.

The decreased training volume and slightly decreased relative intensity compared to the previous investigation (14) were selected due to the use of a free weight modality. It was believed that the increased role of synergistic muscles, proprioceptive input, and kinesthetic awareness with free weights would make this modality more difficult to tolerate. Maximal strength (1 RM) in the parallel barbell squat demonstrated significant increases during the normal training phase (M ± SD; pre = 139.5 ± 29.9 kg, postnormal = 154.6 ± 27.7 kg), but failed to significantly increase further during the overtraining phase (postovertraining = 161.0 ± 27.2 kg). One RM strength instead demonstrated a marked plateau. Although the use of heavy sets of one can be an effective method for short-term 1 RM increases (69), this did not occur in this study. So did overtraining actually occur? There were no strength decrements, and according to some definitions of overtraining a plateau in performance results constitutes a state of overtraining (3, 23, 45, 68). Regardless of how overtraining is defined, muscular strength did not optimally improve. Additionally, 1 RM strength remained stagnant three weeks after the completion of the study despite a greatly reduced training load, indicating the plateau in strength was not transient in nature. Despite the differences in this study with free weights compared to the previous study using a machine modality, there were still some interesting similarities in results. Isokinetic squat strength was attenuated at a bar velocity most similar to the training velocity (0.20 m/sec). In addition, sprint time was slower by 0.1 sec at 9.1 m, although times at 36.6 m did not change. Apparently only the initial sprint start and acceleration phase was affected, while later phases of a sprint were unaffected. Once again, although strength decrements were not evident in the weight room, performance on tasks important for many sports was impaired (i.e., slow velocity strength, sprint starts) due to the training protocol. The reader should be cautioned here that the impaired sprint performance does not mean that strength training in general is detrimental to sprint speed, but rather that excessive amounts of high relative intensity resistance exercise may impair sprint capabilities. The important implications from the free weight study were (1) training capacity is less for free weights compared to machines, (2) performance on sprint tasks may be adversely affected if the strength training protocol is improperly prescribed, and (3) future overtraining research using this free weight modality may need to use greater relative intensities (e.g., 100% 1 RM) or volumes to elicit a definitive decrease in training-specific muscular strength. In relation to the classical general adaptation syndrome proposed by Selye (64), it appears that the exercise protocol used to study high relative intensity resistance exercise addresses the overtraining problem early in its development, before irreversible permanent damage is done (see figure 6.2).

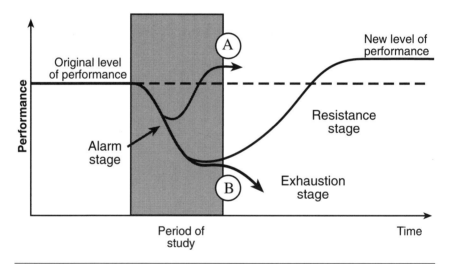

Figure 6.2   High relative intensity resistance exercise overtraining protocol as it relates to Selye's general adaptation syndrome. A = response to a normal training session, B = response to an overtraining stimulus.
Based on data from references 17-21, 56, 57, and 63.

Physiological Responses to High-Intensity Resistance Exercise Overtraining

Based on the initial study using a machine modality (14), a follow-up investigation was performed to gain greater insight on the responses of various physiological systems to an intensity-dependent overtraining stress (18-22, 57, 58). Since the initial study had failed to demonstrate decreases in strength performance, the protocol was modified to include training 6 d/wk using 10 × 1-100% 1 RM loads and 2-min interset rest intervals using the squat simulating machine (19). Whenever a lift was unsuccessful, the following lifts were decreased by 4.5 kg. The training session was completed when 10 successful lifts had been performed, resulting in 10 sets of 1 repetition at a load as close as possible to the individual's 1 RM at that particular moment. Due to the increased relative intensity, the training period was decreased to only two week's duration. The use of this extremely stressful protocol was selected to help insure the development of strength decrements indicative of a state of overtraining. Although typical resistance exercise training programs do not usually use such an extreme protocol, this type of training has been used by some coaches for the sport of weightlifting (67), as well as other activities. Thus, this protocol served an important role for the study of overtraining due to excessively high relative intensity.

The modified protocol proved to be effective in eliciting strength decrements as indicated by a >10% decrease in 1 RM strength (19). Such a strength decrease would be undoubtedly of great concern to any seriously training athlete or coach. Between 2-8 weeks passed before any overtrained subject was able to resume a normal lower body strength training program, again indicating that such a strength decrement was not a transient condition. As was previously observed (14), low velocity knee extension isokinetic strength (0.53 rad/sec) was attenuated, even though the movement was not specific to the training modality (19). Unlike the previous study, high-velocity knee extension strength (5.24 rad/sec) also decreased, as well as isometric strength (0 rad/sec). A number of previously suggested symptoms of overtraining (3, 23, 26, 45, 67, 68, 72) were not apparent with this protocol. These include no change in self-reported sleep patterns, resting heart rate, or body composition (19). The results of this study indicate that overtraining definitively occurred, and that the decreases in maximal strength were evident not only in the training-specific modality (e.g., squat machine), but also with single joint isokinetic assessments across a wide range of the velocity spectrum. It is also critical to note that although overtraining resulted, many of the often-cited symptoms were not present. This indicates that not all forms of overtraining are necessarily identical.

With the presence of a significant decrease in strength (19), the design of the study permitted an awareness concerning the site(s) of physiological maladaptation. Using a combination of voluntary and involuntary isometric quadriceps force measures, it was possible to determine if strength decreases were due to maladaptation in the peripheral musculature or were due to a decreased central drive. The anterior compartment of the thigh was maximally activated via cutaneous stimulation of the femoral nerve, with such stimulation bypassing central regulation of muscular force production. Since both voluntary and stimulated force exhibited significant strength decreases, some type of maladaptation of the peripheral muscles is suggested (19).

Peripheral maladaptation was also suggested by the decreased lactate response to a standardized lifting protocol (i.e., maximal repetitions at 70% 1 RM) (19). Such a response has been observed with increased volumes of weightlifting (15-17, 74), and may simply be an indicator of improved levels of fitness. However, the high relative intensity training protocol used (19) would not appear to elicit much of a training adaptation for the anaerobic glycolysis system. It is possible that the lower lactate response was due to a decrease in the total work performed (as measured in Joules) by the end of the study during the standardized lifting protocol. Since the resistance for this session was always set at 70% 1 RM loads, this decrease in work was due solely to the attenuated 1 RM strength, while the total number of repetitions performed was unchanged. Nevertheless, as might be expected, absolute muscular endurance decreased, while relative muscular endurance was unaffected (19). Dietary records indicated that the overtrained subjects actually ingested greater amounts of carbohydrates during the study (19). Such an unexpected response may have been necessary to maintain muscle glycogen levels as has been necessary during other types of stressful training (5). Whether the lower circulating levels of lactate were due to altered muscle glycogen levels is not known, although in a similar study recently performed, muscle biopsies did not find evidence of altered glycogen levels due to the same experimental protocol (personal communication, Lucille Smith and Jeff Wilson).

Muscle damage did not appear to be a contributing factor to the >10% decreases in 1 RM strength as indicated by circulating levels of creatine kinase, and by a daily training questionnaire inquiring about muscle soreness (19). Concentrations of approximately 600 IU/L were reported for the overtrained subjects (19), which are similar to levels resulting from a typical weight training session (41). While it is likely that some sort of muscle remodeling occurs in response to the high-intensity weightlifting protocol, it can not be classified as muscle damage such as has been reported in the resistance exercise literature (8). When more traditional weight training protocols are used with untrained males, at least four weeks are required before significant muscular adaptations occur such as fiber type transitions and altered protein expression (66). As a result, it is not likely that muscle tissue remodeling contributed greatly to the impaired strength performance during the two week high-intensity overtraining protocol using previously trained subjects.

Joint-Centered Overtraining Mechanism

During the course of the two weeks of high-intensity overtraining, an interesting situation developed with one of the experimental subjects who developed medically diagnosed overuse syndrome of the knees (21). The individual reported pain in both knees, and exhibited slight bilateral effusion. The gross structural integrity of the knees appeared to be intact as indicated by negative results for the anterior-posterior drawer, the Lachman, and the McMurray tests. This subject experienced the greatest decrease in 1 RM strength (>35%), and when performing the training program, was forced to slow down the speed of movement on the lifting task. There appeared to be a critical point in the range of motion on the squat machine when performing at maximal capacity that would produce a shutdown of the involved musculature, and the lift would be unsuccessful. Several interesting factors were evident for this subject. Only maximal strength efforts were adversely affected. The maximal number of repetitions performed at 70% 1 RM actually increased considerably, suggesting a force threshold for the shutdown phenomenon to occur. Movements requiring lower force levels were not affected. Additionally, only dynamic measures of quadriceps force were attenuated. Both voluntary and stimulated isometric force actually increased for this individual. These data indicate that peripheral muscle maladaptation did not occur for this subject. It is speculated that while the shutdown phenomenon may have served to protect the knee joints, it may also have served to protect the peripheral musculature from the maladaptation observed for the rest of the overtrained subjects. It is likely that afferent inhibitory activity from the affected joints is responsible for the results observed with this subject (7, 21, 59, 65). Such a protective mechanism results in a different etiology for the development of an overtraining syndrome (21, 59). This case study also suggests that gross ligamentous and tendinous structural damage is not necessary for dysfunction of the involved joint(s). A more in-depth discussion of biomedical issues related to overtraining may be found in chapter 10.

Endocrine Responses

Numerous studies have investigated the endocrine responses to various types of stressful training. It appears that increased volumes of resistance exercise (4, 12, 15-17, 27-29, 39, 67, 68, 71) will often produce a hormonal profile somewhat similar to what is observed for high volumes of endurance exercise (1-3, 6, 12, 23, 47, 48, 71). Such similarities do not appear to exist for high relative intensity resistance exercise overtraining (12, 18, 58).

The most commonly implicated hormones are testosterone and cortisol. These two hormones have been used to represent the anabolic-catabolic status of an individual (1, 12, 45, 68), and are often expressed as a ratio (e.g., testosterone/cortisol). While this may be a simplistic approach to some complex physiological systems, there is considerable evidence that increased training stresses can result in attenuated resting concentrations of circulating testosterone and augmented resting levels of cortisol. Seemingly, the acute endocrine responses to an exercise stimulus are also critical and are quite dependent on the resistance exercise stimulus (40, 41, 43, 52), but few data are available on acute endocrine responses to overtraining (12, 18, 58). Since the endocrine profile can be perturbed for several hours after a single resistance exercise training session (52), such acute responses can have a considerable impact on the resulting physiological adaptations.

In the case of high relative intensity resistance exercise overtraining, resting concentrations of total testosterone and free testosterone did not decrease, and cortisol did not increase (18). This is contrary to reports of other types of overtraining where resting levels are affected (3, 12, 23, 45, 47, 67, 68, 71). When exercise-induced concentrations are measured, total testosterone actually increased slightly for the overtrained subjects, while cortisol decreased (18). The net result is that the hormonal response to this type of overtraining appears to be distinctly different from other forms of overtraining. This was also evident for the hormonal ratios, where no change was observed for the testosterone/cortisol ratio at rest, while a slight, but significant increase occurred after exercise (18). The ability of binding proteins to interact with testosterone appeared unaffected since the percent unbound testosterone remained constant throughout the study (18). Such a lack of affect on testosterone and cortisol concentrations is not entirely unexpected, since lifting protocols using numerous maximal sets of one repetition provide little stimulus for acute hormonal responses (28).

Altered pituitary activity has been implicated in overtraining of endurance athletes (2). Pituitary activity, as indicated by temporal measures of leutinizing hormone and adrenocorticotropic hormone, were not altered with high relative intensity resistance exercise overtraining (58). Although further research would be needed to determine pulsatility frequency and magnitude of these hormones, the preliminary evidence does not suggest pituitary regulation of testosterone or cortisol is affected by this type of overtraining. It has been well documented that growth hormone is very sensitive to the acute characteristics of a resistance exercise protocol (41, 73), and as a result, may be influenced by resistance exercise overtraining. However, both resting and acute concentrations of immunoreactive growth hormone have not shown a differential response to high relative intensity resistance exercise overtraining (18), further evidence that the pituitary gland is not the site of physiological maladaptation to such a stimulus. In general, the endocrine adaptations to high relative intensity resistance exercise overtraining are distinctly different from overtraining with endurance activities (see chapters 2 and 3), or high volumes of resistance exercise (see chapter 5).

Catecholamine Responses

One of the most critical physiological systems for the study of overtraining may well be the sympathetic nervous system (12, 23, 45, 68). Circulating catecholamines can potentially regulate numerous physiological systems and, therefore, may be potent mediators of an overtraining syndrome. Overtrained endurance athletes have exhibited altered sympathetic activity as indicated by attenuated nocturnal concentrations of urinary epinephrine and norepinephrine (46-49). Furthermore, resistance exercise results in altered catecholamine concentrations acutely (52) and at rest (56), thus presenting the interesting question of how sympathetic activity is affected by high relative intensity resistance exercise overtraining. It is known that exercise intensity is a critical determinant for the acute catecholamine response (44), and it is likely that a heavy resistance exercise protocol would result in a large sympathetic response. No data is currently available, to the author's knowledge, concerning catecholamine responses to volume overtraining with resistance exercise.

With high relative intensity resistance exercise overtraining, resting concentrations of both epinephrine and norepinephrine were unaffected, but acute concentrations exhibited considerable increases (12, 20). The acute catecholamine values observed were among the highest reported in the exercise literature (12, 20, 44, 52), probably due to the intensity of the exercise stimulus, and the timing of the postexercise blood sampling (12, 20). In retrospect, it is likely that some of the acute epinephrine response may have been masked by an anticipatory response, since the preexercise concentrations were higher than would be expected (12, 20). As a result, the acute epinephrine response to the resistance exercise protocol is probably greater than was actually observed. The acute norepinephrine response was elevated even though the total amount of work performed in the exercise stimulus was decreased due to diminished 1 RM levels (19). Since the contractility of skeletal muscle is regulated in part by β-adrenergic receptors (9), the impaired muscle performance in the presence of elevated circulating catecholamines suggests altered functioning of the β-adrenergic system. In general, the elevated catecholamine response to the resistance exercise stimulus during high relative intensity resistance exercise overtraining is evidence of the sympathetic overtraining syndrome (12, 20, 23, 45, 47, 68) described in detail in chapter 2. It is also quite different from the attenuated catecholamine levels (i.e., parasympathetic overtraining) reported for overtrained endurance athletes (45-48), again indicating the unique differences of training and overtraining with different modalities and protocols (35).

Much of the evidence to this point allows speculation that the sympathetic nervous system may be a key factor in high relative intensity resistance exercise overtraining. Many of the symptoms observed in response to such a stimulus can be directly or indirectly regulated by sympathetic activity. For example, as previously mentioned, muscle contractility is influenced by β-adrenergic activity (9). Circulating concentrations of testosterone (34, 62) and cortisol (32) as well are partially under the influence of the sympathetic system, which may explain the slight increases in acute total testosterone levels and acute total testosterone/cortisol values. Lactate concentrations may also be affected since the metabolic characteristics of skeletal muscle can be influenced by the sympathetic system (33, 60, 61). The result is that few physiological systems escape the influence of the sympathetic system, and preservation of this system may allow resistance to high relative intensity resistance exercise overtraining.

Besides direct neural innervation, most physiological systems are also under sympathetic regulation by epinephrine secreted from the adrenal chromaffin cells (70). These secretory characteristics can be modulated by resistance exercise (56), and provide a likely mechanism for the development of high relative intensity resistance exercise overtraining. Epinephrine, however, is not the only substance secreted by the adrenal chromaffin cells. In addition, an enkephalin-containing polypeptide called proenkephalin peptide F is co-secreted with epinephrine (30, 38, 50). The release of peptide F is related to exercise intensity (44), and exhibits differential release patterns in highly trained endurance athletes (42). Since peptide F and epinephrine are co-released in the adrenal medulla, alterations in the circulating concentrations of both can provide insight on the stimulus-secretion coupling mechanisms of the adrenal chromaffin cells. High relative intensity resistance exercise overtraining produces no changes in peptide F concentrations at rest or acutely (57). Considering the large increases in acute epinephrine in response to high relative intensity resistance exercise overtraining (20), this suggests that secretion of epinephrine and peptide F are regulated differentially. With highly trained runners, different release patterns were theorized by Kraemer and colleagues to permit optimal concentrations of epinephrine to occur when most needed during exercise (42), and it is speculated here that a similar situation exists with high relative intensity resistance exercise overtraining. In an attempt to preserve strength performance, epinephrine is preferentially released from the adrenal medulla at the expense of peptide F. Unfortunately, there may be a down side to such a response, since peptide F has been implicated in the function of the immune system (31). Whether this is a contributing factor to increased rates of illness and infection during overtraining remains to be studied.

The role of the sympathetic nervous system appears to provide fertile ground for future overtraining research. It is interesting to note that normally resistance trained subjects exhibit significant relationships between immediately postexercise circulating concentrations of catecholamines and muscular strength performance measures ($r = .79$ to .96), while overtrained subjects do not (12, 20). While such data require further investigation, it is noteworthy that sympathetic activity may have a profound influence on muscular performances as well as on other physiological systems. The interactions between circulating catecholamines and diminished skeletal muscle performance suggest alterations of adrenergic-receptor activity (9, 12, 20, 60, 61). Whether such changes are in receptor number, receptor density, receptor affinity, or are due to postreceptor activity remain to be studied.

Psychological Responses

Few data are available on the psychological responses to high relative intensity resistance exercise overtraining. Initial study of this type of overtraining did not find any changes in mood-state (19) which is often reported as a marker of overtraining (53). The test instrument, however, was an abridged form of the profile of mood-states tool, and it has been suggested that this is not an appropriate instrument for these purposes (55). A relationship has been reported between hypothalamic-pituitary activity and clinical depression (36), and although overtraining may not result in clinical depression, a relationship between endocrine activity and psychological status may partly explain why athletes overtraining with endurance exercises or with high volumes of resistance exercise often exhibit altered mood-states (23, 45, 51). With the present protocol for the study of high relative intensity overtraining, no decreases were observed for circulating hormonal levels (18), despite the fact that training logs indicated a significantly decreased desire to complete the training sessions (19). This would suggest that altered attitudes are not necessarily related to endocrine status. In general, the nature of the psychological data with this overtraining protocol are not very convincing at this time (55), and provide another area for important future research.

An interesting aspect of high relative intensity resistance exercise overtraining is the role of self-efficacy. It appears that although perceptions of difficulty of the lifting task did not change during the two-week overtraining task, perceptions of anticipated lifting success were significantly decreased by day eight of the protocol (22). Despite the fact that all lifts were performed at a known relative intensity (at or near 100% 1 RM), the subjects exhibited a lower confidence for successfully completing the lift. As a result, it may be that these attitudes contributed to the impaired strength performances of the overtrained subjects.

Summary

It appears that maximal strength capabilities during high relative intensity resistance exercise overtraining are vigorously protected and preserved by an individual, and may be simply a survival mechanism of the organism. As a result, it is quite difficult to elicit statistically significant decreases in 1 RM strength in a laboratory setting. A contributing factor is that motivational characteristics of experimental subjects are most likely not the same as those of elite, highly trained athletes. As a result, much unpublished and published pilot work was necessary to develop a protocol that would eventually produce attenuated 1 RM strength (14, 18-20). Once the modality was changed from weight machines to free weights, a completely new protocol needed to be developed (75), and is still in the process of refinement. Regardless of the training modality, it is believed that the role of familiarization is extremely critical so that performance changes can be attributed solely to the overtraining stimulus (10, 63). It should also be noted here that many of the impaired performances reported for the high relative intensity overtraining studies are much greater than would be tolerated by any intelligent coach or athlete. However, it is only through the study of extreme conditions such as the high relative intensity resistance exercise overtraining protocol (14, 18-22, 57, 58), or high volumes of resistance exercise (15-17, 74), or endurance exercise (45, 48, 49), that an understanding of the physiological mechanisms underlying overtraining can be understood.

Figure 6.3 is a simple illustration of several of the physiological factors that may or may not contribute to high relative intensity resistance exercise overtraining. In general, some sort of maladaptation of the peripheral musculature is suggested, with such responses most likely associated with the adrenergic regulation of skeletal muscle. The lone exception to this was when an overuse syndrome of the knees developed, seeming to result in inhibitory input to the working muscles. Circulating hormones did not exhibit previously suggested profiles for overtraining. Instead, resting and acute endocrine concentrations were mostly unchanged, or exhibited slight increases in acute total testosterone and slight decreases in acute cortisol, the opposite of what is seen with other types of overtraining. Up to now, few data exist to support the existence of a sympathetic overtraining syndrome, but it is believed that the augmented catecholamine response to exercise while muscular strength is being attenuated may be indicative of the onset of a

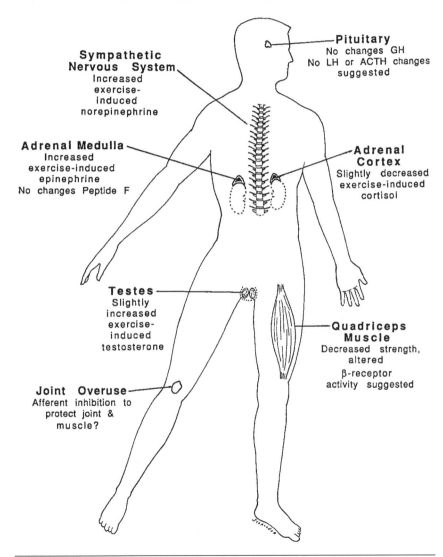

Figure 6.3 Summary of physiological responses to the high relative intensity resistance exercise overtraining protocol.
Art by Roy C. Schroeder, PhD.
Based on data from references 17-21, 56, and 57.

sympathetic overtraining syndrome. Overall, it can be concluded that high relative intensity resistance exercise overtraining results in a different physiological profile than does endurance overtraining, or even high volume resistance exercise overtraining.

Future Study

A number of areas can be suggested for future study with a high relative intensity-based overtraining protocol.

1. Further study of different exercise modalities and training protocols that may result in dissimilar performance and physiological responses (10, 19, 24, 41, 43, 69, 72, 75).
2. Further study of the gross and fine motor control characteristics that may contribute to impaired physical performances (14, 19, 63, 67, 75).
3. Study of the cellular and molecular adaptations of skeletal muscle to overtraining. Adaptations and responses of muscle to normal (66) and stressful (8) forms of resistance exercise, as well as stressful endurance exercise (5), have been investigated, but not in direct response to resistance exercise overtraining.
4. Study of nutritional supplementation protocols that may delay the onset of a high relative intensity resistance exercise overtraining syndrome. Although glycogen depletion may be a contributing factor to other forms of stressful training (5), preliminary data do not indicate a similar role with high relative intensity resistance exercise overtraining. Branched-chain amino acid (BCAA) supplementation has been suggested as being beneficial during stressful training (54), even though this has not demonstrated an ergogenic effect with increased volumes of resistance exercise (16, 17). Use of BCAAs has not been studied with a high relative intensity resistance exercise overtraining protocol. The potential of orally ingested creatine monohydrate for performance enhancement has received much research interest recently (25), and may have potential for providing tolerance to overtraining protocols.
5. Study of pharmaceutical interventions that could influence the onset of a state of overtraining (e.g., adrenergic analogs, anabolic steroids).
6. Further study of the psychological profiles that precede and accompany a state of high relative intensity resistance exercise overtraining. This area of research may provide some of the most practical information for the coach and athlete.
7. More research is needed to track the recovery characteristics of individuals who have experienced a state of overtraining, thus providing insight concerning effective means of recovery. Although attempts have been made to study individuals as they recover from a stressful training stimulus (15, 17, 19, 39, 75), closer scrutiny is required to fully understand the necessary recovery process.
8. Study of gender differences as they relate to overtraining. Important physiological differences exist between the genders that could conceivably influence an overtraining response.

Acknowledgment

Thanks to Roy C. Schroeder, PhD, for contributing the artwork used in this chapter.

References

1. Adlercreutz, H., M. Härkönen, K. Kuoppasalmi, H. Näveri, I. Huhtaniemi, H. Tikkanen, K. Remes, A. Dessypris, J. Karvonen. 1986. Effect of training on plasma anabolic and catabolic steroid hormones and their response during physical exercise. *International Journal of Sports Medicine* 7: S27-28.
2. Barron, J.L., T.D. Noakes, W. Levy, C. Smith, R.P. Millar. 1985. Hypothalamic dysfunction in overtrained athletes. *Journal of Clinical Endocrinology and Metabolism* 60: 803-806.
3. Budgett, R. Overtraining syndrome. 1990. *British Journal of Sports Medicine* 24: 231-236.
4. Busso, T., K. Häkkinen, A. Pakarinen, C. Carasso, J.R. Lacour, P.V. Komi, H. Kauhanen. 1990. A systems model of training responses and its relationship to hormonal responses in elite weight-lifters. *European Journal of Applied Physiology* 61: 48-54.
5. Costill, D.L., M.G. Flynn, J.P. Kirwan, J.A. Houmard, J.B. Mitchell, R. Thomas, S.H. Park. 1988. Effects of repeated days of intensified training on muscle glycogen and swimming performance. *Medicine and Science in Sports and Exercise* 20: 249-254.
6. Cumming, D.C., G.D. Wheeler, E.M. McColl. 1989. The effect of exercise on reproductive function in men. *Sports Medicine* 7: 1-17.
7. DeAndrade, J.R., C. Grant, A.S. Dixon. 1965. Joint distention and reflex muscle inhibition in the knee. *Journal of Bone and Joint Surgery* 47: 313-322.
8. Ebbeling, C.B., P.M. Clarkson. 1989. Exercise-induced muscle damage and adaptation. *Sports Medicine* 7: 207-234.
9. Fellenius, E., R. Hedberg, E. Holmberg, B. Waldeck. 1980. Functional and metabolic effects of terbutaline and propranolol in fast- and slow-contracting skeletal muscle. *Acta Physiologica Scandinavica* 109: 89-95.
10. Fleck, S.J., W.J. Kraemer. 1997. *Designing resistance exercise programs* (2nd ed.) Champaign, IL: Human Kinetics.
11. Fry, A.C. 1992. The effects of acute training status on reliability of integrated electromyographic activity and "efficiency of electrical activity" during isometric contractions: a case study. *Electromyography and Clinical Neurophysiology* 32: 565-570.
12. Fry, A.C., W.J. Kraemer. 1997. Resistance exercise overtraining and overreaching: neuroendocrine responses. *Sports Medicine,* 23(2): 106-129.
13. Fry, A.C., T.A. Aro, J.A. Bauer, W.J. Kraemer. 1991. A kinematic comparison of three barbell squat variations and a squat simulating machine. *Journal of Applied Sport Science Research* 5: 162.
14. Fry, A.C., W.J. Kraemer, J.M. Lynch, N.T. Triplett, L.P. Koziris. 1994. Does short-term near-maximal intensity machine resistance exercise induce overtraining? *Journal of Applied Sport Science Research* 8: 188-191.

15. Fry, A.C., W.J. Kraemer, M.H. Stone, B.J. Warren, S.J. Fleck, J.T. Kearney, S.E. Gordon. 1994. Endocrine responses to overreaching before and after 1 year of weightlifting. *Canadian Journal of Applied Physiology* 19: 400-410.

16. Fry, A.C., W.J. Kraemer, M.H. Stone, B.J. Warren, J.T. Kearney, S.J. Fleck, C.A. Weseman. 1993. Endocrine and performance responses to high volume training and amino acid supplementation in elite junior weightlifters. *International Journal of Sports Nutrition* 3: 306-322.

17. Fry, A.C., W.J. Kraemer, M.H. Stone, J.T. Kearney, S.J. Fleck, J. Thrush, S.E. Gordon, N.T. Triplett. 1992. Endocrine and performance responses during one month of periodized weightlifting with amino acid supplementation. *Journal of Applied Sport Science Research* 6: 183.

18. Fry, A.C., W.J. Kraemer, F. van Borselen, J.M. Lynch, J.L. Marsit, N.T. Triplett, L.P. Koziris. 1993. Endocrine responses to short-term intensity-specific resistance exercise overtraining. *Journal of Strength and Conditioning Research* 7: 179.

19. Fry, A.C., W.J. Kraemer, F. van Borselen, J.M. Lynch, J.L. Marsit, E.P. Roy, N.T. Triplett, H.G. Knuttgen. 1994. Performance decrements with high-intensity resistance exercise overtraining. *Medicine and Science in Sports and Exercise* 26: 1165-1173.

20. Fry, A.C., W.J. Kraemer, F. van Borselen, J.M. Lynch, N.T. Triplett, L.P. Koziris, S.J. Fleck. 1994. Catecholamine responses to short-term high-intensity resistance exercise overtraining. *Journal of Applied Physiology* 77: 941-946.

21. Fry, A.C., J.M. Barnes, W.J. Kraemer, J.M. Lynch. 1996. Overuse syndrome of the knees with high-intensity resistance exercise overtraining: a case study. *Medicine and Science in Sports and Exercise* 28: S128.

22. Fry, M.D., A.C. Fry, W.J. Kraemer. 1996. Self-efficacy responses to short-term high intensity resistance exercise overtraining. *International Conference on Overtraining and Overreaching in Sport: Physiological, Psychological, and Biomedical Considerations*. Memphis, TN.

23. Fry, R.W., A.R. Morton, D. Keast. 1991. Overtraining in athletes: an update. *Sports Medicine* 12: 32-65.

24. Fry, R.W., A.R. Morton, D. Keast. 1992. Periodisation and the prevention of overtraining. *Canadian Journal of Applied Sport Science* 17: 241-248.

25. Greenhaff, P.L. 1995. Creatine and its application as an ergogenic aid. *International Journal of Sports Nutrition* 5: S100-110.

26. Hagerman, F.C. 1992. Failing to adapt to training. *FISA Coach* 3: 1-4.

27. Häkkinen, K., A. Pakarinen. 1991. Serum hormones in male strength athletes during intensive short-term strength training. *European Journal of Applied Physiology* 63: 194-199.

28. Häkkinen, K., A. Pakarinen. 1993. Acute hormonal responses to two different fatiguing heavy-resistance protocols in male athletes. *Journal of Applied Physiology* 74: 882-887.

29. Häkkinen, K., A. Pakarinen, M. Alén, H. Kauhanen, P.V. Komi. 1988. Daily hormonal and neuromuscular responses to intensive strength training in 1 week. *International Journal of Sports Medicine* 9: 422-428.

30. Hanbauer, I., G.D. Kelly, L. Sainani, H.Y.T. Yang. 1982. [Met5]-enkephalin-like peptides of the adrenal medulla: release by nerve stimulation and functional implications. *Peptides* 3: 469-473.

31. Hiddinga, H.J., D.D. Isaak, R.V. Lewis. 1994. Enkephalin-containing peptides processed from the proenkephalin significantly enhances the

antibody-forming cell responses to antigens. *Journal of Immunology* 152: 3748-3759.

32. Holzwarth, M.A., L.A. Cunningham, N. Kleitman. 1987. The role of adrenal nerves in the regulation of adrenocortical function. *Annals of the New York Academy of Sciences* 512: 449-464.

33. Jeukendrup, A.E., M.K. Hesselink. 1994. Overtraining: what do lactate curves tell us? *British Journal of Sports Medicine* 28: 239-240.

34. Jezova, D., M. Vigas. 1981. Testosterone response to exercise during blockade and stimulation of adrenergic receptors in man. *Hormone Research* 15: 141-147.

35. Jost, J., M. Weiss, H. Weicker. 1989. Comparison of sympatho-adrenergic regulation at rest and of the adrenoceptor system in swimmers, long-distance runners, weight lifters, wrestlers, and untrained men. *European Journal of Applied Physiology* 58: 596-604.

36. Kalin, N.H., G.W. Dawson. 1986. Neuroendocrine dysfunction in depression: hypothalamic-anterior pituitary systems. *Trends in Neuroscience* 9: 261-266.

37. Kame, V.D., D.R. Pendergast, B. Termin. 1990. Physiologic responses to high intensity training in competitive university swimmers. *Journal of Swimming Research* 6: 5-8.

38. Kilpatrick, D.L., R.V. Lewis, S. Stein, S. Udenfriend. 1980. Release of enkephalins and enkephalin-containing polypeptides from perfused beef adrenal glands. *Proceedings of the National Academy of Sciences: USA* 77: 7473-7475.

39. Koziris, L.P., A.C. Fry, W.J. Kraemer, M.H. Stone, J.T. Kearney, S.J. Fleck, J. Thrush, S.E. Gordon, N.T. Triplett. 1992. Hormonal and competitive performance responses to an overreaching training stimulus in elite junior weightlifters. *Journal of Applied Sport Science Research* 6: 186.

40. Kraemer, W.J., A.C. Fry, B.J. Warren, M.H. Stone, S.J. Fleck, J.T. Kearney, B.P. Conroy, C.M. Maresh, C.A. Weseman, N.T. Triplett, S.E. Gordon. 1992. Acute hormonal responses in elite junior weightlifters. *International Journal of Sports Medicine* 13: 103-109.

41. Kraemer, W.J., L. Marchitelli, S.E. Gordon, E.A. Harman, J.E. Dziados, R. Mello, P.N. Frykman, D. McCurry, S.J. Fleck. 1990. Hormonal and growth factor responses to heavy resistance exercise protocols. *Journal of Applied Physiology* 69: 1442-1450.

42. Kraemer, W.J., B. Noble, B. Culver, R.V. Lewis. 1985. Changes in plasma proenkephalin peptide F and catecholamine levels during graded exercise in men. *Proceedings of the National Academy of Sciences: USA* 82: 6349-6351.

43. Kraemer, W.J., J.F. Patton, S.E. Gordon, E.A. Harman, M.R. Deschenes, K. Reynolds, R.U. Newton, N.T. Triplett, J.E. Dziados. 1995. Compatibility of high-intensity strength and endurance training on hormonal and skeletal muscle adaptations. *Journal of Applied Physiology* 78: 976-989.

44. Kraemer, W.J., J.F. Patton, H.G. Knuttgen, C.J. Hannan, T. Kettler, S.E. Gordon, J.E. Dziados, A.C. Fry, P.N. Frykman, E.A. Harman. 1991. Effects of high-intensity cycle exercise on sympathoadrenal–medullary response patterns. *Journal of Applied Physiology* 70: 8-14.

45. Kuipers, H., H.A. Keizer. 1988. Overtraining in elite athletes: review and directions for the future. *Sports Medicine* 6: 79-92.

46. Lehmann, M., P. Baumgartl, C. Weisenack, A. Seidel, H. Baumann, S. Fischer, U. Spori, G. Gendrisch, R. Kaminski, J. Keul. 1992. Training-overtraining:

influence of a defined increase in training volume vs. training intensity on performance, catecholamines and some metabolic parameters in experienced middle- and long-distance runners. *European Journal of Applied Physiology* 64: 169-177.

47. Lehmann, M., C. Foster, J. Keul. 1993. Overtraining in endurance athletes: a brief review. *Medicine and Science in Sports and Exercise* 25: 854-862.

48. Lehmann, M., U. Gastmann, K.G. Petersen, N. Bachl, A.N. Khalaf, S. Fischer, J. Keul. 1992. Training-overtraining: performance, and hormone levels, after a defined increase in training volume versus intensity in experienced middle- and long-distance runners. *British Journal of Sports Medicine* 26: 233-242.

49. Lehmann, M., W. Schnee, R. Scheu, W. Stockhausen, N. Bachl. 1992. Deceased nocturnal catecholamine excretion: parameter for an overtraining syndrome in athletes? *International Journal of Sports Medicine* 13: 236-242.

50. Livett, A.R., D.M. Dean, L.G. Whelan, S. Udenfriend, J. Rossier. 1981. Co-release of enkephalin and catecholamines from cultured adrenal chromaffin cells. *Nature* 289: 317-319.

51. Lombardo, J. 1993. The efficacy and mechanisms of action of anabolic steroids. In *Anabolic steroids in sport and exercise*, ed. C.E. Yesalis, 89-106. Champaign, IL: Human Kinetics.

52. McMillan, J.L., M.H. Stone, J. Sartin, D. Marple, R. Keith, D. Lewis, C. Brown. 1993. The 20-hour hormonal response to a single session of weight-training. *Journal of Strength and Conditioning Research* 7: 9-21.

53. Morgan, W.P., D.R. Brown, J.S. Raglin, P.J. O'Connor, K.A. Ellickson. 1987. Psychological monitoring of overtraining and staleness. *British Journal of Sports Medicine* 21: 107-114.

54. Newsholme, E.A., E. Blomstrand, N. McAndrew, M. Parry-Billings. 1992. Biochemical causes of fatigue and overtraining. In *Endurance in sports*, eds. R.J. Shephard, P.-O. Åstrand, 351-364. London: Blackwell Scientific.

55. O'Connor, P.J. In press. Overtraining and staleness. In *Physical activity and mental health,* ed. W.P. Morgan. Bristol Penn, PA: Taylor and Francis.

56. Péronnet, F., G. Thibault, H. Perrault, D. Cousineau. 1986. Sympathetic response to maximal bicycle exercise before and after leg strength training. *European Journal of Applied Physiology* 55: 1-4.

57. Ramsey, L.T., A.C. Fry, W.J. Kraemer, S.J. Fleck, R.S. Staron. 1995. Plasma proenkephalin peptide F responses to short-term high-intensity resistance exercise overtraining. *Southeastern American College of Sports Medicine Conference*. Lexington, KY.

58. Ramsey, L.T., A.C. Fry, W.J. Kraemer, J.M. Lynch. 1996. Pituitary responses to high intensity resistance exercise overtraining. *Southeastern American College of Sports Medicine Conference*. Chattanooga, TN.

59. Renstrom, P. 1988. Overuse injuries. In *The Olympic book of sports medicine*, eds. A. Dirix, H.G. Knuttgen, K. Tittel, 446-468. Oxford: Blackwell Scientific.

60. Richter, E.A., N.B. Ruderman, H. Galbo. 1982. Alpha and beta adrenergic effects on metabolism in contracting, perfused muscle. *Acta Physiologica Scandinavica* 116: 215-222.

61. Richter, E.A., N.B. Ruderman, H. Gavros, E.R. Belur, H. Galbo. 1982. Muscle glycogenolysis during exercise: dual control by epinephrine and contractions. *American Journal of Physiology* 242: E25-32.

62. Robaire, B., S.F. Bayly. 1989. Testicular signaling: incoming and outgoing messages. *Annals of the New York Academy of Sciences* 564: 250-260.

63. Rutherford, O.M., D.A. Jones. 1986. The role of learning and coordination in strength training. *European Journal of Applied Physiology* 55: 100-105.
64. Selye H. 1956. *The stress of life*. New York: McGraw-Hill.
65. Sherman, K.S., D.T. Shakespeare, M. Stokes, A. Young. 1983. Inhibition of voluntary quadriceps activity after menisectomy. *Clinical Science* 64: 70.
66. Staron, R.S., D.L. Karapondo, W.J. Kraemer, A.C. Fry, S.E. Gordon, J.E. Falkel, F.C. Hagerman, R.S. Hikida. 1994. Skeletal muscle adaptations during early phase of heavy-resistance training in men and women. *Journal of Applied Physiology* 76: 1247-1255.
67. Stone, M.H., A.C. Fry, J. Thrush, S.J. Fleck, W.J. Kraemer, J.T. Kearney, and J. Marsh. 1993. Overtraining and weightlifting. In *Proceedings of the 1993 weightlifting symposium*, 133-141. Budapest, Hungary: International Weightlifting Federation.
68. Stone, M.H., R.E. Keith, J.T. Kearney, S.J. Fleck, G.D. Wilson, N.T. Triplett. 1991. Overtraining: a review of the signs and symptoms and possible causes. *Journal of Applied Sport Science Research* 5: 35-50.
69. Stone, M.H., H. O'Bryant, J. Garhammer. 1981. A hypothetical model for strength training. *Journal of Sports Medicine and Physical Fitness* 21: 342-351.
70. Unsworth, C.D., O.H. Viveros. 1987. Neuropeptides of the adrenal medulla. In *Stimulus-secretion coupling in chromaffin cells*, vol I., eds. K. Rosenheck, P.I. Lelkes, 87-109. Boca Raton, Florida: CRC Press.
71. Urhausen, A., H. Gabriel, W. Kindermann. 1995. Blood hormones as markers of training stress and overtraining. *Sports Medicine* 20: 251-276.
72. van Borselen, F., N.H. Vos, A.C. Fry, W.J. Kraemer. 1992. The role of anaerobic exercise in overtraining. *National Strength and Conditioning Association Journal* 14: 74-79.
73. Van Helder, W.P., M.W. Radomski, R.C. Goode. 1984. Growth hormone responses during intermittent weight lifting exercise in men. *European Journal of Applied Physiology* 53: 31-34.
74. Warren, B.J., M.H. Stone, J.T. Kearney, S.J. Fleck, R.L. Johnson, G.D. Wilson, W.J. Kraemer. 1992. Performance measures, blood lactate, and plasma ammonia as indicators of overwork in elite junior weightlifters. *International Journal of Sports Medicine* 13: 372-376.
75. Webber, J.L., A.C. Fry, L.W. Weiss, Y. Li, M.P. Ferreira, C.N. Alexander. 1996. Impaired performances with high intensity free weight resistance exercise. Paper presented at *International Conference on Overtraining and Overreaching in Sport: Physiological, Psychological, and Biomedical Considerations*. Memphis, TN.

Medical Aspects of Overreaching and Overtraining

CHAPTER 7

Cardiovascular and Hematologic Alterations

Pamela S. Douglas, MD, and Mary L. O'Toole, PhD

Introduction

Very little in the scientific literature discusses the effects of overtraining or overreaching on the cardiovascular system. Much has been written regarding cardiac adaptations to chronic exercise training, and in general, the resultant changes are assumed to be beneficial rather than harmful. These changes include modest increases in chamber thickness and cavity size (physiologic hypertrophy), preserved systolic function, and enhanced diastolic filling (10, 18, 23, 37, 65). Investigations of the acute effects of a single bout of prolonged exercise are fewer, although several studies have documented transient reductions in cardiac function following very prolonged aerobic exercise, sometimes termed cardiac fatigue (11, 21, 41, 48, 55, 64). While this may in itself constitute overreaching as far as the cardiovascular system is concerned, no studies have definitively documented the mechanism(s) associated with this decrement in function, although some have been postulated, in part because it is difficult to reproduce cardiac fatigue in the absence of musculoskeletal fatigue and marked changes in fluid and electrolytes and thermoregulation. Indeed, the complex and multiple physiologic changes associated with prolonged exercise are important considerations in documenting the existence and extent of cardiac fatigue itself. Thus, while transient cardiac dysfunction may provide evidence for overreaching, it is difficult to assess the contribution of cardiac dysfunction, if any, to generalized fatigue or decreased exercise performance. Similarly, no studies have documented a progression from overreaching to overtraining as manifest by further cardiac dysfunction.

Cardiovascular Considerations

Cardiac Drift

Early efforts to describe the cardiovascular responses to prolonged exercise
($\geq$ 1 hour) noted preservation of cardiac output despite a gradually increasing heart
rate and decreasing stroke volume over time, termed cardiac drift (19, 20, 26, 49)
(see figure 7.1). The magnitude of these changes appears related to the intensity of
exercise (relative workload) and environmental stress. Heat, in particular, is thought
to be important because thermoregulatory responses produce cutaneous vasodila-
tation, which depletes central blood volume and thereby reduces blood available to
perfuse working muscle. Indeed, the increase in heart rate has been correlated with
the increase in body temperature during prolonged exercise (19), although this one
factor does not appear to fully account for the observed cardiac changes (49). Other
factors have been postulated to contribute to these changes. Dehydration and the

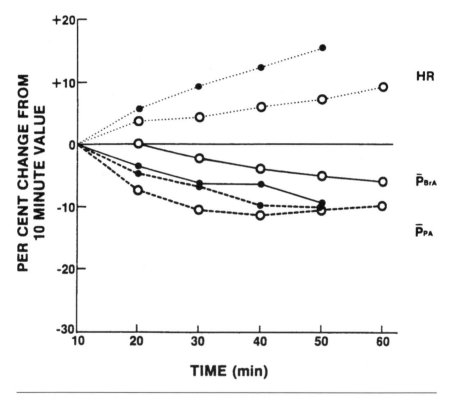

Figure 7.1 Responses of heart rate (HR, dotted line), mean brachial artery
pressure ($\bar{P}_{BrA}$, continuous line), and mean pulmonary artery pressure ($\bar{P}_{PA}$,
dashed line) during prolonged exercise. Data are presented as percent change
from a value obtained 10 minutes after initiation of exercise. Exercise was
performed in the sitting position at two intensities: solid circles represent higher
intensity work than open circles.
Reprinted from Raven and Stevens (Cooper Publishing Group) 1988.

redistribution of blood volume that occurs on initiation of exercise may both play a role but are probably of minor importance (19, 20, 53, 56). Metabolic drift, or the gradual increase in oxygen uptake and subsequent increase in cardiac output required to perform the same external workload (8, 27), may increase heart rate through increased VO_2 and central sympathetic stimulation (38) primarily intended to augment fatiguing skeletal muscle. The importance of such a feedback mechanism in prolonged exercise has not been well demonstrated.

Cardiac Fatigue

The earliest descriptions of cardiac fatigue or myocardial dysfunction following prolonged exercise (8, 20, 54) document an increased heart size. More sophisticated studies have shown reductions in left ventricular systolic performance after prolonged submaximal exercise (11, 21, 41, 48, 55, 64), or reductions in the ability to perform maximal exercise (54, 63). Compared to baseline testing, Upton et al. (63) found a reduction in maximal cardiac output, stroke volume, VO_2max, and work time when testing was performed immediately after 2 h of submaximal exercise. (See figure 7.2a and b.) Similar findings have been reported by Saltin and Stenberg (54).

Niemela et al. (41) reported a reduction in the extent and rate of echocardiographic left ventricular (LV) systolic shortening following a 24-hour run. Because of the increase in end-systolic chamber size and its relation to the decrease in function, and the decrease in blood pressure (afterload), these changes were interpreted as representing a decrease in contractility. All changes returned to baseline 2-3 days after exercise (see figure 7.3). Similar findings have been reported by others following an Ironman triathlon (2.4 mi swim, 112 mi bike, 26.2 mi run) (10) and a marathon run (48). In the former case, the decrease in fractional shortening resolved within 24-48 h, while LV size (preload) remained reduced, providing further evidence that the decrease in function is not simply due to altered loading conditions.

Untrained or sedentary individuals also display a similar decrease in cardiac performance after prolonged exercise. A study by Ketelhut et al. (31) carefully examined cardiac function at matched heart rates before and after five and 60 minutes of cycling. Decreases in cardiac output, ejection fraction, and fractional shortening were felt to contribute to reduced blood pressure and provide evidence of a fall in cardiac pump function. It is of note that the duration of exercise required to produce a similar magnitude of left ventricular dysfunction is much less in the sedentary individual (21, 31, 55, 64) than in the athlete, suggesting that one of the effects of exercise training is raising the threshold required for myocardial fatigue.

Although most studies have concentrated on systolic performance, two have examined diastolic function, and with slightly differing results. Niemela et al. (42) noted a decrease in M-mode echo derived cavity expansion, while Douglas et al. (11) found no change in this index or in the rate of wall thinning. However, the Douglas group did note an alteration in the transmitral LV filling pattern (increased atrial systolic flow velocity), which could not be explained by heart rate or preload alone.

As noted above, a limitation of all these studies is the necessity of measuring cardiac performance in intact human volunteers under markedly changing physiologic conditions. No perfect study method exists, so the attribution of a decrease in a systolic index to reduced contractility cannot be definitively proven.

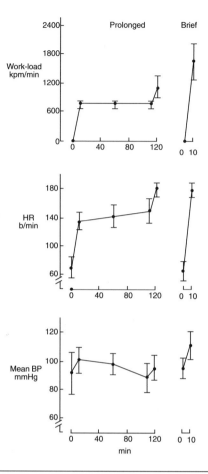

Figure 7.2a   Heart rate (HR), blood pressure (BP), and workload during brief and prolonged exercise. A gradual increase in heart rate during prolonged exercise is associated with a decrease in mean blood pressure. The workload is higher during maximal effort at the conclusion of the brief exercise than at the conclusion of prolonged exercise.

Reprinted from Upton (Excerpta Medica Inc.) 1980.

The issue of whether cardiac fatigue represents a mild form of cardiac damage remains open. Although no elevation in cardiac muscle-specific enzymes has been found following a 24 h run (41, 42) or an Ironman triathlon (11), other studies suggest that such measures may be insensitive. Osbakken and Locko (44) found anterior perfusion defects on thallium scanning performed after 45 min of maximal treadmill testing in marathon runners. Carrio et al. (4) noted myocardial uptake of antimyosin antibodies in 70% of runners following a 6 h run. Douglas et al. (12) noted septal regional wall motion abnormalities following completion of an Ironman race in the same region noted to be abnormal by Osbakken and Locko (44). In a subsequent study, the septal region also displayed altered ultrasonic myocardial tissue characteristics (5, 14).

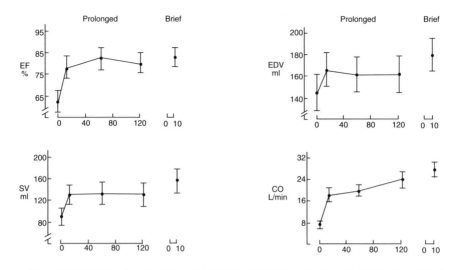

Figure 7.2b The cardiovascular response to brief and prolonged exercise. The increase in stroke volume (SV) during exercise is the result of an increase in both ejection fraction (EF) and end-diastolic volume (EDV). Cardiac output (CO) during maximal effort was higher during the brief than during the prolonged exercise period, primarily as a result of a larger end-diastolic volume.

Reprinted from Upton (Excerpta Medica Inc.) 1980.

A variety of mechanisms has been postulated to account for cardiac fatigue. The obvious possibility of ischemia has been discussed above. Metabolic abnormalities, including high levels of circulating free fatty acids (55), oxidative stress (57), or an as yet unspecified derangement have all been postulated. Among the most compelling possibilities is a reduction in cardiac beta-adrenergic responsiveness, which has been reported in dogs following as little as 1 h of exercise (24). Studies in sedentary humans found similar results (21). The amount of intravenously administered isoproterenol required to raise the heart rate by 15 bpm was increased more than 2-fold after prolonged submaximal exercise (average duration 95 min). The reduced responsiveness to catechols correlated closely with the degree of cardiac fatigue, as measured by a reduction in ejection fraction. Similar findings have been noted following completion of an Ironman triathlon (13). The possibility that cardiac fatigue is due to catechol excess also draws support from the transient cardiomyopathy associated with pheochromocytoma (30, 60), and from reports that norepinephrine produces cardiac dysfunction only when infused in a setting of increased myocardial work (1).

A final possible contributor to reduced cardiac function is the sustained tachycardia of exercise; in this case, the clinical analogy is with the transient decreases in LV function associated with incessant supraventricular tachycardia(s) (6, 45) or rapid pacing (58). It is likely that given the complex physiologic events associated with prolonged exercise, no single etiology or mechanism will be identified, and that associated factors such as blood volume, thermoregulation, autonomic nervous system and baroreceptor activity, and hydration will never be fully excluded as contributors.

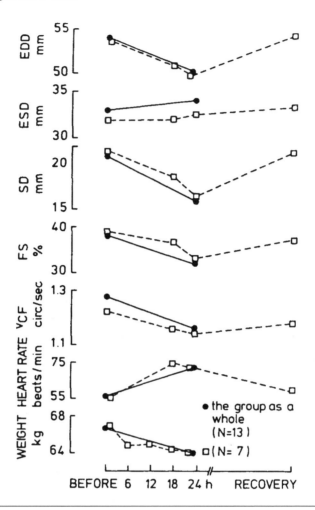

Figure 7.3   Echocardiographic data and body weight before, during, and after a 24-hour run. The measurements during recovery were made 2-3 days after the race. SD = stroke dimension, FS = fractional shortening, ESD = end-systolic dimension, EDD = end-diastolic dimension. ^{v}CF = velocity of circumferential shortening. Reprinted from Niemela (American Heart Association) 1984.

Overwork

Day-to-day variations in cardiovascular factors have been suggested as potential markers for overreaching and overtraining. Overwork has been postulated to result in increased resting, exercise, and recovery heart rates (43, 62). Anecdotally, it is not uncommon for athletes to monitor morning heart rates as a signal of impending overreaching or overtraining. In a prospective study, 12 experienced marathoners were monitored while competing in a 20-day road race. Total mileage was 312 mi with an average 17.3 mi/d (twice normal training distances). Within 30 minutes of awakening, resting heart rates were measured daily. Mean heart rates increased for all subjects from day eight to day 20 (M =12 bpm; min/max = 1-20 bpm) (16). Most of the runners complained of persistent leg soreness and all complained of chronic fatigue, but average running speed did not vary significantly from day to day. By strict definition, these athletes were not overreached or overtrained, but perhaps would have been classified as such if performance requirements had been higher. For example, typical daily running pace was 8 minutes 30 seconds per mile, while best marathon paces were an average of one minute per mile faster (approximately 15%). Lower resting heart rates and normal or faster recovery heart rates have also been reported to occur in overtrained athletes and have been attributed to exhaustion of the neuroendocrine system (32) or a decrease in catecholamine sensitivity (36). Some evidence exists in support of the decreased catecholamine sensitivity hypothesis. Lehmann et al. (35) reported a reduced heart rate response to submaximal exercise performance accompanied by higher plasma noradrenaline levels. At maximal exercise, both unchanged (35) and reduced (61) catecholamine levels have been observed.

Blood pressure alterations have also been suggested to accompany overexercise (62). Diastolic and systolic blood pressures have been reported to increase (6 and 10 mmHg, respectively) from pre- to post-season in a group of college swimmers (39). Increased resting blood pressures have also been reported in football players, weightlifters, and cyclists during the course of their seasons (52). It is unclear how many, if any, of these athletes could be classified as overreached or overtrained. Verma et al. (66) reported a slower return of blood pressure to resting levels following exercise in overtrained athletes. Increased postural drop in blood pressure has also been reported to occur in overtrained athletes (52). Conversely, the marathoners followed by Dressendorfer et al. (16) did not exhibit significant changes in resting blood pressures.

Hematologic Considerations

Responses to Strenuous Training

Many hematologic variables change as a part of the normal response to strenuous training. For example, pseudoanemia or dilutional anemia is an extremely common finding in strenuously training athletes. This syndrome is characterized by suboptimal levels of hemoglobin (<16 g/dl for men; <14 g/dl for women), and is attributed to a training-induced increase in plasma volume unmatched by a proportional increase in hemoglobin (2, 47). However, a great deal of difficulty exists in distinguishing between these physiologic changes and abnormal changes that can be associated with the overtraining syndrome. Dressendorfer et al. (15) studied hematologic variables in a group of well-trained marathon runners during a 20-day road race covering 312 miles. Distances run were approximately twice their normal daily training mileages. Hematologic changes that occurred included marked reductions in red blood cell count, hemoglobin concentration, and hematocrit, all evident within the first few days of the race. The increased running mileage had no effect on mean corpuscular volume or on mean corpuscular hemoglobin concentration. The total white blood cell concentration increased significantly on the second day, but was within the normal range. Neutrophils increased and lymphocytes declined during the race (15). Interestingly, performance of the runners was maintained despite symptoms that could be interpreted as indicative of overreaching or overtraining.

Iron status of strenuously training athletes has been the subject of considerable study and debate (7). While dilutional anemia (previously discussed) appears to be related to training intensity and duration, iron deficiency appears to be related to exercise mode as well. Low serum ferritin levels have been reported in middle- and long-distance runners, but not in equally well-trained swimmers, cyclists, and rowers (9, 17). During long distance (mean finish times = 11-13 h) triathlon racing (overreaching), a 50-60% decrease in serum iron has been reported to occur (22, 50). However, during multiday overreaching, Dressendorfer et al. (15) reported that serum iron levels more than doubled by day two (149 μg/dL-326 μg/dL) and remained high for the remaining 20 days of running. Haymes et al. (28) tracked iron status parameters in highly trained cross-country skiers for eight months. They noted that total iron binding capacity increased as the season progressed, suggesting an increased need for iron. Changes in iron status were not reported in relation to performance nor were any of the athletes reported to suffer from overtraining-related performance decrements.

Overreaching

Hematologic responses may be variable and not always related to decreased performance in studies that induced short-term overreaching. R.W. Fry et al. (25) trained five subjects excessively with twice-daily interval training sessions for 10 days and were able to induce performance decrements in treadmill running. Although performance decreased, several hematologic variables expected to change in association with strenuous training did not change. The authors point out that the diagnosis of overtraining should not be excluded because of failure of blood variables to change. They also noted that following five days of rest, serum ferritin was significantly reduced from pretraining levels and suggested that specific blood variable changes may be useful in identifying an etiology of overtraining in individual athletes.

Overtraining

A few studies have addressed changes in hematologic variables or iron status in athletes clinically diagnosed as overtrained. Rowbottom et al. (51) examined multiple blood parameters traditionally associated with chronic exercise training, but were unable to find markers of diagnostic value in a group of 10 athletes who had been clinically diagnosed as overtrained. In a prospective study, Hooper et al. (29) examined hematologic variables to find a marker for monitoring overtraining during a six-month swim season. Creatine phosphokinase and erythrocyte, total leukocyte counts, as well as hemoglobin and hematocrit were measured five times in a group of 14 national-class swimmers. Three became overtrained during the season, but no significant differences in CK, total leukocytes or other differential counts, hemoglobin, or hematocrit were seen between these overtrained athletes and those who did not exhibit signs of overtraining. Neutrophil number, however, was significantly higher for the overtrained swimmers during the taper. The reason for this is not clear and may be related to residual effects from previous strenuous workouts, since neutrophil count can be elevated for 24 h postexercise (40).

Previously, neutrophil count has been reported to be either increased (68) or decreased (36) in overtrained athletes. Shephard et al. (59) have suggested that leukocytosis and granulocytosis in response to acute exercise is mainly a reflection of changes in blood volume and thus difficult to relate to overtraining. Wishnitzer et al. (67) studied hematologic variables in 18 normally trained competitive distance runners and compared them with 18 overtrained distance runners. No significant differences were found between the two groups of runners in level of hemoglobin, mean corpuscular volume, serum iron, iron saturation, and ferritin. Decreased bone marrow hemosiderin was seen in both groups, but overtrained runners had a more pronounced decrease in bone marrow cellularity. The authors concluded that competitive distance runners may not suffer from overt iron deficiency, but that a decreased bone marrow cellularity may be indicative of overtraining.

Increased plasma urea and creatine kinase (CK) levels have previously been associated with overtraining (3). However, Lehmann et al. (33, 34) reported no change in urea, creatinine, or uric acid in middle- and long-distance runners who exhibited performance decrements as a result of markedly increased training volumes over a 4-week period. In a prospective study in which excessive training loads induced overtraining in race horses, plasma urea levels also remained within normal limits and CK did not increase (3). These findings suggest that overtraining is not associated with protein breakdown or disruption of sarcolemmal integrity, and that neither plasma urea nor CK levels are useful as markers for overtraining. However, there is accumulating evidence that glutamine status is affected by overtraining and may contribute to impaired immune function. Parry-Billings et al. (46) studied 40 athletes from various sports who had been clinically diagnosed as overtrained. Overtraining symptoms had been present in all athletes for at least three weeks. Resting plasma concentrations of glutamine were significantly lower in overtrained athletes compared with matched controls.

Other Considerations

The foregoing discussion has been confined to the effects of prolonged aerobic exercise. Although the acute hemodynamic effects of resistive exercise are well described and striking (36), little is known regarding persistent, postexercise cardiac or hematologic effects, or manifestations of overtraining or overreactivity. Similarly, the previous discussion has focused solely on the healthy, young individual, whether trained or sedentary. In particular, the cardiovascular effects of exercise are obviously quite different in those with cardiovascular disease or taking medications, or in people with other diseases that affect the cardiovascular system such as diabetes or severe anemia. In general, however, exercise tolerance in such cases is typically limited by the underlying pathology rather than by overreaching. The effects of healthy aging on cardiovascular or hematologic manifestations of overreaching are unknown.

References

1. Bosso, F.J., F.D. Allman, C.F. Pilati. 1994. Myocardial work load is a major determinant of norepinephrine-induced left ventricular dysfunction. *American Journal of Physiology* 266 (*Heart Circulatory Physiology* 35): H531-539.

2. Brotherhood, J., B. Brozovic, L.G.C. Pugh. 1975. Haematological status of middle- and long-distance runners. *Clinical Science and Molecular Medicine* 48: 139-145.

3. Bruin, G., H. Kuipers, H.A. Keizer, G.J. Van der Vusse. 1994. Adaptation and overtraining in horses subjected to increasing training loads. *Journal of Applied Physiology* 76: 1908-1913.

4. Carrio, I., R. Serra-Grima, L. Berna, M. Estorch, C. Martinez-Dunker, J. Ordonez. 1990. Transient alterations in cardiac performance after six-hour race. *American Journal of Cardiology* 65: 1471.

5. Chafizadeh, E., S. Katz, M. O'Toole, S. Howdell, A. D'Sa, P.S. Douglas. 1995. Altered global and regional myocardial backscatter characteristics in cardiac fatigue following prolonged exercise. *Journal of American College of Cardiology* 25: 2174A.

6. Chen, S.A., C.J. Yang, C.E. Chiang, C.P. Hsia, W.P. Tsang, D.C. Wang, C.T. Ting, S.P. Wang, B.N. Chiane, M.S. Chang. 1992. Reversibility of left ventricular dysfunction after successful catheter ablation of supraventricular reentrant tachycardia. *American Heart Journal* 124: 1512.

7. Clement, D.B., L.L. Sawchek. 1984. Iron status and sports performance. *Sports Medicine* 1: 65-74.

8. Davies, C.T.M., M.W. Thompson. 1986. Physiologic responses to prolonged exercise in man. *Journal of Applied Physiology* 61: 611-617.

9. Dickson, D.N., R.S. Wilkinson, T.D. Noakes. 1982. Effects of ultra-marathon training and racing on hematologic parameters and serum ferritin levels in well-trained athletes. *International Journal of Sports Medicine* 3: 111-117.

10. Douglas, P.S., M.L. O'Toole, W.D.B. Hiller, N. Reicher. 1986. Left ventricular structure and function by echocardiography in ultraendurance athletes. *American Journal of Cardiology* 58: 805-809.

11. Douglas, P.S., M.L. O'Toole, W.D.B. Hiller, K. Hackney, N. Reicher. 1987. Cardiac fatigue after prolonged exercise. *Circulation* 76: 1206-1213.

12. Douglas, P.S., M.L. O'Toole, J. Woolard. 1990. Regional wall motion abnormalities after prolonged exercise in the normal left ventricle. *Circulation* 82: 2108-2114.

13. Douglas, P.S., S.E. Katz, M.L. O'Toole. 1994. Adrenergic desensitization accompanies left ventricular dysfunction following prolonged exercise. Paper presented at the annual meeting of the American Heart Association.

14. Douglas, P.S., A. D'Sa, S. Katz, S. Howell. 1994. Ultrasonic integrated backscatter: a new method for imaging and analysis. *Circulation* 88: I326.

15. Dressendorfer, R.H., C.E. Wade, E.A. Amsterdam. 1981. Development of pseudoanemia in marathon runners during a 20-day road race. *Journal of the American Medical Association* 246: 1215-1218.

16. Dressendorfer, R.H., C.E. Wade, J.H. Scaff. 1985. Increased morning heart rate in runners: a valid sign of overtraining? *Physician and Sportsmedicine* 13: 77-86.

17. Dufaux, B., A. Hoederath, I. Streitberger. 1981. Serum ferritin, transferrin, haptoglobin, and iron in middle- and long-distance runners, elite rowers, and professional racing cyclists. *International Journal of Sports Medicine* 2: 43-46.

18. Ehsani, A.A., G.W. Heath, J.M. Hagberg, K. Schechtman. 1981. Noninvasive assessment of changes in left ventricular function induced by graded isometric exercise in healthy subjects. *Chest* 80: 51-55.

19. Ekelund, L.G. 1967. Circulatory and respiratory adaptations during prolonged exercise. *Acta Physiologica Scandinavica* 70: 5-38.

20. Ekelund, L.G., A. Holmgren, C.O. Ovenfors. 1967. Heart volume during prolonged exercise in the supine and sitting position. *Acta Physiologica Scandinavica* 70: 88-98.

21. Eysmann, S.B., E. Gervino, D.E. Vatner, S.E. Katz, L. Decker, P.S. Douglas. 1996. Prolonged exercise alters beta-adrenergic responsiveness in healthy sedentary man. *Journal of Applied Physiology* 80: 616-622.

22. Farber, H., J. Arbetter, E. Schaefer, G. Dallal, R. Grimaldi, N. Hill. 1987. Acute metabolic effects of an endurance triathlon. *Annals of Sports Medicine* 3: 131-138.

23. Fisman, E.Z., A.G. Frank, E. Ben-Ari, G. Kessler, A. Pines, Y. Drorv, J.J. Kellerman. 1990. Altered left ventricular volume and ejection fraction response to supine dynamic exercise in athletes. *Journal of the American College of Cardiology* 15: 582-588.

24. Friedman, D.B., G.A. Ordway, R.S. Williams. 1987. Exercise-induced functional desensitization of canine cardiac beta-adrenergic receptors. *Journal of Applied Physiology* 62: 1721-1723.

25. Fry, R.W., A.R. Morton, P. Garcia-Webb, G.P. Crawford, D. Keast. 1992. Biological responses to overload training in endurance sports. *European Journal of Applied Physiology* 64: 335-344.

26. Gliner, J.A., P.B. Raven, S.M. Horvath, B.L. Drinkwater, J.C. Sutton. 1975. Man's physiological responses to long-term work during thermal and pollutant stress. *Journal of Applied Physiology* 39: 628-632.

27. Hartley, L.H. 1977. Central circulatory function during prolonged exercise. *Annals of New York Academy of Science* 301: 189-194.

28. Haymes, E.M., J.L. Puhl, T.E. Temples. 1986. Training for cross-country skiing and iron status. *Medicine and Science in Sports and Exercise* 18: 162-167.

29. Hooper, S.L., L.T. Mackinnon, A. Howard, R.D. Gordon, A.W. Bachman. 1995. Markers for monitoring overtraining and recovery. *Medicine and Science in Sports and Exercise* 27: 106-112.

30. Imperato-McGinley, J., T. Gautier, K. Ehlers, M.A. Zullo, D.S. Goldstein, E.D. Vaughan. 1987. Reversibility of catecholamine-induced dilated cardiomyopathy in a child with pheochromocytoma. *New England Journal of Medicine* 316: 793-797.

31. Ketelhut, R., C.J. Losem, F.H. Messerli. 1994. Is a decrease in arterial pressure during long-term aerobic exercise caused by a fall in cardiac pump function? *American Heart Journal* 127: 567-571.

32. Kuipers, H., H.A. Keizer. 1988. Overtraining in elite athletes. *Sports Medicine* 6: 79-92.

33. Lehmann, M., H.H. Dickhuth, G. Gendrisch, W. Lazar, M. Thum, R. Kaminski, J.F. Aramend, E. Peterke, W. Wieland, J. Keul. 1991. Training-overtraining: a prospective, experimental study with experienced middle- and long-distance runners. *International Journal of Sports Medicine* 12: 444-452.

34. Lehmann, M., P. Baumgartl, C. Wiesenack, A. Seidel, H. Baumann, S. Fischer, U. Spori, G. Gendrischet, R. Kaminski, J. Keul. 1992. Training-overtraining: influence of a defined increase in training volume vs training intensity on performance, catecholamines and some metabolic parameters in experienced middle- and long-distance runners. *European Journal of Applied Physiology* 64: 169-177.

35. Lehmann, M., C. Foster, J. Keul. 1993. Overtraining in endurance athletes: a brief review. *Medicine and Science in Sports and Exercise* 25: 854-862.

36. Lentini, A.C., R.S. McKelvie, N. McCartney, C.W. Tomlinson, J.D. MacDougal. 1993. Left ventricular response in healthy young men during heavy-intensity weight-lifting exercise. *Journal of Applied Physiology* 75: 2703-2710.
37. Levine, B.D., L.D. Lane, J.C. Buckey, D.B. Friedman, C.G. Blomqvist. 1991. Left ventricular pressure-volume and Frank-Starling relations in endurance athletes: implications for orthostatic tolerance and exercise performance. *Circulation* 84: 1016-1023.
38. McCloskey, D.I., J.H. Mitchell. 1972. Reflex cardiovascular and respiratory responses originating in exercising muscle. *Journal of Physiology* (London) 224: 173-186.
39. Morgan, W.P., D.L. Costill, M.G. Flynn, J.J. Raglin, P.J. O'Connor. 1988. Mood disturbance following increased training in swimmers. *Medicine and Science in Sports and Exercise* 20: 408-414.
40. Nieman, D.C., L.S. Berk, M. Simpson-Westerberg, K. Arabatzis, S. Youngberg, S.A. Tan, J.W. Lee, W.C. Eby. 1989. Effects of long-endurance running on immune system parameters and lymphocyte function in experienced marathoners. *International Journal of Sports Medicine* 10: 317-323.
41. Niemela, K.O., I.J. Palatsi, M.J. Ikaheimo, J.T. Takkunen, J.J. Vuori. 1984. Evidence of impaired left ventricular performance after an uninterrupted competitive 24 hour run. *Circulation* 70: 350-356.
42. Niemela, K., I. Palatsi, M. Ikaheimo, J. Airaksinen, J. Takkunen. 1987. Impaired left ventricular diastolic function in athletes after utterly strenuous prolonged exercise. *International Journal of Sports Medicine* 8: 61-65.
43. Noakes, T. 1986. *Lore of running*. Cape Town: Oxford University Press.
44. Osbakken, M., R. Locko. 1984. Scintigraphic determination of ventricular function and coronary perfusion in long-distance runners. *American Heart Journal* 108: 296.
45. Packer, D.L., G.H. Bardy, S.J. Worley, M.S. Smith, F.R. Cobb, R.E. Coleman, J.J. Galagher, L.D. German. 1986. Tachycardia-induced cardiomyopathy: a reversible form of left ventricular dysfunction. *American Journal of Cardiology* 57: 563-570.
46. Parry-Billings, M., R. Budgett, Y. Koutedakis, E. Blomstand, S. Brooks, C. Williams, P.C. Calder, S. Pilling, R. Baigre, E.A. Newsholme. 1992. Plasma amino acid concentrations in the overtraining syndrome: possible effects on the immune system. *Medicine and Science in Sports and Exercise* 24: 1353-1358.
47. Pate, R. 1983. Sports anemia: a review of the current research literature. *Physician and Sportsmedicine* 11: 115-126.
48. Perrault, H., F. Peronnet, R. Lebeau, R.A. Nadeau. 1986. Echocardiographic assessment of left ventricular performance before and after marathon running. *American Heart Journal* 112: 1026-1031.
49. Raven P.B., G.H.J. Stevens. 1988. Cardiovascular function and prolonged exercise. In *Perspectives in exercise science and sports medicine.* vol. 1, *Prolonged exercise*, eds. D.R. Lamb, R. Murray, 43-47. Indianapolis: Benchmark Press.
50. Rogers, G., C. Goodman, D. Mitchell, J. Hattingh. 1986. The response of runners to arduous triathlon competition. *European Journal of Applied Physiology* 55: 405-409.
51. Rowbottom, D.G., D. Keast, C. Goodman, A.R. Morton. 1995. The hematological, biochemical and immunological profile of athletes suffering from

the overtraining syndrome. *European Journal of Applied Physiology* 70: 502-509.

52. Ryan, A. 1983. Overtraining in athletes: a roundtable. *Physician and Sportsmedicine* 11: 93-110.
53. Saltin, B. 1964. Circulatory responses to submaximal and maximal exercise after thermal dehydration. *Journal of Applied Physiology* 19: 1125-1132.
54. Saltin, B., J. Stenberg. 1964. Circulatory responses to prolonged severe exercise. *Journal of Applied Physiology* 19: 833-838.
55. Seals, D.R., M.A. Rogers, J.M. Hagberg, C. Yamamoto, P.E. Cryer, A.A. Ehsani. 1988. Left ventricular dysfunction after prolonged strenuous exercise in healthy subjects. *American Journal of Cardiology* 61: 875-879.
56. Senay, L.C., J.M. Pivarnick. 1985. Fluid shifts during exercise. *Exercise and Sports Science Reviews* 13: 335-387.
57. Seward, S.W., K.S. Seiler, J.W. Starnes. 1995. Intrinsic myocardial function and oxidative stress after exhaustive exercise. *Journal of Applied Physiology* 79: 251-255.
58. Shannon, R.P., K. Komamura, B.S. Stambler, M. Bigaud, W.T. Manders, S.F. Vatner. 1991. Alterations in myocardial contractility in conscious dogs with dilated cardiomyopathy. *American Journal of Physiology* 260 (*Heart and Circulatory Physiology* 29): H1903-1911.
59. Shephard, R.J., R.J. Verde, S.G. Thomas, P. Shek. 1991. Physical activity and the immune system. *Canadian Journal of Sport Science* 16: 169-185.
60. Shub, C., L. Cueto-Garcia, S.G. Sheps, D.M. Ilstrup, A.J. Tajik. 1986. Echocardiographic findings in pheochromocytoma. *American Journal of Cardiology* 57: 971-975.
61. Stachenfeld, N., G.W. Glelm, P.M. Zabetakis, K.H. Briggs, J.A. Nichols. 1990. Markers of training status during increase in training time. *Medicine and Science in Sports and Exercise* 22: S96.
62. Stone, M.H., R.E. Keith, J.T. Kearney, S.J. Fleck, G.D. Wilson, N.T. Triplett. 1991. Overtraining: a review of the signs, symptoms and possible causes. *Journal of Applied Sport Science Research* 5: 35-50.
63. Upton, M.T., S.K. Rerych, J.R. Roeback. 1980. Effect of brief and prolonged exercise on left ventricular function. *American Journal of Cardiology* 45: 1154-1160.
64. Vanoverschelde, J.L.J., L.T. Younis, J.A. Melin. 1991. Prolonged exercise induces left ventricular dysfunction in healthy subjects. *Journal of Applied Physiology* 70: 1356-1363.
65. Vanoverschelde, J.J., B. Essamri, R. Vanbutsele, A. D'Hondt, J.R. Detry, J.A. Melin. 1993. Contribution of left ventricular diastolic function to exercise capacity in normal subjects. *Journal of Applied Physiology* 74: 2225-2233.
66. Verma, S.K., S.R. Mahindroo, D.K. Kansal. 1978. Effect of four weeks of hard physical training on certain physiological and morphological parameters of basketball. *Journal of Sports Medicine* 18: 379-384.
67. Wishnitzer, R., A. Eliraz, N. Hurvitz, A. Vorst, M. Sternfeld, A. Beregy. 1990. Decreased bone marrow cellularity and hemosiderin in normal and overtrained runners. *Harefuah* 118: 74-78.
68. Wolfe, W. 1961. A contribution to the question of overtraining. In *Health and fitness in the modern world*, ed. L.A. Larson, 291-301. Chicago: The Athletic Institute.

CHAPTER 8

Neuroendocrine Aspects of Overtraining

Hans A. Keizer, MD, PhD

Introduction

In modern sport practice, the principle of overload training is used in order to improve performance level of athletes. By definition, overload training will transiently disturb the homeostasis of many processes in the body. Improvement in performance, though, will only be achieved when the athlete implements enough time with either rest or low training loads. If done correctly, the same exercise load will not severely disturb homeostasis anymore. However, if the balance between exercise-induced fatigue and time devoted to recovery is incorrect, adaptation will fail and a situation of chronic fatigue will occur (36, 71). The symptoms associated with overtraining, such as changes in emotional behavior, sleep disturbances, and hormonal dysregulations (36, 71), are indicative of changes in the regulatory and coordinative function of the hypothalamus. Therefore, the purpose of the present chapter is to review and discuss some of the available data on hypothalamic-pituitary involvement in exercise stress and relevant data on the involvement of the neuroendocrine system in exercise stress and overtraining, and link them to those obtained from the stress literature.

The Response of the Hypothalamic-Pituitary Axis to Acute and Chronic Exercise Stress

Any stress, including acute physical exercise of moderate to high intensity, is able to elicit marked changes in stress hormone secretions. These increments in plasma stress hormone levels are necessary to augment the muscle enzyme activities (and consequently energy release and expenditure) manifold (20). In the recovery phase, termination of the stress response must occur in order to adapt to a higher level or to habituate or adapt to the same stress. Now we may describe an acute bout of intensive exercise, overload training (i.e., repeated bouts of acute exercise on consecutive days), overreaching, and finally overtraining, in terms of transient or waning states of stress, which inevitably involves the neuroendocrine system.

The exercise overload stimulus then, considered to be a homologic stressor, will cause a transient (up to 1-7 days) period wherein the same exercise stress is less tolerated (i.e., impaired stress tolerance). If this stressor is precipitated upon other (heterologic) stressors (such as social, psychological), the athlete may recover more slowly and eventually slip into a state of overreaching or overtraining. In this next section, we will discuss the effects of acute exercise and normal, adaptive training on several hypothalamic-pituitary systems.

The Hypothalamic-Pituitary-Growth Hormone (HPGH) Axis

Growth hormone (GH) secretion is age dependent (35). It is maximal during puberty and early adolescence; thereafter GH secretion gradually declines to much lower levels (95). Under basal conditions, GH is secreted in an intermittent, pulsatile pattern by the anterior pituitary and also shows a diurnal rhythm (96). Daytime levels of growth hormone were reported to be very low (actually below the detection limits of the assays), with no pulsatations. With sleep onset and coinciding with the initial periods of slow-wave sleep, distinct bursts of GH secretion occur (111).

The pituitary release of GH into the bloodstream is increased by hypothalamic growth hormone releasing hormone (GHRH) and decreased by somatostatin or somatotropin release-inhibiting hormone (SRIH). The rhythmic pulsatations of GHRH and SRIH each have a periodicity of 3-4 hr, and are 180 degrees out of phase (112). The amount of GHRH released determines the GH amplitude, while SRIH determines the frequency and duration of the GH bursts (70). Finally, peripheral afferent signals might be important modulators of GH secretion or action since no effect of growth hormone on spaceflight-induced muscle atrophy could be found (53).

Many hormones, neuropeptides, and neurotransmitters influence GH secretion either by direct action on the somatotrope or by altering GHRH and SRIH release. GH secretion is enhanced by estrogens (27), testosterone, progesterone, and thyroid hormones (100). High levels of corticoids suppress GH secretion, but low levels of corticoids sensitize the pituitary response to GHRH (124). Gonadal steroids and thyroid hormone stimulate GH synthesis in the somatotrope (32, 51). The neuropeptides that influence GH secretion include GH itself, somatostatin exerting a short-loop negative feedback, and IGF-1 exerting a long-loop negative feedback.

Actions of GH.   GH is an important metabolic hormone that stimulates net protein anabolism, lipolysis, and linear bone growth. The cellular effects of GH are mediated directly and in many cases indirectly by their autocrine action upon the target cells. In this situation GH induces the release of insulin-like growth factor 1 and 2 (IGF_1, IGF_2). GH stimulates IGF_1 production in, among others, the liver, fibroblasts, and skeletal muscle. GH has insulin-like effects (increased glucose and amino acid uptake by target cells) and anti-insulin effects. In healthy subjects GH administration stimulates muscle protein synthesis almost instantaneously (38) but not whole body protein synthesis (37). Muscle mass and function are increased

(108), whereas fat mass is decreased (28) after GH substitution in GH-deficient (GHD) patients. GH also exerts very important effects on mood-state. Low plasma GH levels impair cognitive function and cause mood disturbances (83) that disappear after GH substitution (54).

Acute Exercise and GH Secretion.   Acute exercise increases GH secretion in an intensity-dependent manner (88), albeit a wide interindividual variation exists (98). That means the more anaerobic the work, the more GH increases (118), a fact also observed after heavy resistance training (69). During long-term endurance exercise, plasma GH reaches peak values before the end of the exercise bout and levels off thereafter or eventually declines. After exercise, GH levels normalize within 1-2 hours.

Chronic Exercise and GH Secretion.   In evaluating hormone responses to acute and chronic exercise, we have to consider changes occurring in the recovery phase, eventually during sleep. Indeed, regular endurance exercise and training also increase the nocturnal pulsatile secretion, as has been shown in hamsters (11) and man (46, 125). These changes are linked to training intensity, as has been shown by Weltman and colleagues (125) in female runners (see figure 8.1). Very demanding exercise, however, may depress the nocturnal GH secretion (65), which then resembles the pattern observed in obese subjects, depressive patients, and in the elderly (50).

The Hypothalamic-Pituitary-Thyroid (HPT) Axis

The thyroid gland is stimulated to produce and release thyroxine (T4) and the metabolically more active 3,3',4-triiodothyronine (T3) by thyroid stimulating hormone (TSH) released from the anterior pituitary. TSH production is stimulated by

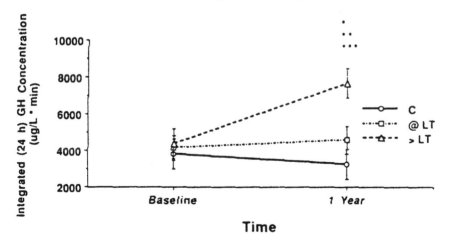

Figure 8.1   Effects of 1 yr of run training on 24 h integrated serum GH concentrations. C = control, no training; @LT = training below the lactate threshold; >LT = training at or slightly above lactate threshold.
Reprinted from Weltman et al. 1991.

thyrotropin releasing hormone (TRH), synthesized in the paraventricular nuclei of the hypothalamus. Thyroid hormones (T4 and T3) exert negative feedback at the hypothalamus and at the pituitary to regulate TRH and TSH levels (87). In addition, the thyroid hormones stimulate the secretion of somatostatin, thus adding a hypothlamic level of inhibition of TRH secretion (99).

Like other anterior pituitary hormones, serum TSH displays both episodic (pulsatile) and circadian variations. In healthy subjects, the mean TSH pulse frequency is approximately 9-12 per 24 hours (40). The 24-hour secretion pattern shows a distinct circadian rhythm, with increases in TSH pulse amplitude at night. These nightly pulses are blunted in hypothyroidic patients' secretion (99). Thyroid function is influenced by a variety of other hormones, including ACTH and glucocorticoids. Pharmacological doses of glucocorticoids decrease thyroid function by suppressing TSH secretion (99).

Actions of Thyroid Hormones.   Thyroid hormones have important effects on almost all tissue, including skeletal muscle. They modulate calorigenesis, and protein, carbohydrate, and lipid metabolism. Increased thyroid hormone-stimulated calorigenesis is reflected in increased oxygen consumption of most tissues. Protein synthesis (for example, of lysozymal enzymes) in skeletal muscle is enhanced at moderate increments of plasma T4. Optimal doses of thyroid hormones are required for the elicitation of the full growth potency of GH (99). Thyroid hormones influence all aspects of carbohydrate and lipid metabolism. These effects appear to be biphasic (i.e., they are dose dependent). High doses increase the glycogenolytic and hyperglycemic actions of epinephrine. Low doses increase glycogen synthesis in the presence of insulin (99).

The effect of thyroid hormones on lipid metabolism involves synthesis, mobilization, and degradation. Degradation is more pronounced than synthesis. The effect of thyroid hormones on lipolysis may be brought about either directly, by activation of the cyclic AMP system, or indirectly by sensitizing target cells to other lipolytic agents (99). Thyroid hormones influence muscle contractility by controlling the expression of gene coding for myosin isoforms (56, 58), Na-K ATPase pumps, and Ca-ATPase canals of the sarcoplasmatic reticulum (56). Probably according to these mechanisms, physical performance is severely decreased both in hypo- and hyperthyroidism. Indeed, in resting skeletal muscle of patients with hypothyroidism a rise in the inorganic phosphate to ATP ratio and a defective pyruvate oxidation was found, whereas in working muscle the rate of phosphocreatine decrease and the fall in pH was greater than in healthy control subjects (57).

Acute Exercise and Thyroid Secretion.   Acute exercise is able to stimulate the hypothalamic-pituitary-thyroid axis almost instantaneously (85) as observed from an increase in thyroid stimulating hormone (TSH) (see figure 8.2, panel A). This will result in an increased conversion of T4 to T3. Hackney et al. (46) evaluated the change in nocturnal concentrations of T4 after exercise (90 min, 70% VO_2max) compared with a control day and found a significant augmentation of T4. More recently, Hackney and Gulledge (45) found only a transiently increased plasma T3 level directly after acute exercise.

Chronic Exercise and Thyroid Secretion.   Boyden and Parameter (15) investigated the effects of endurance training on the thyroid axis of 17 women. The training was increased from zero to 50 miles/week within a year. The authors found an increased resting TSH level which paralleled the increase in mileage. However,

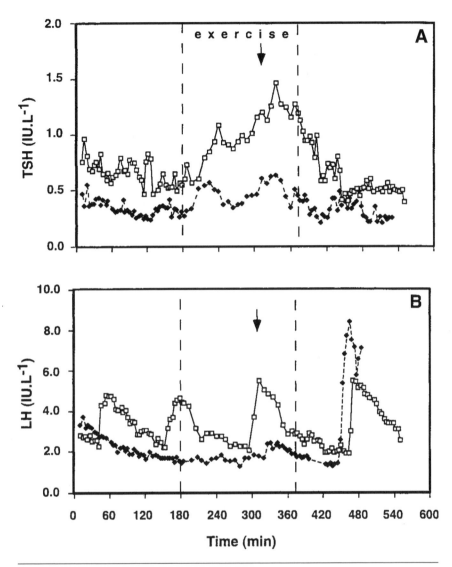

Figure 8.2   TSH (panel A) and LH (panel B) secretion pattern of a female athlete in her eighth day of two consecutive menstrual cycles, before, during, and after aerobic exercise to exhaustion. In the first cycle (open squares), the subject was tested after five days of light aerobic run training. In the second cycle, she was retested after five days of very distressing training (filled squares). The arrow indicates the end of exercise.

the TRH-stimulated TSH response was exaggerated at 30 miles/week, but was diminished at 50 miles/week. This might indicate that the amount of training stress changes the conversion of T4 to T3. Loucks and coworkers (76, 77) tried to clarify the low T3 syndrome commonly found in amenorrheic athletes. In women with normal menstrual cycles they did not find a relationship with the exercise per se, but there was a relation between the energy cost of exercise and dietary intake. If dietary intake was too low, the subjects responded with a reduced T3 level after only four days of low energy intake. Changes in T4 could not be distinguished. Alén et al. (2) followed 11 weightlifters before, during, and after a one-year period of strength training. They did not find pathological changes during that year. But, during the precompetition period, a gradual increase in serum T3, T4, and free T4 was found. In the final two weeks before competition where the training intensity was increased, but the volume was decreased, thyroid hormone values returned to baseline.

The Hypothalamic-Pituitary-Adrenal (HPA) Axis

The HPA axis, together with the autonomic nerve system, is the most important stress system of the body. Once a stressor (physical exercise, psychic stress) has exceeded a certain threshold, a systemic reaction takes place that includes the brain and peripheral components, the HPA axis, and the sympathetic nervous system. The central stress system includes the CRH neurons located in the hypothalamic paraventricular nucleus and other brain areas, the locus ceruleus, and other central sympathetic systems in the brain stem (22).

The secretions of the corticotroph include adrenocorticotropic hormone (ACTH) and β-endorphin. In man, β-lipotrophin (β-LPH) is co-released with ACTH after cleavage from pro-opiomelanocortin (POMC) in the corticotrophs of the anterior pituitary. Thereafter, β-LPH is partially cleaved to g-LPH and β-endorphin. Because the majority of β-LPH remains intact, the circulating levels of β-endorphin are extremely low.

Although CRH is the principal regulator of ACTH and β-endorphin secretion, other factors also exert important regulator influences on the corticotroph. Experimental evidence obtained from in vitro and in vivo experiments suggests that AVP, oxytocin, and angiotensin II may also stimulate the corticotroph (94). Even lactate is able to activate the HPA axis, although ACTH and cortisol responses are significantly less than observed during exercise (22).

CRH, ACTH, and consequently, cortisol are secreted in a pulsatile and diurnal pattern. The ACTH pulse frequency is rather constant during night and day; the pulse amplitude, however, varies. The secretory bursts increase in amplitude after 3-5 hours of sleep, reach maximum the last few hours before and about one hour after awakening, decline thereafter, and are minimal in the evening (119). Accordingly, plasma cortisol values are highest at awakening, are low in the afternoon and evening to reach nadir values during the first non-REM/REM sleep cycles, but rise markedly during the second half of sleep (12). Meals (especially those with high protein content) at lunch, and sometimes at dinnertime, cause additional secretory episodes (110).

Results from a study of Kjaer et al. (66) indicated that afferent nervous activity from working muscles stimulated the secretion of ACTH and β-endorphin. The authors also found that this was not true for GH, catecholamines, and insulin. They argued that the direct, feed forward, stimulation of neuroendocrine centers from motor centers in the brain is more important than the nervous feedback in these cases.

Actions of CRH, ACTH, β-Endorphin, and Cortisol.   The actions of CRH, ACTH, β-endorphin, and cortisol are manifold, and are too vast to describe in detail in this chapter. Therefore the author has chosen only to discuss briefly those actions that might be relevant to the topic of this book.

The action of CRH goes well beyond its role as a mere releasing factor. CRH is considered as an integrator of the (neuro)endocrine, autonomic, and behavioral responses to stress. It is probably through CRH that GH, LH, FSH, and ACTH and β-endorphin secretions are modulated (4). For instance, CRH modulates growth hormone secretion (101), suppresses the secretion of gonadotrophins and consequently the gonadal hormones (93), activates the autonomic nerve system, but inhibits the parasympathetic nervous system (16, 17). This results in, among others, increased heart rate and blood pressure. Probably by means of the autonomic nervous system, CRH has also profound, and dose-dependent effects on mood-state as has been shown by animal experiments. These effects range from increased arousal to anxiety, from fear to depression (91). In addition, the CRH molecule itself has also profound effects on peripheral tissues, including the sites of inflammation, where it stimulates cytokine production (interleukin 1) (23).

The primary action of ACTH on the adrenal cortex is to increase cortisol secretion by increasing its synthesis, but the hormone also increases the LDL-receptor, which affects steroid synthesis (89). Extra adrenal effects of ACTH involve the increase in glucose and amino acid uptake by skeletal muscle and stimulation of lipolysis in the adipocyte (72). Although under normal conditions the quantitative contribution to energy release is unknown, we hypothesize that the trained cell, which is much more sensitive for ACTH and other hormones, responds on the relatively large increase in ACTH levels with intensive exercise.

The actions of β-endorphin can also be divided into central and peripheral. Opioids raise the threshold of pain, influence extrapyramidal motor activity, and are involved in behavior and mood-state (100). The HPA axis is, unlike GH secretion, subjected to inhibitory opioid control (42). In addition, high levels of opioids also suppress adrenomedullary function in man (13). Another peripheral effect of opioids that might impinge upon physical performance is brought about by their influence on ventilation and cardiac function. Grossman et al. (42) showed that blocking the μ-receptors with naloxone caused (1) a significant increase in ventilation, (2) an elevation of plasma lactate, GH, cortisol, epinephrine, and norepinephrine, and (3) an increase in perceived exertion.

The effects of glucocorticoids can be divided into anabolic and catabolic effects. The anabolic effect involves the stimulation of gluconeogenesis and glycogen synthesis in the liver. The catabolic effects involve a decreased glucose uptake in peripheral tissues (89) and an increased protein degradation in, among others, skeletal muscle. In part, this is caused by the competition of glucocorticoids with androgen receptors in skeletal muscle (80). Furthermore, glucocorticoids acutely increase lipolysis in fat cells (33).

Acute Exercise and HPA Hormone Secretion.   Acute exercise of sufficient intensity (at least 70% VO_2max) significantly increases plasma ACTH, cortisol, and β-endorphin levels in an intensity-dependent manner (34, 60, 68). With very high intensities (i.e., 120% VO_2max), ACTH was already increased after 1 min exercise and cortisol increments were significant after a time delay (i.e., after 15 min of recovery), suggesting an increased ACTH-induced steroidogenesis (19). Increasingly elevated plasma β-endorphin levels are found after 100 m, 1500 m and 10,000 m running events (92), indicating that both exercise intensity and duration are important.

Chronic Exercise and HPA Hormone Secretion.   After physical training the ACTH response (the absolute values) to a standard submaximal exercise is significantly suppressed both in rats (122, 123) and man (61, 68), albeit at the same relative workload, ACTH levels were relatively more increased after training (61). From cross-sectional studies evidence emerged that peak β-endorphin levels are higher after training (34). Prospective training studies, however, were unable to confirm that training augments plasma ß-endorphin response to acute exercise (52, 120).

In response to CRH stimulation, highly trained subjects had blunted ACTH and cortisol responses, resembling a state of hypercortisolism (78, 79). Recently, however, Chrousos (22) argued that in these cases other non-CRH factors play a more important role in the activation of the HPA axis by exercise. He suggested that vasopressin produced by the parvocellular component of the paraventricular nucleus might be a good candidate.

Work in rats showed that exercise training attenuated ACTH responses for heterologic stressors as well, suggesting that cross-adaptation occurred (90, 122). Whether this occurs in athletes as well is not known.

The Hypothalamic-Pituitary-Gonadal (HPG) Axis

The gonads are stimulated to synthesize and secrete androgens, estrogens, and progestins by the concerted action of luteinizing hormone (LH) and follicle stimulating hormone (FSH). The anterior pituitary gonadotrophs are under control of gonadotropin releasing hormone (GnRH). The HPG axis is regulated by the feedback effects of gonadal hormones (estrogens, progestagens, and androgens). Since steroid receptors are found in many parts of the brain (100), gonadal hormones regulate sexuality, but influence mood-state as well. Gonadotropins are secreted in a rhythmic, pulsatile fashion. This is, for the greater part, caused by the hypothalamic pulse generator GnRH (67). Following the rhythmic GnRH discharges, LH is secreted in bursts approximately every 90-120 min. GnRH rhythmicity varies to some extent according to the endocrine milieu (67). Testosterone (T) and progesterone decrease the frequency. In response to estrogens, normal women release LH, whereas men do not. Endogenous opiates inhibit hypothalamic GnRH release in the late follicular and mid-luteal phases of the menstrual cycle, but not in the early follicular phase (97, 104). This indicates that the effect of β-endorphin upon gonadotropin secretion depends on the steroid milieu.

Actions of Gonadal Hormones.   Gonadal hormones exert many actions upon the body. From experiments in male and female rats, it appeared that testosterone increases skeletal muscle glycogen content and glucose uptake (8, 43, 117). During submaximal exercise a glycogen-sparing effect of testosterone was found in exercising female rats (116). Recent work of Van Breda (115) showed that the effect of testosterone is brought about by its aromatization to estradiol, which supports the findings of an estradiol-induced glycogen sparing in exercising oophorectomized rats (39).

Androgens compete both in vivo and in vitro with receptor proteins that bind glucocorticoids (80, 81), thereby antagonizing the glucocorticoid muscle wasting. In rats, Dohm and Lewis (31) found that the exercise-induced decline in plasma T levels was accompanied by increased excretion of products of muscle catabolism. Studies on such effects of gonadal hormones in human subjects are completely lacking. Although Allenberg and colleagues (3) showed that testosterone might have some influence on glycogenesis, their data are not (yet) confirmed by other researchers. Therefore, it is unknown whether the rather moderate decreases in testosterone and estradiol-17β levels as shown in sport practice have any effect on muscle metabolism.

One of the major effects of gonadal hormones is their influence on mood-state. For example, Coppen et al. (24) showed that hysterectomized premenopausal women improved mood and vigor after estrogen replacement, whereas Schiff and colleagues (109) observed that the administration of estrogens was also associated with a shorter mean sleep latency, a longer period of rapid eye movement sleep, and a positive correlation between psychological intactness (as clinically ranked) and latency to sleep onset.

Acute Exercise and HPG Hormone Secretion.   Acute exercise in female athletes has been found to increase plasma estradiol, progesterone, and testosterone, probably irrespective of the exercise duration and phase of the menstrual cycle (55, 60). In men, acute exercise of different modalities (running, weight training, cycling) is able to increase plasma T levels (see Cumming and Wheeler (26) for review) with the exception of a maximal tethered swim of 14 minutes, where a consistent decrease in plasma T levels has been observed (25). Unlike in female athletes, in male athletes T levels during exercise prolonged >60 minutes may increase during the first hour, but decrease when the exercise continues (26). During the recovery phase, this decline may continue and result in plasma T levels well beyond the preexercise values (25-50%).

Prolonged exercise to exhaustion may have pronounced effects on nocturnal hormone secretions and consequently, the anabolic influence of sleep. Kern and colleagues (65) found in triathletes that 120-150 km bicycle exercise (causing an "overwhelming distress") was able to decrease plasma testosterone during the whole night, but increase cortisol levels during the first half and decrease cortisol secretion during the second half of sleep, as compared to the nonexercise control or the low distressing exercise day. However, plasma gonadal hormone changes are in most cases very transitory; i.e., return to a normal range occurs within hours to a couple of days after exhaustive exercise.

The mechanisms of the acute exercise-induced changes in plasma gonadal hormones seem to differ somewhat between males and females. For estradiol and progesterone, these changes are explained by a decreased metabolic clearance rate (63), since either no changes in LH secretion could be observed, or the increase of gonadal hormones anticipated the increase in LH (55). For testosterone this mechanism is also operative, but in the female athlete an increased adrenal secretion is probably involved as well (60). The decreases in testosterone in the male athlete may be explained (among others) by a catecholamine-induced decrease in testicular blood flow (see Cumming and Wheeler (26) for review).

Chronic Exercise and HPG Hormone Secretion.   Effects of strenuous exercise, particularly endurance training, on the HPG axis have been extensively studied in women (see Bonen (9), Keizer and Rogol (64) for recent reviews). Oligomenorrhea, amenorrhea, and defects in the luteal phase and anovulatory cycles occur, albeit with a much lower incidence than originally thought (10, 103). In contrast, it appears that long-term vigorous physical training is able to induce slight alterations in the HPG axis without inducing pathological changes in the menstrual cycle in women with a proven normal menstrual cycle before they were engaged in training (14, 60, 61). Especially during the adaptation phase to training, i.e., in the first four months of training, changes in LH pulsatility are observed (103).

The available data on the effects of strenuous training on plasma testosterone levels in male athletes are somewhat conflicting. Retrospective and prospective studies, either using isolated or serial blood sampling, have shown that vigorous exercise training (running, cycling, rowing) may result in decreased resting free and total testosterone concentrations (44, 47, 48, 82, 126). However, in a strictly standardized prospective study of almost two years, we were unable to confirm these results (59). In this study we retested our subjects after the recovery training week, which might have normalized testosterone levels (see figure 8.3)

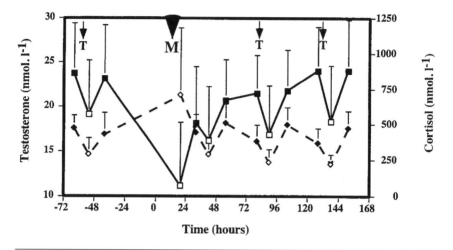

Figure 8.3   Plasma testosterone (solid line) and cortisol levels (dashed line) in 25 male runners before and after a marathon run (M). The subjects performed an incremental treadmill test to exhaustion, before and after the marathon (arrows). The hours indicate clock hours before and after the marathon, beginning on the marathon day (0 = 12.00 P.M.). Filled squares = samples taken between 8 and 10.00 A.M.; open squares = samples taken between 17.00 and 18.00 P.M. T = incremental treadmill test to exhaustion.
Based on data from Keizer et al. 1987.

Involvement of the Neuroendocrine System in Overreaching or Overtraining

From the previous paragraphs it might be clear that the neuroendocrine system is involved in all kinds of stress, including physical exercise. Dysregulation of the HPGH, HPT, HPA, and HPG axes has indeed profound influence upon mood-state, muscle metabolism, and muscle contraction, given that a severe hormonal deficit occurs. However, is there evidence that training alone is able to deregulate the system, and are the rather moderate alterations as have been observed in athletes able to elicit similar changes? In this paragraph an attempt is made to review the available literature and to dissociate between various factors that are able to dysregulate the neuroendocrine system.

The first evidence for the involvement of the neuroendocrine system in over-training was provided by the results of a study from Barron and colleagues (7).

Although based on only six subjects (male marathon runners) who were diagnosed as suffering from the overtraining syndrome, their study offered us for the first time some clues for understanding this syndrome. The authors found a blunted GH, ACTH, and cortisol response after an insulin-induced hypoglycemia. Interestingly, responses to TRH and LHRH were normal, which proved (according to the authors) that the dysfunction was at the level of the hypothalamus. The exact contribution of training volume or intensity to the overtraining syndrome is, unfortunately, unknown, since precise information on training variables and possible other stressors is lacking in the study of Barron and colleagues.

Adlercreutz and colleagues (1) found that a decrease in the ratio of free testosterone (FT) to cortisol (C) correlated with an impairment in performance in runners. They hypothesized that a decreased FT/C ratio hampers the anabolic processes after exercise, and thus delays recovery. Although some investigators reported similar results (102), others were unable to confirm the validity of the FT/C ratio as an early marker of overtraining in various types of endurance and strength athletes (49, 73, 114, 121, 128). This discrepancy may be explained by the resistance against stress of the HPG axis (103), the fact that more subtle changes occur during sleep and will be missed if one relies on only one blood sample (46, 65), and by not taking into account the normal diurnal rhythm, since plasma testosterone levels are about 25% lower in the late afternoon and evening as compared to the values around noon (41). Thus, we agree with Urhausen and colleagues (113), that decrements in gonadal hormones more likely indicate the actual physiological strain of the previous training bout, rather than overtraining.

Besides the methodological problems of blood sampling to detect parameters for overtraining, the validity of cross-sectional studies can be doubted. In addition, precise information about the training intensity, volume, and frequency is found in none of these studies. Recently, though, Lehmann et al. (74) tried in a prospective study to dissociate between run training intensity and volume as causative factors for overtraining. In a period of three weeks, they increased either the amount of high-intensity run training 2.5-fold in one group or, in another group, the weekly training volume about 2-fold. They found a decrease in resting and exercise plasma cortisol levels after the increased volume training only, whereas plasma T and free T did not change. The CRH-induced plasma ACTH and cortisol responses were interesting. Compared to the overtrained condition, baseline ACTH levels and response on CRH were lower and cortisol response higher. The authors also postulated that akin to ACTH, the resting levels of ß-endorphin might be increased, which explains the inhibition of gonadotropin release as discussed in this chapter.

In an attempt to evaluate the response of gonadal hormones, cortisol, and β-endorphins on standardized exhaustive and nonexhaustive training, we conducted a prospective study with 12 male runners (Keizer et al., unpublished data). The subjects were tested three times with an in-between interval of eight days, i.e., testing occurred after a light training period (LiTr), after a very intensive training period (ITr), and again after a light training and recovery period (RecTr). The core part of the training (warm-up and cool-down not included) consisted of one hour of strictly standardized daily running either of low intensity (continuous running at 60% VO_2max in the LiTr and RecTr periods) or with bouts of high intensity (95-110% VO_2 alternated with 60% VO_2max in the ITr period). The results showed an unaltered FT/C ratio after the intensive training period. In fact, the FT/C ratio increased because the plasma C levels significantly decreased. Nevertheless, five out of 12 subjects had symptoms of overreaching (no training desire, impossibility to fulfill the training load, sleep disturbances, lethargy, and constant feelings of tiredness), but only two of them showed a decreased performance after ITr. It was striking that all five runners had severe problems with the light training program in the third period. One of the subjects (JS) was deviant in his reaction on the heavy training. When he arrived at the laboratory he was apparently very tired and fell asleep during the 2 hr preparation for the test. However, his run time to exhaustion was much longer (1.5 min) than after the light training period. In the days of the recovery period, he became more and more tired (despite the very low training load) and his performance was much lower than expected; actually he ran 2.3 min less than after the first light training period. His normal performance level was restored after as much as five weeks of light training, i.e., this subject was by definition overtrained. There were slight decrements in plasma β-endorphin responses at exertion when the whole group was considered, but the overreached athletes showed (to our surprise) a blunted β-endorphin response to standardized treadmill running (see figure 8.4) after the heavy and distressing training week. Exercise-induced β-endorphin levels of JS remained low after the recovery period. Although these results must be considered as preliminary, they might indicate that the feelings of severe fatigue and lack of concentration, as mentioned by the athletes, are associated with decreased endorphin release.

These findings appear to contrast with the findings of Russel et al. (105) who found increased instead of decreased levels of β-endorphin levels at rest in young female swimmers after an almost tripled training volume in a period of three months. The authors hypothesized that these increments in endorphin levels might have inhibited pituitary LH secretion, since all subjects became oligomenorrheic. Unfortunately, the information about the training was scanty. Therefore, we attempted to further evaluate the effects of very intensive, and light, highly standardized training upon the HPG axis in eumenorrheic athletes. We investigated the plasma LH responses after (1) having tripled the training for one week (62), or (2) increasing

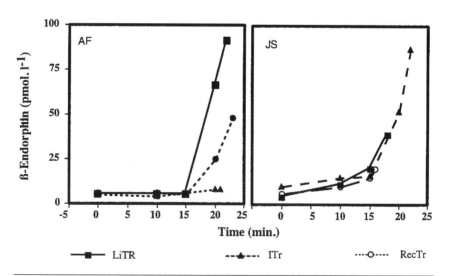

Figure 8.4   Plasma ß-endorphin responses to incremental treadmill running to exhaustion in two male runners after three consecutive training weeks. In the first (LiTR) and third (RecTr) weeks, training intensity was low; in the second week training intensity (ITr) was very high. Training volume was kept constant.

the training intensity (from 60% VO$_2$max to 80-90% VO$_2$max) in two subsequent menstrual cycles without reducing the training volume (Platen, unpublished data). In the first experiment the subjects felt extremely fatigued. The subjects were investigated for LH secretion three hours before and after exercise, either after a period of normal training or after one week of very exhaustive training. It was shown that LH pulse frequency was not altered, but LH amplitude was dramatically decreased in all six subjects (see table 8.1).

In the second experiment, the training from day two to day seven differed in intensity along the two menstrual cycles. In the first menstrual cycle, the training intensity was kept low (blood lactate 2.5 mmol/L) as monitored by the frequent measurement of blood lactate, whereas in the second cycle the subjects trained more anaerobically (blood lactate 6-8 mmol/L). On the eighth day, they ran to exhaustion (65% VO$_2$max) on a motor-driven treadmill. Blood for determination of LH was drawn every 4 min the 3 h before and after exercise, and during exercise every 8 min. Although there was a nonuniform response, defects in LH pulsatility were detected in five of the subjects (see figure 8.2, panel B). In only these five subjects, plasma estradiol levels were lower, whereas a slight, but significant, increase in pooled plasma cortisol levels was observed for the whole group. Whether the decreased LH amplitude is due to a decreased pituitary LH content or a decreased GnRH secretion cannot be concluded from these results. Evidence for the former hypothesis was presented by the results of a study by Boyden et al. (14) who showed a training volume-dependent decrease of LH secretion after a bolus injection with luteinizing hormone releasing hormone (LHRH).

Another possible factor involved in training-induced menstrual cycle alterations is an increased catecholestrogen (CE) formation in the brain. Catecholestrogen formation by 2- and 4-hydroxylation of primary estrogens constitutes a major pathway of estrogen metabolism (6), whereas in rats, CE decreased plasma LH but not FSH levels 60-120 min after injection (5). In teenage swimmers, Russell et al. (105) showed that an increase in the amount of training by 70% was accompanied by a 2.5-fold increase in basal plasma CE (and β-endorphin) levels, whereas all swimmers became oligomenorrheic. According to these authors the physiological

Table 8.1 The Influence of Acute Exhaustive Bicycle-Ergometer Exercise and Training on LH Secretion Pattern in Six Eumenorrheic Women

Condition	Pulses/3 hr		Pulse/Amplitude (IU/L)	
	Before	After	Before	After
Normal training	2.86	2.66	2.90	1.66[a]
(± SD)	0.88	0.99	1.57	0.67
Exhaustive training	2.77	1.59[a]	0.65[b]	0.81
(± SD)	1.85	1.23	0.29	0.59

[a]Significantly different (p < .02) between, before, and after exercise.
[b]Significantly different between N and E.

role of the formation of CE may be a reduction of the degradation rate of CE during exercise, thus increasing the availability of these important substrates in the hypothalamus. The possible role of CE in the incidence of exercise-induced menstrual cycle disturbances has been discussed further by De Cree (29), whereas very recently further evidence for a possible role of CE has been found (30). In the latter study, eumenorrheic women performed a strictly standardized training on days 2-7 of three consecutive menstrual cycles. The training was either low intensity (menstrual cycle 2) or very high intensity (menstrual cycle 3). The results showed a very pronounced decrease in total unconjugated estrogens and a decreased estrogen/2-hydroxy-catecholestrogen ratio.

The mechanisms whereby the secretions of the various hypothalamic-pituitary axes will be changed are difficult to reconcile. Again, the hypothesis of Adlercreutz and colleagues (1) seems to be attractive, but remains questionable. It can not be denied that increments in circulating stress hormones affect the recovery rate and the duration of the recovery phase (106, 107). Glucocorticoids and sex steroids exert catabolic and anabolic effects, respectively, on many tissues, among them skeletal muscle. Prolonged exposure to glucocorticoids will down regulate cortisol receptors in the putamen, which disinhibits the HPA axis (106, 107) and inhibits GH release (75), probably via an increased CRH tone (101). However, as we have discussed, the effect of a period of strenuous training on plasma stress levels of cortisol, β-endorphins, and gonadal hormones is far from uniform among athletes. There is evidence, however, that combinations of stressors can be detrimental. For example, an energy deficit combined with strenuous exercise can account for the decrements in plasma gonadal hormone levels both in rats (43), and man (18, 127).

Summarizing the effects of strenuous exercise and training on the neuroendocrine system and the involvement of this system in overreaching and overtraining is quite difficult, since the results of research devoted to this subject matter are far from unanimous; not unreasonably, since well-conducted prospective studies for obvious reasons are scanty. However, mood-state changes, sleep disturbances, changes in recovery rate, prolonged feelings of fatigue, and changes in reproductive state all indicate maladaptation in certain parts in the brain with consequent changes in hypothalamic effector output. These changes are caused by exercise-induced changes in neurotransmitter metabolism as described by Chaouloff (21). Very recently, Meeusen and colleagues (84) reported, by using the micro-dialysis technique, clear exercise-induced increases in the extracellular content of NE and DA, but not of GABA in the striatum of the conscious rat. Training caused a dramatic decrease in basal levels of NE and DA, but the percentile increase during exercise remained unaltered. This may indicate an increased receptor sensitivity for these neurotransmitters, which may explain the attenuated plasma ACTH responses after training. Conversely, in depressed patients, and maybe also in overtrained athletes, either a decreased sensitivity for these neurotransmitters may have occurred, or production rate may be impaired. In turn, this may explain the decreased GH, LH, β-endorphin, and TSH levels, since in general these neurotransmitters are excitatory. Additionally, lower β-endorphin levels may be involved in lower stimulation of GH release (86). Since these effects apparently take place during sleep, extended periods of blood sampling are mandatory in order to obtain more insight in training- and overtraining-induced alterations of the neuroendocrine system.

In conclusion, we hypothesize the following in regard to the neuroendocrine system and overtraining. First, overtraining originates at the level of the hypothalamus and higher brain centers. Changes in neurotransmitter content and receptor sensitivity might play an important role in adaptation to training as well as in maladaptation. Second, blunted responses of stress hormones to acute exercise explain the feelings of fatigue, decreased pain perception, and mood-state; it is not known whether this situation precedes other disturbances of the neuroendocrine system. Third, there is no evidence that the C/T ratio is a good early marker for overreaching or overtraining. Fourth, the mild hypercortisolism found during sleep in periods of strenuous training is most probably caused by a higher CRH tone. Finally, it is possible that overreaching and overtraining are characterized by an array of events over time. In order to detect the timely sequence, prolonged observation periods and a prospective study design are necessary to unravel the mechanisms of neuroendocrine dearrangements.

References

1. Adlercreutz, H., M. Härkönen, K. Kuoppasalmi, H. Näveri, I. Huhtaniemi, H. Tikkanen, K. Remes, A. Dessypris, J. Karvonen. 1986. Effect of training on plasma anabolic and catabolic steroid hormones and their response during physical training. *International Journal of Sports Medicine* 7: S27-28.
2. Alén, M., A. Pakarinen, K. Häkkinen. 1993. Effects of prolonged training on serum thyrotropin and thyroid hormones in elite strength athletes. *Journal of Sports Science* 11: 493-497.
3. Allenberg, K., N. Holmquist, S.G. Johnson, P. Bennet, J. Nielsen, H. Galbo, N.H. Secher. 1983. Effect of exercise and testosterone on the active form of glycogen synthase in human skeletal muscle. In *Biochemistry of exercise*, eds. H.G. Knuttgen, J.A. Vogel, J. Poortmans, 625-630. Champaign, IL: Human Kinetics.
4. Almeida, O.F., A.H. Hassan, F. Holsboer. 1993. Intrahypothalamic neuroendocrine actions of corticotropin-releasing factor. *Ciba Foundation Symposium* 172: 151-169.
5. Ball, P., G. Emons, R. Knuppen. 1982. Importance of catecholestrogens in the regulation of the ovarian cycle. *Archives of Gynecology* 231: 315-320.
6. Ball, P., R. Knuppen. 1980. Catecholestrogens (2- and 4-hydroxyoestrogens): chemistry, biogenesis, metabolism, occurrence and physiological significance. *Acta Endocrinology (Copenhagen)* 93 Supplement 232: S1-128.
7. Barron, J.L., T.D. Noakes, W. Levy, C. Smith, R.P. Millar. 1985. Hypothalamic dysfunction in overtrained athletes. *Journal of Endocrinology and Metabolism* 60: 803-806.
8. Bergamini, E. 1974. Different mechanisms in testosterone action on glycogen metabolism in rat perineal and skeletal muscles. *Endocrinology* 96: 77-84.
9. Bonen, A. 1994. Exercise-induced menstrual cycle changes: a functional, temporary adaptation to metabolic stress. *Sports Medicine* 17: 373-392.
10. Bonen, A., S.M. Shaw. 1995. Recreational exercise participation and aerobic fitness in men and women: analysis of data from a national survey. *Journal of Sports Science* 13: 297-303.
11. Borer, K.T., D.R. Nicoski, V. Owens. 1986. Alteration of pulsatile growth hormone secretion by growth-inducing exercise: involvement of endogenous opiates and somatostatin. *Endocrinology* 118: 844-850.

12. Born, J., W. Kern, K. Bieber, G. Fehm-Wolfsdorf, M. Schiebe, H.L. Fehm. 1986. Night-time plasma cortisol secretion is associated with specific sleep stages. *Biology of Psychiatry* 21: 1415-1424.

13. Bouloux, P.G.M., A. Grossman, S. Al-Damluji, T. Bailey, G.M. Besser. 1986. Enhancement of the sympathoadrenal response to cold-pressor test by naloxone in man. *Clinical Science* 69: 365-368.

14. Boyden, T.W., R.W. Parameter, P.R. Stanforth, T.C. Rotkis, J.H. Wilmore. 1984. Impaired gonadotropin responses to gonadotropin-releasing hormone stimulation in endurance-trained women. *Fertility and Sterility* 41: 359-363.

15. Boyden, T.W., R.W. Parameter. 1985. Exercise and the thyroid. In *Exercise endocrinology*, eds. K. Fotherby, S.B. Pal. Berlin: Walter de Gruyter.

16. Brown, M.R., L.A. Fisher, J. Rivier, J. Spiess, C. Rivier, W. Vale. 1982. Corticotropin-releasing factor: effects on the sympathetic nervous system and oxygen consumption. *Life Science* 30: 207-210.

17. Brown, M.R., L.A. Fisher, J. Spiess, C. Rivier, J. Rivier, W. Vale. 1982. Corticotropin-releasing factor: actions on the sympathetic nervous system and metabolism. *Endocrinology* 111: 928-931.

18. Bullen, B.A., G.S. Skrinar, I.Z. Beitins, D.B. Carr, S.M. Reppert, C.O. Dotson, M.M. Fencl, E.V. Gervino, J.W. McArthur. 1984. Endurance training effects on plasma hormonal responsiveness and sex hormone excretion. *Journal of Applied Physiology* 56: 1453-1463.

19. Buono, M.J., J.E. Yeager, J.A. Hodgdon. 1986. Plasma adrenocorticotropin and cortisol responses to brief high-intensity exercise in humans. *Journal of Applied Physiology* 61: 1337-1339.

20. Challis, J.R.A., B. Crabtree, E.A. Newsholme. 1987. Hormonal regulation of the rate of the glycogen/glucose-1-phosphate cycle in skeletal muscle. *Biochemistry Journal* 163: 205-210.

21. Chaouloff, F. 1989. Physical exercise and brain monoamines: a review. *Acta Physiologica Scandinavica* 137: 1-13.

22. Chrousos, G.P. 1992. Regulation and dysregulation of the hypothalamic-pituitary-adrenal axis. In *Endocrinology and metabolism clinics of North America*, ed. J.D. Veldhuis. Philadelphia: W.B. Saunders.

23. Chrousos, G.P., P.W. Gold. 1992. The concepts of stress and stress system disorders: overview of behavioral and physical homeostasis. *Journal of the American Medical Association* 267: 1244-1252.

24. Coppen, A., M. Bishop, R.J. Beard, G.J. Barnard, W.P. Collins. 1981. Hysterectomy, hormones, behaviour: a prospective study. *Lancet* 1: 126-128.

25. Cumming, D.C., S.R. Wall, H.A. Quinney, A.N. Belcastro. 1987. Decrease in serum testosterone levels with maximal intensity swimming exercise in trained male and female swimmers. *Endocrine Research* 13: 31-41.

26. Cumming, D.C., G.D. Wheeler. 1994. Exercise, training, and the male reproductive system. In *Physical activity, fitness, and health*, eds. C. Bouchard, R.J. Shephard, T. Stephens, 980-992. Champaign, IL: Human Kinetics.

27. Dawson-Hughes, B., D. Stern, J. Goldman, S. Reichlin. 1986. Regulation of growth hormone and somatomedin-C secretion in postmenopausal women: effect of physiological estrogen replacement. *Journal of Clinical Endocrinology and Metabolism* 63: 424-432.

28. De Boer, H., G.J. Blok, E.A. Van der Veen. 1995. Clinical aspects of growth hormone deficiency in adults. *Endocrinology Reviews* 16: 63-86.

29. De Cree, C. 1990. The possible involvement of endogenous opioid peptides and catecholestrogens in provoking menstrual irregularities in women athletes. *International Journal of Sports Medicine* 11: 329-438.

30. De Cree, C., H.A. Keizer, P. Ball, K. Mannheimer, G. Van Kranenburg, P. Geurten. In press. Menstrual cycle-related plasma 2-hydroxy-catecholestrogen responses to short-term submaximal and maximal exercise. *Journal of Applied Physiology.*

31. Dohm, G.L., T.M. Louis. 1978. Changes in androstenedione, testosterone and protein metabolism as a result of exercise. *Proceedings of the Society of Experimental Biology and Medicine* 158: 622-625.

32. Evans, R.M. 1988. The steroid and thyroid receptor superfamily. *Science* 240: 889-895.

33. Fain, J.N., R. Saperstein. 1970. The involvement of RNA synthesis and cyclic AMP in the activation of fat cell lipolysis by growth hormone and glucocorticoids. *Hormone and Metabolism Research* 2: 20-27.

34. Farrell, P.A., M. Kjaer, F.W. Bach, H. Galbo. 1987. Beta-endorphin and adrenocorticotropin response to supramaximal treadmill exercise in trained and untrained males. *Acta Physiologica Scandinavica* 130: 619-625.

35. Finkelstein, J.W., H.P. Roffwarg, R.M. Boyar, J. Kream, L. Hellman. 1972. Age related change in twenty four hour spontaneous secretion of growth hormone. *Journal of Clinical Endocrinology and Metabolism* 38: 519-524.

36. Fry, R.W., A.R. Morton, D. Keast. 1991. Overtraining in athletes: an update. *Sports Medicine* 12: 32-65.

37. Fryburg, D.A., E.J. Barrett. 1993. Growth hormone acutely stimulates skeletal muscle but not whole-body protein synthesis in humans. *Metabolism* 42: 1223-1227.

38. Fryburg, D.A., R.A. Gelfland, E.J. Barret. 1991. Growth hormone actually stimulates forearm muscle protein synthesis in normal humans. *American Journal of Physiology (Endocrinology and Metabolism)* 260: E499-504.

39. Gorski, J., B. Staniciewicz, R. Brycka, K. Kiczka. 1976. The effect of estradiol on carbohydrate utilization during prolonged exercise in rats. *Acta Physiology (Poland)* 27: 361-367.

40. Greenspan, S.L., A. Klibanski, D. Schoenfeld, E.C. Ridgway. 1986. Pulsatile secretion of thyrotropin in man. *Journal of Clinical Endocrinology and Metabolism* 63: 661-668.

41. Griffin, J.E., J.D. Wilson. 1992. Disorders of the testis and the male reproductive tract. In *Williams textbook of endocrinology*, eds. J.D. Wilson, D.W. Foster, 799-852. Philadelphia: W.B. Saunders.

42. Grossman, A., P. Bouloux, P. Price, P.L. Drury, K.S. Lam, T. Turner, J. Thomas, G.M. Besser, J. Sutton. 1984. The role of opioid peptides in the hormonal responses to acute exercise in man. *Clinical Science* 67: 483-491.

43. Guezennec, C.Y., P. Ferre, B. Serrurier, D. Merino, P.C. Pesquies. 1982. Effects of prolonged physical exercise and fasting upon plasma testosterone level in rats. *European Journal of Applied Physiology* 49: 159-168.

44. Hackney, A.C. 1989. Endurance training and testosterone levels. *Sports Medicine* 8: 117-127.

45. Hackney, A.C., T. Gulledge. 1994. Thyroid hormone responses during an 8-hour period following aerobic and anaerobic exercise. *Physiology Research* 43: 1-5.

46. Hackney, A.C., R.J. Ness, A. Schrieber. 1989. Effects of endurance exercise on nocturnal hormone concentrations in males. *Chronobiology International* 6: 341-346.

47. Hackney, A.C., W.E. Sinning, B.C. Bruot. 1990. Hypothalamic-pituitary-testicular axis function in endurance trained males. *International Journal of Sports Medicine* 11: 298-303.

48. Hackney, A.C., W.E. Sinning, B.C. Bruot. 1988. Reproductive hormonal profiles of endurance-trained and untrained males. *Medicine and Science in Sports and Exercise* 20: 60-65.

49. Häkkinen, K., A. Pakarinen, M. Alén, H. Kauhanen, P.V. Komi. 1987. Relationships between training volume, physical performance capacity, serum hormone concentration during prolonged training in elite weight lifters. *International Journal of Sports Medicine* 8: S61-65.

50. Hartman, M.L., A. Iranmanesh, M.O. Thorner, J.D. Veldhuis. 1993. Evaluation of pulsatile patterns of growth hormone release in humans: a brief review. *American Journal of Human Biology* 5: 603-614.

51. Hodin, R.A., M.A. Lazar, B.I. Wintman, D.S. Darling, R.J. Koening, P.R. Larsen, D.D. Moore, W.W. Chain. 1989. Identification of a thyroid hormone receptor that is pituitary-specific. *Science* 244: 76-79.

52. Howlett, T.A., S. Tomlin, L. Ngahfoong, L.H. Rees, B.A. Bullen, G.S. Skrinar, J.W. McArthur. 1984. Release of beta endorphin and met-enkephalin during exercise in normal women: response to training. *British Medical Journal of Clinical Research and Education* 288: 1950-1952.

53. Jiang, B., R. R. Roy, C. Navarro, V. R. Edgerton. 1993. Absence of a growth hormone effect on rat soleus atrophy during a 4-day spaceflight. *Journal of Applied Physiology* 74: 527-531.

54. Jorgensen, J.O., L. Thuesen, J. Muller, P. Ovesen, N.E. Skakkebaek, J.S. Christiansen. 1994. Three years of growth hormone treatment in growth hormone-deficient adults: near normalization of body composition and physical performance. *European of Journal of Endocrinology* 130: 224-228.

55. Jurkowski, J.E., N. L. Jones, W. C. Walker. 1978. Ovarian hormone response to exercise. *Journal of Applied Physiology* 44: 109-114.

56. Kaminsky, P., M. Klein, M. Duc. 1993. Control of muscular bioenergetics by the thyroid hormones. *Presse Medicale* 22: 774-778.

57. Kaminsky, P., B. Robin Lherbier, F. Brunotte, J.M. Escanye, P. Walker, M. Klein, J. Robert, M. Duc. 1992. Energetic metabolism in hypothyroid skeletal muscle, as studied by phosphorus magnetic resonance spectroscopy. *Journal of Clinical Endocrinology and Metabolism* 74: 124-129.

58. Katzeff, H.L., K.M. Ojamaa, I. Klein. 1994. Effects of exercise on protein synthesis and myosin heavy chain gene expression in hypothyroid rats. *American Journal of Physiology (Endocrinology and Metabolism)* 267: E63-E67.

59. Keizer, H., G.M.E. Janssen, P. Menheere, G. Kranenburg. 1989. Changes in basal plasma testosterone, cortisol, dehydroepiandrosterone sulfate in previously untrained males and females preparing for a marathon. *International Journal of Sports Medicine* 10: 139-145.

60. Keizer, H.A., H. Kuipers, J. de Haan, L. Habets. 1987. Multiple hormonal responses to physical exercise in eumenorrheic trained and untrained women. *International Journal of Sports Medicine* 8: 139-150.

61. Keizer, H.A., H. Kuipers, J. de Haan, G.M. Janssen, E. Beckers, L. Habets, G. van Kranenburg, P. Geurten. 1987. Effect of a 3-month endurance training

program on metabolic and multiple hormonal responses to exercise. *International Journal of Sports Medicine* 3: 154-160.

62. Keizer, H.A., P. Platen, P.P.C.A. Menheere, R. Biwer, C. Peters, R. Tietz, H. Wust. 1989. The hypothalamic pituitary axis under exercise stress: the effects of aerobic and anaerobic training. In *Hormones and sport*, eds. Z. Laron, A.D. Rogol, 101-115. New York: Raven Press.

63. Keizer, H.A., J. Poortman, G.S. Bunnik. 1980. Influence of physical exercise on sex-hormone metabolism. *Journal of Applied Physiology* 48: 765-769.

64. Keizer, H.A., A.D. Rogol. 1990. Physical exercise and menstrual cycle alterations: what are the mechanisms? *Sports Medicine* 10: 218-235.

65. Kern, W., B. Perras, R. Wodick, H.L. Fehm, J. Born. 1995. Hormonal secretion during nighttime sleep indicating stress of daytime exercise. *Journal of Applied Physiology* 79: 1461-1468.

66. Kjaer, M., N.H. Secher, F.W. Bach, S. Sheikh, H. Galbo. 1989. Hormonal and metabolic responses to exercise in humans: effect of sensory nervous blockade. *American Journal of Physiology (Endocrinology and Metabolism)* 257: E95-101.

67. Knobil, E. 1988. The neuroendocrine control of ovulation. *Human Reproduction* 3: 469-472.

68. Kraemer, W.J., S.J. Fleck, R. Callister, M. Shealy, G.A. Dudley, C.M. Maresh, L. Marchitelli, C. Cruthirds, T. Murray, J.E. Falkel. 1989. Training responses of plasma beta-endorphin, adrenocorticotropin, and cortisol. *Medicine and Science in Sports and Exercise* 21: 146-153.

69. Kraemer, W.J., S.J. Fleck, J.E. Dziados, E.A. Harman, L.J. Marchitelli, S.E. Gordon, R. Mello, P.N. Frykman, L.P. Koziris, N.T. Triplett. 1993. Changes in hormonal concentrations after different heavy-resistance exercise protocols in women. *Journal of Applied Physiology* 75: 594-604.

70. Kraicer, J., M.S. Sheppard, J. Luke, B. Lussier, B.C. Moor, J.S. Cowan. 1988. Effect of withdrawal of somatostatin and growth hormone (GH)-releasing factor on GH release in vitro. *Endocrinology* 122: 1810-1815.

71. Kuipers, H., H.A. Keizer. 1988. Overtraining in elite athletes. *Sports Medicine* 6: 79-92.

72. Lebovitz, H.E., K. Bryan, L.A. Frohman. 1965. Acute effects of corticotropin and related peptides on carbohydrate and lipid metabolism. *Annals of the New York Academy of Science* 131: 274-287.

73. Lehmann, M., U. Gastmann, K.G. Petersen, N. Bachl, A. Seidel, A.N. Khalaf, S. Fischer, J. Keul. 1992. Training-overtraining: performance, and hormone levels, after a defined increase in training volume versus intensity in experienced middle- and long-distance runners. *British Journal of Sports Medicine* 26: 233-242.

74. Lehmann, M., K. Knizia, U. Gastmann, K.G. Petersen, A.N. Khalaf, S. Bauer, L. Kerp, J. Keul. 1993. Influence of 6-week, 6 days per week, training on pituitary function in recreational athletes. *British Journal of Sports Medicine* 27: 186-192.

75. Lima, L., V. Arce, M.J. Diaz, J.A. Tresguerres, J. Devesa. 1993. Glucocorticoids may inhibit growth hormone release by enhancing beta-adrenergic responsiveness in hypothalamic somatostatin neurons. *Journal of Clinical Endocrinology and Metabolism* 76: 439-444.

76. Loucks, A.B., R. Callister. 1993. Induction and prevention of low-T3 syndrome in exercising women. *American Journal of Physiology* 264: R924-R930.

77. Loucks, A.B., E.M. Heath. 1994. Induction of low-T3 syndrome in exercising women occurs at a threshold of energy availability. *American Journal of Physiology* 266: R817-R823.
78. Loucks, A.B., J.F. Mortola, L. Girton, S.S.C. Yen. 1989. Alterations in the hypothalamic-pituitary-ovarian axes in athletic women. *Journal of Clinical Endocrinology and Metabolism* 68: 402-411.
79. Luger, A., P. Deuster, S.B. Kyle. 1987. Acute hypothalamic-pituitary-adrenal responses to the stress of treadmill exercise: physiological adaptations to physical training. *New England Journal of Medicine* 316: 1309-1315.
80. Mayer, M., F. Rosen. 1975. Interaction of anabolic steroids with glucocorticoid receptor sites in rat muscle cytosol. *American Journal of Physiology* 229: 1381-1386.
81. Mayer, M., F. Rosen. 1977. Interaction of glucocorticoids and androgens with skeletal muscle. *Metabolism* 26: 937-962.
82. McColl, E.M., G.D. Wheeler, P. Gomes, Y. Bhambhani, D.C. Cumming. 1989. The effects of acute exercise on pulsatile LH release in high-mileage male runners. *Clinical Endocrinology (Oxford)* 31: 617-621.
83. McGauley, G.A., R.C. Cuneo, F. Salomon, P.H. Sönksen. 1990. Psychological well-being before and after growth hormone treatment in adults with growth hormone deficiency. *Hormone Research* 33: 52-54.
84. Meeusen, R., I. Smolders, S. Sarre, K. De Meirleir, H. Keizer, M. Serneel, G. Ebinger, Y. Michotte. 1997. Endurance training effects on striatal neurotransmitter release: an 'in vivo' microdialysis study. *Acta Physiologica Scandinavica* 159: 335-342.
85. Menheere, P.P.C.A. 1989. Sensitive estimations of glycoprotein hormones: clinical and scientific applications. PhD diss. University of Limburg, Maastricht, The Netherlands.
86. Moretti, C., A. Fabbri, L. Gnessi, M. Cappa, A. Calzolari, F. Fraioli, A. Grossman, G.M. Besser. 1983. Naloxone inhibits exercise-induced release of PRL and GH in athletes. *Clinical Endocrinology (Oxford)* 18: 135-138.
87. Morley, J.E., G.L. Brammer, B. Sharp, T. Yamada, A. Yuwiler, J.M. Hershman. 1981. Neurotransmitter control of hypothalamic-pituitary-thyroid function in rats. *European Journal of Pharmacology* 70: 263-271.
88. Näveri, H., K. Kuoppasalmi, M. Härkönen. 1985. Metabolic and hormonal changes in moderate and intense long-term running exercises. *International Journal of Sports Medicine* 6: 276-281.
89. Orth, D.N., W.J. Kovacs, C. Rowan Debold. 1992. The adrenal cortex. In *Williams textbook of endocrinology*, eds. J.D. Wilson, D.W. Foster, 489-620. Philadelphia: W.B. Saunders.
90. Overton, J.M., K.C. Kregel, G. Davis-Gorman, D.R. Seals, C.M. Tipton, L.A. Fisher. 1991. Effects of exercise training on responses to central injection of CRF and noise stress. *Physiology of Behavior* 49: 93-98.
91. Owens, M.J., C.B. Nemeroff. 1993. The role of corticotropin-releasing factor in the pathophysiology of affective and anxiety disorders: laboratory and clinical studies. *Ciba Foundation Symposium* 172: 296-308.
92. Petraglia, F., C. Barletta, F. Facchinetti, F. Spinazzola, A. Monzani, D. Scavo, A.R. Genazzani. 1988. Response of circulating adrenocorticotropin, beta-endorphin, beta-lipotropin and cortisol to athletic competition. *Acta Endocrinology (Copenhagen)* 118: 332-336.
93. Petraglia, F., S. Sutton, W. Vale, P. Plotsky. 1987. Corticotropin-releasing factor decreases plasma luteinizing hormone levels in female rats by inhibit-

ing gonadotropin-releasing hormone release into hypophysial-portal circulation. *Endocrinology* 120: 1083-1088.

94. Plotsky, P.M., T.O. Bruhn, W. Vale. 1985. Evidence for multifactor regulation of the adrenocorticotropin secretory response to hemodynamic stimuli. *Endocrinology* 116: 633-639.

95. Printz, P.N., E.D. Weitzman, G.R. Cunningham, I. Karacan. 1983. Plasma growth hormone during sleep in young and aged men. *Journal of Gerontology* 38: 519-524.

96. Quabbe, H.J., E. Schilling, H. Helge. 1966. Pattern of growth hormone secretion during a 24-hour fast in normal adults. *Journal of Clinical Endocrinology and Metabolism* 26: 1173-1177.

97. Quigley, M.E., K.L. Sheehan, R.F. Casper, S.S. Yen. 1980. Evidence for increased dopaminergic and opioid activity in patients with hypothalamic hypogonadotropic amenorrhea. *Journal of Clinical Endocrinology and Metabolism.* 50: 949-954.

98. Raynaud, J., A. Capderou, J.P. Martineaud, J. Bordachar, J. Durand. 1983. Intersubject viability in growth hormone time course during different types of work. *Journal of Applied Physiology* 55: 1682-1687.

99. Reed Larsen, P., S.H. Ingbar. 1992. The thyroid gland. In *Williams textbook of endocrinology*, eds. J.D. Wilson, D.W. Foster, 357-487. Philadelphia: W.B. Saunders.

100. Reichlin, S. 1992. Neuroendocrinology. In *Williams textbook of endocrinology*, eds. J.D. Wilson, D.W. Foster, 135-219. Philadelphia: W.B. Saunders.

101. Rivier, C., W. Vale. 1985. Involvement of corticotropin-releasing factor and somatostatin in stress-induced inhibition of growth hormone secretion in the rat. *Endocrinology* 117: 2478-2482.

102. Roberts, A.C., R.D. McClure, R.I. Weiner, G.A. Brooks. 1993. Overtraining affects male reproductive status. *Fertility and Sterility* 60: 686-692.

103. Rogol, A.D., A. Weltman, J. Y. Weltman, R.L. Seip, D.B. Snead, S. Levine, E.M. Haskvitz, D.L. Thompson, R. Schurrer, E. Dowlin. 1992. Durability of the reproductive axis in eumenorrheic women during 1 yr of endurance training. *Journal of Applied Physiology* 72: 1571-1580.

104. Ropert, J.F., M.E. Quigley, S.S. Yen. 1981. Endogenous opiates modulate pulsatile luteinizing hormone release in humans. *Journal of Clinical Endocrinology and Metabolism* 52: 583-585.

105. Russell, J.B., D.E. Mitchell, P.I. Musey, D.C. Collins. 1984. The role of beta-endorphins and catechol estrogens on the hypothalamic-pituitary axis in female athletes. *Fertility and Sterility* 42: 690-695.

106. Sapolsky, R.M., L.C. Krey, B.S. McEwen. 1986. The neuroendocrinology of stress and aging: the glucocorticoid cascade hypothesis. *Endocrinology Reviews* 7: 284-301.

107. Sapolsky, R.M., L.C. Krey, B.S. McEwen. 1984. Stress down-regulates corticosterone receptors in a site-specific manner in the brain. *Endocrinology* 114: 287-292.

108. Sartorio, A., M.V. Narici. 1994. Growth hormone (GH) treatment in GH-deficient adults: effects on muscle size, strength and neural activation. *Clinical Physiology* 14: 527-537.

109. Schiff, I., Q. Regestein, D. Tulchinsky, K.J. Ryan. 1979. Effects of estrogens on sleep and psychological state of hypogonadal women. *Journal of the American Medical Association* 242: 2405-2414.

110. Slag, M.F., M. Ahmad, M.C. Gannon, F.Q. Nuttall. 1981. Meal stimulation of cortisol secretion: a protein induced effect. *Metabolism* 30: 1104-1108.

111. Takahashi, Y., D.M. Kipnis, W.H. Daughaday. 1968. Growth hormone secretion during sleep. *Journal of Clinical Investigations* 47: 2079-2090.

112. Tannenbaum, G.S., N. Ling. 1984. The interrelationship of growth hormone (GH)-releasing factor and somatostatin in generation of the ultradian rhythm of GH secretion. *Endocrinology* 115: 1952-1957.

113. Urhausen, A., H. Gabriel, W. Kindermann. 1995. Blood hormones as markers of training stress and overtraining. *Sports Medicine* 20: 251-276.

114. Urhausen, A., T. Kullmer, W. Kindermann. 1987. A 7-week follow-up study on the behaviour of testosterone and cortisol during competition in rowers. *European Journal of Applied Physiology* 56: 528-533.

115. Van Breda, E. 1994. The effect of testosterone on skeletal muscle energy metabolism in diabetic and non-diabetic endurance trained rats. PhD diss. University of Limburg, Maastricht, The Netherlands.

116. Van Breda, E., H.A. Keizer, P. Geurten, G. Van Kranenburg, H. Kuipers. 1992. Testosterone enhances glycogen synthesis after submaximal exercise in muscles of trained female rats by an increased synthase activity. *Medicine and Science in Sports and Exercise.* 24: S15, 1992.

117. Van Breda, E., H.A. Keizer, P. Geurten, G. van Kranenburg, P.P. Menheere, H. Kuipers, J.F. Glatz. 1993. Modulation of glycogen metabolism of rat skeletal muscles by endurance training and testosterone treatment. *Pflugers Arch* 424: 294-300.

118. Vanhelder, W.P., R.C. Goode, M.W. Radomski. 1984. Effect of anaerobic and aerobic exercise of equal duration and work expenditure on plasma growth hormone levels. *European Journal of Applied Physiology* 52: 255-257.

119. Veldhuis, J.D., A. Iranmanesh, G. Lizarralde, M. Johnson. 1989. Amplitude modulation of a burst-like mode of cortisol secretion subserves the circadian glucocorticoid rhythm. *American Journal of Physiology (Endocrinology and Metabolism)* 257: E6-14.

120. Vervoorn, C., H. Keizer, H. Koppeschaar, P. Platen, W. Erich, J. Thijssen, W. De Vries. 1991. Changes in plasma beta-endorphin and prolactin after a short term intensive training program. *Medicine and Science in Sports and Exercise* 23: S108, 1991.

121. Vervoorn, C., L.J.M. Vermulst, A.M. Boelens-Quist, H.P.F. Koppeschaar, W.B.M. Erich, J.H.H. Thijssen, W.R. de Vries. 1992. Seasonal changes in performance and free testosterone: cortisol ratio of elite female rowers. *European Journal of Applied Physiology* 64: 14-21.

122. Watanabe, T., A. Morimoto, Y. Sakata, N. Tan, K. Morimoto, N. Murakami. 1992. Running training attenuates the ACTH responses in rats to swimming and cage-switch stress. *Journal of Applied Physiology* 73: 2452-2456.

123. Watanabe, T., A. Morimoto, Y. Sakata, M. Wada, N. Murakami. 1991. The effect of chronic exercise on the pituitary-adrenocortical response in conscious rats. *Journal of Physiology (London)* 439: 691-699.

124. Wehrenberg, W.B., A. Baird, N. Ling. 1983. Potent interaction between glucocorticoids and growth hormone-releasing factor in vivo. *Science* 221: 556-558.

125. Weltman, A., J.Y. Weltman, R. Schurrer, W.S. Evans, J.D. Veldhuis, A.D. Rogol. 1991. Endurance training amplifies the pulsatile release of growth hormone: effects of training intensity. *Journal of Applied Physiology* 72: 2188-2196.

126. Wheeler, G.D., M. Singh, W.D. Pierce, W.F. Epling, D.C. Cumming. 1991. Endurance training decreases serum testosterone levels in men without change in luteinizing hormone pulsatile release. *Journal of Clinical Endocrinology and Metabolism* 72: 422-425.

127. Williams, N.I., J.C. Young, J.W. McArthur, B. Bullen, G.S. Skrinar, B. Turnbull. 1995. Strenuous exercise with caloric restriction: effect on luteinizing hormone secretion. *Medicine and Science in Sports and Exercise* 27: 1390-1398.

128. Wittert, G.A., J.H. Livesey, E.A. Espiner. 1996. Adaptation of the hypothalamopituitary adrenal axis to chronic exercise stress in humans. *Medicine and Science in Sports and Exercise* 28: 1015-1019.

CHAPTER 9

Musculoskeletal and Orthopedic Considerations

W. Ben Kibler, MD, and T. Jeff Chandler, EdD

Introduction

The concept of overtraining is most frequently applied to biochemical or psychological alterations that occur as a result of athletic activity. This chapter will focus on the alterations that may exist in the musculoskeletal system and may be the musculoskeletal manifestations of the overtraining syndrome. These alterations have their genesis in the same imbalance between injury and recovery that has been identified as the overtraining phenomenon (53, 63, 77). Physical exercise disrupts body homeostasis, and the body must recover. Therefore, two responses must be evaluated: the musculoskeletal system's response to an appropriate or inappropriate training load, and the effect this response or adaptation has on injury or injury potential.

Musculoskeletal overtraining may occur as a result of inappropriate physiological, biomechanical, or anatomical stresses. Appropriate stresses to the musculoskeletal system cause positive adaptation to occur (1). Inappropriate volume or intensity of exercise may cause a maladaptive cellular or tissue response due to an imbalance between load and recovery. These maladaptive responses occur to some extent in most all sports; however, they can certainly become a part of the overtraining syndrome. The maladaptive responses may be objectively documented as distinct musculoskeletal injuries, such as alterations in muscle strength, flexibility, or balance, changes in joint range of motion, or stress reactions in bone.

The exact mechanisms underlying musculoskeletal overtraining are not completely understood, but accumulating evidence indicates that disruptions in cellular homeostasis appear to be basic to the process. Tissue effects arise from these cellular disruptions. These changes may be seen as a possible contributing factor to the musculoskeletal aspects of the overtraining syndrome.

Cellular Aspects of Musculoskeletal Overtraining

The cellular aspects of overtraining have been discussed in the scientific literature. These changes include an elevated resting creatine kinase activity, elevated resting cortisol levels, decreased muscle glycogen and testosterone, alteration in size and shape of mitochondria with decreased mitochondrial enzymes, and alterations in calcium release and uptake (3, 7, 91). Electron microscopy of affected connective tissue indicates intracytoplasmic calcifications, longitudinal collagen fiber splitting, collagen fiber kinking, and abnormal fiber cross links.

Some researchers have associated overtraining with an increase in tissue catabolism (54). The premise of this theory is that the catabolic state induced by excessive training may be directly related to the maladaptations that occur. Muscle damage does occur with high intensity exercise and has been thought to occur from a number of possible mechanisms (91). These include an imbalance of calcium influx resulting in reduced mitochondrial function and production of adenosine triphosphatase, an increase in intracellular pH causing a destabilization of lysosomal membranes, and macrophage induction of various lymphokines and monokines that enhance protease activity. Vailas et al. (91) concluded that overtraining does result from a progressive increase in basal proteolytic activity of muscle tissue. However, more recent thought has been directed toward the tensile load theory of overload injury.

Microtraumatic damage to muscle fibers has been established as the cause of the delayed onset muscle soreness (DOMS) that occurs 24-48 h after strenuous exercise (4). It is also known that microscopic muscle damage is more pronounced after strenuous exercise involving intense eccentric muscle contractions (4, 15, 55, 57, 70, 71, 83). Delayed onset muscle soreness is thought to be primarily due to direct structural damage to the muscle (5). Direct myofibrillar damage, defined as an alteration in muscle fiber structure or function observed immediately after exercise, has been shown to occur in response to both eccentric exercise and fatigue (79). Crenshaw et al. (13) demonstrated that DOMS was positively correlated with intramuscular fluid pressure in the vastus lateralis muscle indicating intercellular swelling. Although evidence suggests the damage seen in DOMS is not permanent (4, 19, 20, 37), this model used acute eccentric overload to produce the changes. Chronic repetitive eccentric loads at submaximal levels have been shown to produce continued and irreversible changes in rabbit (23) myotendinous junctions. Although DOMS and the resulting acute cellular disruptions occur in all athletes, the overtrained athlete is particularly susceptible to maladaptation and injury as the result of chronic overloads and disruptions.

The mechanisms by which tensile load causes cellular changes is not clear. Recent work suggests that a direct mechanical tension effect, rather than an indirect hormonal or vascular effect, is the primary causative mechanism for the cellular changes (85). This study evaluated the effects of rest periods on muscle microinjury induced from eccentric contractions. If the cause of the microinjury was metabolic in nature, rest periods would alleviate some of the muscle soreness. If the cause of the microinjury was from tensile stress, rest periods would make no difference in the degree of muscle damage. Changes in soreness, strength, resting elbow angle, and muscle girth were monitored after 10 eccentric contractions of the elbow flexors. There were no differences between the continuous contractions and contractions with 15 sec of rest. Loss of strength and soreness occurred even with the groups that rested five and 10 min between repetitions. The results were most consistent with a materials failure of tissue due to tensile load. Leiber (58) has shown that in muscular tissue, this cellular damage is due to excessive strain in the contracting fiber, not the absolute force developed in the fiber or the muscle. The anatomic site of myofibrillar injury is the attachment of the myofibril to the extrasarcolemic cytoskeleton (21). The exact mechanism of this disruption is "muscle fiber strain that occurs during lengthening of an activated muscle" (58). Implied in this description is that the load is important only as a relative load compared to the muscle's ability to protect itself against strain. Normal loads on weakened muscles, a relative force overload, are as capable of causing strain as supernormal loads on normal muscles, an absolute force overload.

Muscular Adaptations Related to Intense Participation in Sports

Adaptations to the musculoskeletal systems of athletes have been shown to occur corresponding with the sites of stress. In muscular tissue, these adaptations have been documented in both strength measurements and flexibility measurements (8, 9). One recent report indicated that range of motion deficits in national-level junior tennis players were correlated with both age and years of tournament play (46). Also, the prepubertal musculoskeletal base apparently does respond in the same manner to tensile loads as does the postpubertal musculoskeletal base in terms of decreases in strength and flexibility (44), indicating the adaptations to muscles used repetitively in sports are not related to hormonal changes that occur with aging.

Adaptations in Range of Motion

Studies on the flexibility of athletes demonstrate that a muscle placed under high tensile load repetitively may become tight, limiting the range of motion and perhaps changing the mechanics of the joint. Kibler et al. (42) reported that upper body athletes were more likely to experience tightness in upper body flexibility measurements, and lower body athletes were more likely to experience tightness in lower body flexibility measurements. The frequency of muscle tightness and injuries in senior division soccer players, mean age 24.6 years, was reported by Ekstrand and Gillquist (16). Soccer players were generally less flexible than a group of age-matched nonplaying controls. All measurements were performed on the lower body and included hip flexion with the knee straight, hip extension, hip abduction, knee flexion lying prone, and ankle dorsiflexion with the knee straight.

Flexibility comparisons of junior elite tennis players to other athletes indicated specific musculoskeletal adaptations in areas of the body under highest stress (8). Tennis players were tighter than other athletes on shoulder internal and external rotation on the dominant and nondominant side, as well as on low back flexibility. These tennis players, mean age 15.4, spent approximately six days per week for a total of 22 h on the tennis court.

In lacrosse, Jackson and Nyland (34) reported that overall upper body and lower body flexibility in club players was average, with areas of tightness in the hamstrings and quadriceps muscle groups. Professional baseball pitchers demonstrate a decrease in internal rotation flexibility of the dominant arm, along with an increase in external rotation flexibility when compared to the dominant arm of position players (6).

In a more recent study, shoulder internal rotation and total rotation have been shown to decrease both as a function of age (age groups: <16, 16-18, >18) and by years of tournament play (yrs of play: <6, 6-9, >9) in a group of elite tennis players (U.S. National Team) (46). Dominant shoulder internal rotation declined and the difference between dominant and nondominant internal rotation increased with both age and years of tournament play (28). Males and females showed the same degree of deficits, indicating that both are susceptible to tensile overload injuries with intense training.

Adaptations in Strength

Strength adaptations to repetitive activities may actually be deleterious to both performance and injury potential. There may be two types of imbalances. Muscle strength imbalances may occur when muscles causing the joint to perform a specific movement (prime movers) are selectively strengthened without subsequent strengthening of other muscles surrounding the joint, including both stabilizing and decelerating muscle groups. This may occur due to selective strengthening during conditioning, or may occur as a result of the plyometric-like stretch and activation that occurs in prime movers in actual athletic activity. Also, muscle weakness may occur in a paired set of muscles as a result of direct injury or repetitive microtrauma over a long period of time. In both cases, the net result is a change in the force couple balance that is necessary to control movement or position. Both muscle strength imbalances and muscle weakness have been demonstrated in athletes participating in repetitive athletic activities.

Muscle strength adaptations have been reported throughout the scientific literature (2, 9, 12, 33, 43, 60, 73). Microscopic muscle damage does occur with intense muscle contractions (4) or with tensile overload to the muscle and tendon unit (26). As mentioned previously, eccentric exercise has been shown to produce considerably more damage to skeletal muscle than other types of exercise (3, 57, 83). This damage may be a factor in loss of strength just as it is a factor in the loss of flexibility. At this point, however, no clinical study has documented a direct correlation of muscle strength loss with muscle damage.

Sport-specific muscle imbalances have been documented in several groups of athletes. An imbalance of shoulder external and internal rotation strength has been reported in tennis players (9), water polo players (60), and baseball pitchers (32). Runners with plantar fasciitis have been shown to have weakness of the plantar flexor muscles (44). Imbalances in rotator cuff strength in adduction/abduction and external/internal rotation were demonstrated in water polo players (60). The adductor muscles in water polo players had been selectively strengthened compared to the abductors, resulting in an adduction/abduction strength ratio of about 2:1. The internal rotators had increased in strength relative to the external rotators, resulting in a decrease in the external/internal strength ratio to about 0.6:1. The muscle imbalances demonstrated in water polo were more apparent at a test speed of 30 degrees per second.

Chandler et al. (9) demonstrated that college-level tennis players increase the strength of the shoulder musculature in internal rotation without subsequent strengthening of the external rotator muscle group creating a functional muscle imbalance that may predispose the athlete to injury. Using the nondominant arm as a control, tennis players demonstrated significantly greater strength in internal rotation in peak torque at 60 degrees per second, peak torque at 300 degrees per second, average power at 60 degrees per second, and average power at 300 degrees per second. In external rotation, the tennis players were significantly stronger on only one of the four measurements, peak torque at 300 degrees per second. By increasing the strength of the internal rotator muscle groups without subsequent strengthening of the external rotators, a muscle imbalance is created that may predispose the athlete to overload injury.

In summary, adaptations in muscular strength and flexibility do exist as a result of athletic play and conditioning. Most of the adaptations are thought to be due to a disparity between tissue load and tissue recovery, that may eventually be related to the overtraining syndrome. Some of these adaptations may result in injury or may increase injury potential. The muscular maladaptations are manifested as decreases in the flexibility or range of motion at a joint, and decreases or imbalances in muscle strength in the muscle groups involved. Along with these adaptations, biomechanical deficits in movement patterns can be identified, probably as a result of deficits in muscle strength and flexibility. These biomechanical variations, however small, may have a dramatic effect on performance and injury risk (84).

Skeletal Adaptation

Skeletal adaptations may first be noted as stress reactions, that are localized areas of bony hyperactivity in areas subject to repetitive stress. The exact mechanism of these reactions is unclear; they probably represent alterations in bone synthesis and turnover, and have relatively few early clinical manifestations. The exact role stress reactions play in injury pathogenesis is also unclear. But their location, in the same anatomical areas that have the highest incidence of overt stress fractures, implies that some may be prodromal. These reactions are usually more diffuse than stress fractures, and may be objectively evaluated by bone scan.

The Effect of Overtraining on Injury Potential

Overtraining

Overtraining adaptations, while usually not extensive enough to cause overt clinical symptoms, do arise from damage to the musculoskeletal system and place the system at some mechanical or performance disadvantage, causing an alteration in mechanics or decrease in efficiency. These adaptations most certainly can occur from overuse of the musculoskeletal system, as well as from intense participation in a specific sport or activity. In all instances, these adaptations can be manifestations of the overtraining syndrome in that the training load results in adaptations that appear to be less than beneficial for athletic performance. Although the recognition of such injuries is increasing, the actual role of strength and flexibility deficits in the cause and prevention of these injuries is not completely clear (65). It appears that these deficits may create abnormal mechanics that predispose the tissues to injury with continued use. The injuries may occur locally or at a distance from the mechanical deficit. These deficits may play a role mainly in influencing the potential for overload injuries.

Overload Injuries

Overload types of injuries to the musculoskeletal system are becoming more recognized (30, 75), but are probably still underreported in the literature. This is because these injuries are due to repetitive microtrauma overload, with more gradual onset of symptoms and less acute athletic function.

According to Micheli (64) overload injuries result not simply from overuse but from improper training techniques and anatomic malalignment. Kibler et al. (41) have suggested that the strength and flexibility of the musculotendinous unit are also factors in such injuries. In baseball pitchers, the overload of pitching has been linked with injury and inflammation of the muscles of the shoulder girdle complex (72). The cycle begins with localized pain that has a secondary effect of inhibiting muscle action. Repeated minor injuries, secondary to contracture, tightness, weakness, imbalance, and inflammation have a cumulative effect resulting in a chronic injury. Jobe and Bradley (35) state that the glenohumeral joint is particularly vulnerable to altered mechanics because the throwing motion puts tremendous stresses on the stabilizing mechanisms.

The risk of repetitive overload injury related to inappropriate overload has been shown to increase with increasing height and weight (56, 81, 90, 94). Taimela et al. (82) state that a high center of gravity or increased limb length may be a factor in these injuries. An increased limb length could produce more leverage and therefore more stress on the joints. The forces encountered on landing and push-off, as well as during other movements, must also be considered as potential causes of such injuries.

Mechanism of Injury Production

Injuries to the musculoskeletal system that can be related to the overtraining syndrome range from minor to severe. Our belief is that some degree of musculoskeletal adaptation occurs in response to each dose of exercise, whether normal (use) or abnormal (abuse), and there can be various degrees of adaptation or maladaptation. The degree of maladaptation depends on the preparation and maintenance of strength and flexibility of the musculoskeletal system, the degree of use of the musculoskeletal system due to the demands placed on it by the particular athletic activity, the degree of abuse to the musculoskeletal system by the same demands, and the degree of recovery prior to the next use or abuse. The extent of these adaptations may also be related to sex, age, and the onset of physical maturity.

These maladaptations may be described in terms of a vicious cycle of overload injuries (see figure 9.1, tables 9.1 and 9.2); where each use or abuse of the musculoskeletal system pushes the athlete further toward overt injury, and each recovery period brings the athlete back from possible overt injury.

In this framework, a relative or absolute force overload would create a tensile overload. If this tensile overload is small enough, the tissue may heal with appropriate rest or treatment, or the body will bypass the area by using other mechanics or other functions. This tissue compromise can more frequently create functional

Table 9.1 Four Categories of Clinical Presentation of Muscle Injury (43)

- Acute injury
- Chronic injury
- Acute exacerbation of chronic injury
- Subclinical injury

Reprinted from Kibler 1990.

Table 9.2 Components of Muscle Injury (43)

- Tissue injury complex
- Clinical symptom complex
- Functional biomechanical deficit
- Functional adaptations
- Tissue overload complex

Reprinted from Kibler 1990.

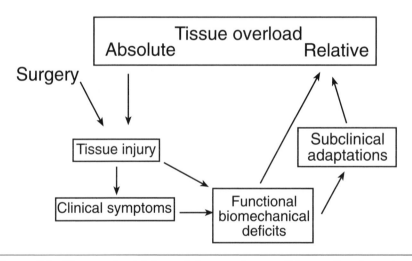

Figure 9.1 The overload injury vicious cycle: possible results of overload injuries to the musculotendinous unit.

biomechanical deficits as a result of alterations in flexibility, strength, strength balance, or skeletal reaction. The athlete attempts to compensate for these deficits by adopting alternate patterns of movement, position, and activity. These patterns are usually less efficient, creating even more overload, thereby closing the circle (46). This can also be described as a cascade-to-overload injury, as it seems that these injuries begin with relatively minor adaptations in terms of strength, flexibility, and biomechanics, with adaptations to these areas becoming more pronounced as the athlete reaches the point of overt injury. Factors related to the ability of the musculoskeletal base to adequately accept the load placed upon it are outlined in figure 9.2. In this construct, the degree of use of the musculoskeletal system to promote optimal performance must also be evaluated in terms of the possibility of maladaptations to the musculoskeletal system, both in terms of overt injury as well as increasing the potential for injury.

Meuwisse (62) has created a model that demonstrates how these changes interact with sport activities to create injuries (see figure 9.3). The intrinsic musculoskeletal changes of inflexibility, strength weakness, or bony reaction create a predisposed athlete. Interactions of the predisposed athlete with the extrinsic demands of the sport creates a susceptible athlete. If the susceptible athlete continues to participate and to be exposed to the extrinsic demands, tissue failure and clinical symptoms may result.

Overload Injuries in Specific Sports

Injuries in soccer players have been shown to be related to strength, range of motion, and degree of rehabilitation (17). Twenty-eight percent of the moderate and

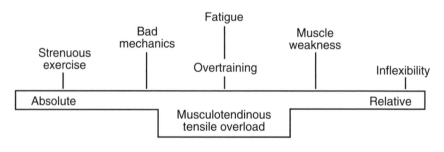

Figure 9.2 Potential causes of tensile overload injuries to soft tissue.

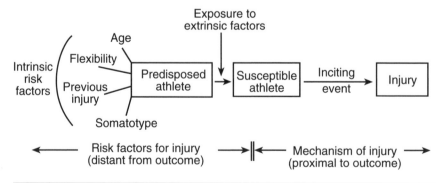

Figure 9.3 Meuwisse's model.

Reprinted, by permission, from W.H. Meuwisse, 1994, "Assessing causation in sport injury: a multifactoral model," *Clinical Journal of Sports Medicine* 4: 166-170.

major knee injuries reported in 180 male soccer players were attributed to inadequate rehabilitation with persistent muscle weakness. Soccer players sustaining noncollision knee injuries had reduced knee extension strength in the injured leg compared to noninjured players. No significant change in the hamstring/quadricep ratio was found in the knee injury group compared to the noninjured players. The soccer players were less flexible than a control group in hip abduction, hip extension, knee flexion, and ankle dorsiflexion. Seventy-seven percent of the players who sustained muscle rupture or tendinitis in the lower extremity had significant muscle tightness, with the remainder demonstrating normal flexibility. All 13 athletes with hip adductor rupture or tendinitis demonstrated decreased hip abduction ROM. No significant differences in ROM were found between players with hamstring strains compared to the other athletes. In club lacrosse players, Jackson and Nyland (34) reported that all overuse injuries except one occurred in athletes who demonstrated a muscle imbalance, inflexibility, or both on the preparticipation physical exam.

Knapik et al. (48) reported on the incidence of injuries in female collegiate athletes in eight weight-bearing varsity sports. These athletes experienced an increased incidence of lower extremity injury if they had (1) a right knee flexor 15% stronger than the left knee flexor tested isokinetically at 180 degrees per second, (2) a right hip extensor 15% more flexible than the left hip extensor, and (3) a knee flexor/knee extensor ratio of less than 0.75 tested isokinetically at 180 degrees per second.

In a review of tennis injuries, Kuland et al. (52) indicated that resistance training exercises were effective in preventing pain from tennis shoulder as described by Priest and Nagel (74). Kuland and colleagues also stated that strength and flexibility exercises for the wrist extensors and flexors were effective in the prevention of tennis elbow pain. Kibler et al. (41) reported on the injury incidence of elite junior tennis players and recreational tennis players. In the elite juniors, overload injuries accounted for 63% of the injuries, sprains, 25%, and fractures, 12%. In recreational players, overload injuries accounted for 62% of the injuries, sprains, 22%, and fractures, 14%.

In swimmers, Dominquez (14) found a significant decrease in shoulder pain in age-group swimmers after they began resistance training. There was evidence presented in this study to support the opinion that lack of strength training was related to the development of shoulder pain. Hawkins and Kennedy (29) utilized isokinetic exercises to increase the strength and endurance of the shoulder muscles in swimmers and other upper body athletes. These exercises reduced the incidence of shoulder problems developed by the athletes.

Injury to the Skeletal System From Overload

Forces occurring to the body during athletic participation may injure the skeletal system (38). By the 1980s, as many as 10% of the injuries seen in sports medicine practices were stress fractures (59). The skeletal system may respond to these stresses with calcium deposition and resorption along the lines of stress. If calcium resorption outpaces deposition, stress fractures may occur. Stress fractures can occur from inappropriate overload, and thus may be a manifestation of the overtraining syndrome. The potential causes of stress fractures can be related to muscle strength imbalances, inflexibilities, muscle weakness, or abnormal mechanical loads on the bone. Weak muscles may be unable to dissipate either absolute or relative force overloads, causing those forces to be transferred to the skeletal system. Fatigued muscles, that may be present in an overtrained state, may react in the same manner as weak muscles, increasing the forces on the skeletal system and producing a stress reaction. Muscle strength imbalances may modify the forces on the skeletal system in a similar manner and may cause an abnormal stress to the skeletal system producing a stress reaction. Excessively strong muscles, either absolutely or relative to weaker antagonistic muscles, may also concentrate stresses in bone and create stress reactions and stress fractures (76, 88).

Inflexibilities of the surrounding muscle groups may also be a factor in stress fractures. By altering the biomechanical pattern of the normal activity, these muscular adaptations may cause stresses to be placed on bone that in a repetitive overload situation might develop into a stress reaction. Other mechanical stresses, such as bony malalignment, improper shoes, abnormal stress due to running surfaces, or injury to other body parts, may also increase the bony load and predispose to stress fracture.

An area that is receiving more attention is the effect of inappropriate overload on the immature skeleton as more youths participate intensely in athletics (61). The amount of load placed on a skeleton that is not adequately prepared to accept the load is of great importance. Traction apophysitic conditions such as Osgood-Schlatter disease and Sever disease, both due to an abnormally tight muscle putting high tensile loads on an attachment that is weak, are becoming very common. Juvenile osteochondritis dissecans of the elbow or knee is thought to be due to abnormal compressive loads, with bony and eventually cartilaginous fracture. Spinal conditions such as spondylolysis or spondylolisthesis, and vertebral end plate problems are also becoming more common. These have been shown to be related to the amount of time spent in gymnastics or running. In all these cases, it appears that the exercise dose of repetitive trauma is too high for the skeletal system to properly adapt, increasing the athlete's chance of a skeletal stress reaction.

Soft Tissue Injuries From Overload

Soft tissue injuries from inappropriate overload may develop in a similar fashion. As muscles contract, tensile stress is applied to the muscle and tendon unit. In repetitive loading conditions, this tissue may become strained. The weakest point of the musculotendinous system is the muscle-tendon junction (23). The structural components of muscle injuries in animal models are well documented in the literature (22, 23, 25, 68).

Muscle strains can occur as acute or chronic conditions. Inappropriate overload can be a causal factor in the acute muscle strain both in the form of abnormal biomechanics and a decrease in the ability of the muscle to protect itself. Muscle strength imbalances or muscle weakness as may be seen in the overtrained athlete may lead to a muscle strain injury. Improper or incomplete rehabilitation of acute muscle injuries can cause chronic problems for athletes (31), and total training volume and intensity is certainly a factor in this process. If intense training begins too soon after the injury performance may be decreased, while the chance of a recurrence of the injury is increased. One example of this is the acute hamstring injury that is common in many sports. If not completely rehabilitated, or if the athlete is allowed to return to stressful activity too early, this injury can become a chronic problem, limiting performance for an extended period of time.

Tendinitis is a common soft tissue injury in sports. Tendinitis can be sudden and seemingly unrelated to repetitive loading, or it can be correlated with musculoskeletal overload as a cascade-to-overt injury. Even in a situation where a hamstring injury is seemingly unrelated to overtraining, hamstring tightness from continued use or a muscle strength imbalance may be a hidden contributor to the injury. Other factors of overtraining such as fatigue and recovery may also play a critical role.

Specific instances of tendinitis such as lateral or medial epicondylitis, plantar fasciitis, and rotator cuff tendinitis fit the model of the cascade-to-overt injury. Tennis elbow, or epicondylitis of the lateral or medial epicondyle of the elbow is a common injury in throwing and racquet sports. Priest and Nagle (74) estimated that 45% of those tennis players who play daily suffered from this condition. Approximately 90% of the instances of tennis elbow is backhand tennis elbow, and is an overload condition to the extensor muscles that extend the wrist on the backhand stroke. Many potential causes of tennis elbow have been identified: improper stroke mechanics, weakness of the wrist extensors, inflexibility of the wrist musculature (27, 69), and posterior shoulder weakness (45). Increased strength of the wrist extensors would allow the muscles to absorb a greater force, lessening the force transferred to the elbow. Flexibility of these muscles would also cause less force to be transmitted to the elbow. Similarly, strong posterior shoulder muscles could move the arm and wrist more rapidly through the hitting zone, decreasing the tensile load on the elbow. Regardless of the actual cause of the initial trauma, achievement of normal strength and flexibility is a major component of both preventive and rehabilitative programs for tennis elbow. Similar scenarios could be constructed for overload injuries to the shoulder, as in rotator cuff tendinitis (47), and to the foot of running athletes, as in plantar fasciitis (10, 44).

Ligamentous injuries can also be linked with the vicious cycle of musculoskeletal adaptations model. Chronic ankle sprains can continue to produce symptoms, often because of incomplete rehabilitation of the initial injury. Lack of strength and flexibility are often contributing factors in the chronic condition, and may possibly have been factors in the initial injury.

Subclinical soft tissue injuries are injuries that may not be recognized as injuries by the athlete or the coach and are often overlooked as a possible causative factor of more severe injuries. However, under the broad definition of injury, these abnormalities may produce symptoms even though the athlete continues to participate. The athlete may modify participation, and at times may cause an injury in a distant part of the kinetic chain, likely due to abnormal biomechanical movement patterns. The athlete, coach, physician, and sport scientist must always remember that repetitive use injuries may be due to abnormal mechanics in a proximal segment of the kinetic chain. Overt injuries that result in soft tissue injury or adaptations may begin as subclinical soft tissue problems. These subclinical adaptations may occur as a result of playing a sport, and may be related to the frequency and intensity of playing the sport. As a result of these maladaptations, musculoskeletal changes occur that predispose the athlete to overt injury, as well as modify the ability of the athlete to perform at the highest possible level.

Specific Musculoskeletal Injuries and Their Evaluation

Many injuries may be seen in the athlete with the overtraining syndrome. Certainly all athletes with these injuries are not suffering from overtraining syndrome, but rather possible inappropriate overload of the involved tissue. The injuries listed below are the most common and serve as examples of repetitive microtrauma that progresses from normal to maladaptations to overt injury.

Bony Injuries

Elbow Valgus Overload.   The throwing motion places tremendous stress on the joint surfaces of the elbow. In sports that necessitate repetitive throwing, chronic valgus stress on the posterior medial aspect of the elbow may cause bone spur formation on the joint surfaces at the elbow. Soft tissue alterations seen in these conditions include pronator and biceps inflexibility, and medial collateral ligament incompetence. An X ray is diagnostic.

Elbow Osteochondritis Dissecans.   Some throwing athletes damage the articular cartilage in the lateral aspect of the elbow by exerting too much load on the surface of the bones. Subchondral fractures and chondral injuries may occur, causing a piece of the joint surface along with a portion of the underlying bone to loosen itself from the surface of the bone. This may be seen by itself, or as a component of the valgus overload. Plain X ray or MRI is diagnostic. Shoulder muscle imbalance may be seen complicating all cases of elbow valgus overload.

Spondylolysis and Spondylolisthesis.   Spondylolysis may develop from continual cyclic loading of the pars interarticularis, especially in the immature or repetitively hyperextended spine. Spondylolisthesis is the slipping of one vertebral element over another after a bilateral spondylolysis occurs.

Lumbar End Plate Injury.   Repetitive impact loading of the immature spine has been shown to cause microfractures and failure of the developing vertebral end plate. The extent of injury has been positively associated with a high dose of training.

Osgood-Schlatter Disease.   Osgood-Schlatter disease is an apophysitis of the patellar tendon at its insertion on the tibial tubercle. In young athletes this tendon is attached to prebone, that is weaker than normal adult bone. Predisposing factors include quadriceps and illiotibial band inflexibility and quadriceps muscle weakness. These all decrease the compliance of the muscle-tendon-bone unit. With the stresses of running and jumping, irritation and microinjury begins and apophysitis eventually develops.

Posterior (Medial) Tibial Stress Reaction.   The posterior tibial muscle originates on the back of the tibia. Overload of this area leads to posterior medial shin pain. Excess training on hard surfaces or increasing running mileage are predisposing factors.

Metatarsal Stress Fracture.   The metatarsal stress fracture is the most common stress fracture seen in the human body (78). The fracture occurs as a cumulative result of the force applied to the foot in running. The first metatarsal bone is thick and strong, and is seldom a candidate for a stress fracture; stress fractures are most common in the other four metatarsal bones. Predisposing factors include high arches, change in running intensity or duration, or posterior muscle inflexibility.

Overt Soft Tissue Injury

Rotator Cuff Tendinitis.   Rotator cuff tendinitis is secondary to eccentric overload or compressive impingement. Associated findings include scapular muscle weakness, internal rotation inflexibility, ligamentous instability, and rotator cuff muscle weakness. Substitute throwing motions are commonly present. Rotator cuff impingement is rarely the sole cause of rotator cuff tendinitis in young athletes.

Medial Epicondylitis.   Medial epicondylitis is an inflammation of the wrist flexors at their origin on the medial epicondyle. This is usually because of an absolute force overload and is, therefore, more common in elite tennis or baseball players, with pain generally more severe in the service or throwing motion. As was mentioned in the discussion of the shoulder, maladaptations are frequently seen in elbow soft tissue injuries.

Lateral Epicondylitis.   Lateral epicondylitis occurs from an overload to the wrist extensors, that originate on the lateral epicondyle of the elbow. Tennis elbow is commonly lateral epicondylitis, and pain is generally more severe on backhand strokes, where the wrist extensors are active to stabilize the wrist. This is usually seen in less accomplished athletes and is often secondary to abnormal mechanics of the backhand motion. Muscle weakness in the posterior shoulder is seen as well as weakness and inflexibility of the wrist and finger extensors.

Abdominal Muscle Strain.   Abdominal muscle strain may be seen in soccer or other kicking athletes, or in throwing athletes, due to kicking or throwing. The abdominal muscles are often tight, especially on the contralateral side of the dominant arm or leg.

Hamstring Strain.   A strain to the knee flexors, the hamstrings, occurs with the muscle fully stretched. The muscle tears at the weakest point, the muscle-tendon junction. Garrett et al. (25) demonstrated that the muscle-tendon junction is oblique and involves almost the entire length of the muscles, so tears may occur anywhere along the muscle. With a severe hamstring strain, there will be a palatable defect in the posterior thigh and the athlete will be unable to flex the knee with any force. Predisposing factors include muscle imbalance. Inadequate evaluation and rehabilitation will predispose the athlete to recurrent injury and scar.

Patellar Tendinitis.   Patellar tendinitis is tendinitis of the patellar tendon and is common in jumping sports such as basketball. The condition has been demonstrated to be almost purely secondary to repetitive microtears. Predisposing factors include quadriceps inflexibility.

Gastrocnemius Strain.   The gastrocnemius is a two-joint muscle with a medial and a lateral head. The weakest portion of the muscle is where the muscle attaches to the Achilles tendon. A gastrocnemius strain is a tearing of the muscle, usually at the musculotendinous attachment.

Achilles Tendinitis.   The Achilles tendon may become involved in an overload situation anywhere along its course. The most common causes include repetitive microtears in a tight, noncompliant muscle and tendon unit, similar to patellar tendinitis. It is common in athletes who participate in running and jumping sports, especially with changes in training.

Plantar Fasciitis.   Plantar fasciitis is an irritation and partial detachment of the fascia and its calcaneal attachment. Kibler et al. (44) demonstrated plantar flexor inflexibilities and muscle strength deficits associated with the clinical presentation of the syndrome.

Rehabilitation of Overload Injuries

Proper rehabilitation of overload soft tissue injuries is important both to the successful return of the athlete to normal activity or sport, as well as in the prevention of a recurrence of the injury. If the injuries indeed were related to inappropriate training load, correcting the causative factor is an important part of the rehabilitation process. The goal of rehabilitation should be to return the athlete as soon as possible to the highest level of function possible, while at the same time reducing the chance of further injury or damage to the injured area. Functional restoration rather than symptom resolution is the desired endpoint.

The traditional goals of rehabilitation of any type of musculoskeletal injury have been (1) to establish an accurate and complete diagnosis of all the anatomic and functional deficits resulting from the injury, (2) to minimize the deleterious local effects of the acute injury, (3) to allow anatomic healing of the injury, (4) to maintain other components of athletic fitness, and (5) to regain previous athletic function (31). The methods by which these goals can be accomplished include (1) adequate tissue protection during healing, (2) proper sequence and use of therapeutic techniques, (3) proper rehabilitation of muscular flexibility, strength, and balance, and (4) proper rehabilitation of all the links of sport-specific or activity-specific criteria for return to play, backed by a complete assessment of the ability of the athlete to function normally in the sport or activity. This also includes assessment of the training program, so that overtraining and subsequent adaptations do not recur.

The rehabilitation process can be broken down into three phases. The acute phase of rehabilitation is concerned with protection of the injured tissues by splinting, bracing, or other support to allow healing and control of the early inflammatory process. The primary goal of this phase is to reduce the inflammatory response and overt clinical symptoms. This phase deals mainly with the tissue injury and clinical symptom complexes of the negative feedback vicious cycle.

The acute phase overlaps with the recovery phase, where efforts are made to return injured tissues to normal flexibility and strength and to integrate them with the entire kinetic chain of the sporting activity. Gradual controlled increases in physical activity are characteristic of the recovery phase, sometimes with the aid of taping, bracing, or other form of support. This phase concentrates on normalizing the tissue overload, functional biomechanical deficit, and subclinical adaptation complexes.

The maintenance phase is the final phase of rehabilitation. During this phase, the primary goal is to promote maximal strength, flexibility, and muscle balance in the injured area as well as the entire kinetic chain, and to ensure these levels are maintained with appropriate sport-specific evaluation to make sure the athlete can participate safely. In this phase, sport-specific criteria for return to play can be developed.

The goal of sports medicine rehabilitation is to restore the athlete safely and rapidly to as normal a level of function as possible. For this to occur, the athlete must progress as soon as possible from one phase of rehabilitation to the next phase. It should be noted that the criteria to move from one phase to the next are based on function, not time. Sport-specific functional testing is then an essential part of moving from one phase of rehabilitation to the next, and finally to full participation. Volume and intensity of training are very carefully controlled in all of these phases.

Prevention of Overtraining Related Injuries

Although it is often claimed that muscle strength will reduce injuries, specific studies that indicate an actual decrease in the injury rate with resistive training are few. Studies are available that show possible mechanisms by which muscle strength may be a factor in injury prevention. Differences may exist in the mechanisms of injury prevention in traumatic and overload injuries.

Studies that show strength increases with resistive training are numerous. A consistent finding of research over the past 40 years is that resistive training results in increased muscular strength (4). Stone (80) states that muscle and connective tissue can undergo adaptations to physical training resulting in greater tissue mass and increased maximal tensile strength. Garrett et al. (24) demonstrated that maximally stimulated muscle absorbs significantly more energy prior to failure than submaximally or nonstimulated muscle. Both tetanically stimulated and wave summation contracted muscle required a greater force to tear than nonstimulated control muscles. There was no statistical difference in force to failure between muscles stimulated at 16 Hz and 64 Hz. This may indicate that a stronger muscle may absorb more energy than a weak muscle before reaching the point of muscle strain. In addition to strengthening muscle tissue to absorb loads, resistive training has also been shown to strengthen other structures around a joint that may be a factor in injury prevention (1, 49, 50, 56, 86, 87, 89).

The effect of physical activity, and specifically resistive training, on bone density may affect the injury rate to bone tissue in both fractures and overload related stresses. Localized bone loss with immobilization does occur (18, 51). Studies from space indicate that activation of the antigravity muscles may play an important role in the prevention of bone mineral loss (92).

In a study of nationally competitive male athletes, those athletes who had the greatest strain placed on the lower limbs had the highest bone mineral content (67). Two groups of controls for this study consisted of 24 age-matched physically active males and 15 age-matched sedentary males. The physically active controls had significantly more bone mineral than the sedentary controls. The competitive athletes had more bone mineral than the controls as a single group. In a study such as this, the possibility exists that the subjects with the greatest bone density chose to be more active in sports.

Bone mass of the humerus in tennis players has been shown to be significantly greater in the dominant arm compared to the nondominant arm (36). Montoye et al. (66) also demonstrated that in the dominant arm both bone width and mineral content of the dominant ulna, radius, and humerus were significantly greater than in the nondominant arm of 61 senior male tennis players. In young male baseball players, the bone mineral content of the humerus but not the ulna and radius was significantly greater on the dominant compared to the nondominant side (93). Due to the lack of control groups or very small control groups, no definitive conclusions can be drawn regarding bone density in the dominant compared to the nondominant arms in these studies.

The effect of muscle strengthening exercises on joint stability has been reported in the literature. As mentioned earlier, strength training has an effect on the ligament strength as well as the ligament-bone junction strength, that would likely have a positive effect on injury prevention. Also, increasing the strength of muscles as dynamic stabilizers would likely play an important role.

The concept of the prevention of overload injuries by preconditioning the musculotendinous unit to handle increasingly higher and sport-specific loads is referred to as prehabilitation (11). If muscle imbalances, strength deficits, and flexibility deficits are indeed created by excessive training, then performing exercises to prevent these imbalances before they occur may be a factor in their prevention. Strength and flexibility exercises for athletes may be prescribed based on the demands of the sport and the musculoskeletal base of the athlete to prevent these deficits from occurring. Using prehabilitation exercises, it may be possible to prevent certain musculoskeletal overload injuries associated with overtraining that are common to a sport or activity. These injuries comprise a significant part of injuries sustained in repetitive use sports, and play a large role in performance in those activities. A sample prehabilitation program for throwing athletes is presented in table 9.3.

Table 9.3 Sample Prehabilitation Exercises for Throwing Athletes

Flexibility exercises

- Wrist flexion, extension, pronation, supination

- Shoulder internal rotation

- Hip rotation stretch

- Calf stretch, hamstring stretch

Strength exercises

- Shoulder external rotation

- Wrist flexion, extension, pronation, supination

- Scapular strengthening

- Abdominal crunches

- Abdominal twists

- Cross step lunges

- Side step lunges

Prehabilitation exercises, to be effective in the modification of injury risk, must be sport specific and should be incorporated into the total conditioning program, beginning with the conditioning of the isolated links of the kinetic chain and progressing to sport-specific activities. Rehabilitation exercises consist of both strength and range of motion exercises for the links of the kinetic chain considered to be at risk in a particular sport and should be continued throughout the "in season" phase of conditioning. These exercises are an effective extension of rehabilitation exercises in cases where the athlete was previously injured, and may have subclinical adaptations present in the injured area.

A preventive conditioning program, in order to be effective, must be designed based on the metabolic and musculoskeletal demands of the sport, as well as the characteristics of the individual athlete's musculoskeletal base. A conditioning program for sports should progress to providing a variety of movement experiences, including lower body speed and movement training. By including movement training in the prehabilitation program, the neuromuscular system of the athlete is exposed to variety, which may be helpful in decreasing the risk of traumatic injuries as well.

Summary

Intense training places a demand on the musculoskeletal system that may lead to overt clinical damage as well as functional and biomechanical adaptations that may be detrimental to sport performance. These adaptations can be seen to some extent in all athletes, but the overtrained athlete is possibly at the highest level of risk. The types of injuries identified range from the overt, that are obvious injuries that will usually prevent athletic performance for some period of time, to the subclinical, that decrease performance but may be seldom recognized. These injuries apparently may be avoided or lessened in severity by a combination of several methods. A thorough preparticipation evaluation is important to detect subtle adaptations in strength and flexibility that can result from inappropriate overload and that may increase the athlete's chances of injury. A sport-specific conditioning program is necessary to give the athlete a strong musculoskeletal base on which to build athletic skills and decrease the risk of maladaptation. In many sports, prehabilitation exercises can be performed for those musculoskeletal areas that are under high stress in a particular sport. Also, a maintenance conditioning program that extends through the season may be important to maintain fitness throughout the season. Following proper principles of conditioning including specificity, recovery, and progression are important.

A complete and accurate diagnosis of the injuries that do occur is necessary so that proper treatment may follow. This can be facilitated by understanding the types of clinical presentations of injuries, and the different anatomical and functional alterations that may be acting to cause or continue the clinical presentation. By following these general guidelines, safe participation in sporting activities as well as performance will be enhanced.

The exact point where "training" becomes "overtraining" is difficult to define, especially prospectively. An exciting area of sports medicine research will be to define the anatomic parameters and exercise doses that may be a factor in the development of the overtraining syndrome, allowing maximal performance with minimal overload risk. At present, retrospective studies indicate that adaptations do occur in muscles, tendons, and bones in response to high training loads, and that some of these adaptations are not beneficial to performance and may be associated with increased injury risk. Musculoskeletal maladaptations and injuries can be a warning signal to the athlete and coach that the volume or intensity of training is too high, and overtraining is a possible causative factor.

References

1. Adams, A. 1966. Effect of exercise upon ligamentous strength. *Research Quarterly* 37: 163-167.
2. Alderink, G.J., D.J. Kuck. 1986. Isokinetic shoulder strength of high school and college-aged pitchers. *Journal of Orthopedic and Sports Physical Therapy* 14: 163-172.
3. Armstrong, R.B., R.W. Ogilvie, J.A. Schwane. 1983. Eccentric exercise induced injury to rat skeletal muscle. *Journal of Applied Physiology* 54: 8093.
4. Armstrong, R.B. 1990. Initial events in exercise-induced muscular injury. *Medicine and Science in Sports and Exercise* 22: 429-435.
5. Atha, J. 1981. Strengthening muscle. *Exercise and Sport Science Reviews* 9: 173.
6. Brown, L.P., S.L. Niehues, A. Harrah, P. Yavorsky, H.P. Highman. 1988. Upper extremity range of motion and isokinetic strength of internal and external rotators in major league baseball players. *American Journal of Sports Medicine* 16: 577-585.
7. Butler, D., A. Siegel. 1989. Alterations in tissue response: conditioning effects at different ages. In *Sports induced inflammation: workshop*, eds. W.B. Leadbetter, J.A. Buckwalter, S.L. Gordon, 713-730. Bethesda, MD: American Academy of Orthopedic Surgeons.
8. Chandler, T.J., W.B. Kibler, T.L. Uhl, B. Wooten, A. Kiser, E. Stone. 1990. Flexibility comparisons of junior elite tennis players to other athletes. *American Journal of Sports Medicine* 18: 134-136.
9. Chandler, T.J., W.B Kibler, A.M. Kiser, B.P. Wooten. 1991. Shoulder strength, power, and endurance in college tennis players. *American Journal of Sports Medicine* 20: 455-457.
10. Chandler T. J., W.B. Kibler. 1993. A biomechanical approach to the prevention, treatment, and rehabilitation of plantar fasciitis. *Sports Medicine* 15: 344-352.
11. Chandler, T.J. 1995. Exercise training for tennis. *Clinics in Sports Medicine*, ed. R. Lehman, 14: 33-46.
12. Cook, E.E., V.L. Gray, E. Savinar-Nogue, J. Medeiros. 1987. Shoulder antagonistic strength ratios: a comparison between college-level baseball pitchers and nonpitchers. *Journal of Orthopedic Sports and Physical Therapy* 8: 451-461.
13. Crenshaw, A.G., L.E. Thornell, J. Friden. 1994. Intramuscular pressure, torque, and swelling for the exercise-induced sore vastus lateralis muscle. *Acta Physiologica Scandinavica* 152: 265-277.
14. Dominquez, R.H. 1978. Shoulder pain in age group swimmers. In *Swimming medicine* IV, eds. B. Eriksson, B. Furlong, 105-109. Baltimore: University Park Press.

15. Ebbling, C.B., P.M. Clarkson. 1989. Exercise induced muscle damage and adaptation. *Sports Medicine* 7: 207-234.

16. Ekstrand, J., J. Gillquist. 1982. The frequency of muscle tightness and injuries in soccer players. *American Journal of Sports Medicine* 10: 75-78.

17. Ekstrand, J., J. Gillquist. 1983. The avoidability of soccer injuries. *International Journal of Sports Medicine* 4: 124-128.

18. Falch, J.A. 1982. The effect of physical activity on the skeleton. *Scandinavian Journal of Social Medicine* 29: S5558.

19. Friden, J., J. Seger, M. Sjostrom, B. Ekblom. 1983. Adaptive response in human skeletal muscle subjected to prolonged eccentric training. *International Journal of Sports Medicine* 4: 170-176.

20. Friden, J. 1984. Muscle soreness after exercise: implications of morphological changes. *International Journal of Sports Medicine* 5: 57-66.

21. Friden, J., R.L. Leiber, L.E. Thornell. 1991. Subtle indications of muscle damage following eccentric contractions. *Acta Physiologica Scandinavica* 142: 523-524.

22. Garrett, W.E, J.C. Califf, F.H. Bassett. 1984. Histochemical correlates of hamstring injuries. *American Journal of Sports Medicine* 12: 98-103.

23. Garrett, W.E., J. Tidball. 1987. Myotendinous junction: structure, on and failure. In *Injury and repair of musculoskeletal soft tissues*, eds. S.L. Woo, J.A. Buckwalter. American Academy of Orthopedic Surgeons and National Institute of Arthritis and Musculoskeletal and Skin Diseases. Chicago: AAOS.

24. Garrett, W.E., M.R. Safran, A.V. Seaber, R.R. Glisson, B.M. Ribbeck. 1987. Biomechanical comparison of stimulated and nonstimulated skeletal muscle pulled to failure. *American Journal of Sports Medicine* 15: 448-454.

25. Garrett, W.E., P.K. Nickolaou, B.M. Ribbeck, R.R. Glisson, A.V. Seaber. 1988. The effect of muscle architecture on the biomechanical failure properties of skeletal muscle under passive extension. *American Journal of Sports Medicine* 16: 712.

26. Garrett, W.E. 1990. Muscle strain injuries: clinical and basic aspects. *Medicine and Science in Sports and Exercise* 22: 436-443.

27. Gruchlow, W., D. Pelteiter. 1979. An epidemiologic study of tennis elbow. *American Journal of Sports Medicine* 7: 234-238.

28. Harryman, D.T., J.A. Sidles, J.M. Clark, K.J. McQuade, T.D. Gibb, F.A. Matsen. 1990. Translation of the humeral head on the glenoid with passive glenohumeral motion. *Journal of Bone and Joint Surgery* 72A: 1334-1343.

29. Hawkins, R., J. Kennedy. 1980. Impingement syndrome in athletes, *American Journal of Sports Medicine* 8: 151-158.

30. Hess, G.P., W.L. Cappiello, R.B. Poole, S.C. Hunter. 1989. Prevention and treatment of overuse tendon injuries. *Sports Medicine* 8: 371-384.

31. Herring, S. A. 1990. Rehabilitation from muscle injury. *Medicine and Science in Sports and Exercise* 22: 453-456.

32. Hinton, R.Y. 1988. Isokinetic evaluation of shoulder rotational strength in high school baseball pitchers. *American Journal of Sports Medicine* 16: 274-279.

33. Ivey, F.M., J.H. Calhoun, K. Rusche, J. Biershenk. 1985. Isokinetic testing of the shoulder strength: normal values. *Archives of Physical Medicine and Rehabilitation* 66: 384-386.

34. Jackson, D.L., J. Nyland. 1990. Club lacrosse: a physiological and injury profile. *Annals of Sports Medicine* 5: 114-117.

35. Jobe, W.F., J.P. Bradley. 1988. Rotator cuff injuries in baseball. *Sports Medicine* 6: 378-387.

36. Jones, H.H., J.D. Priest, W.C. Hayes, C.C. Tichenor, D.A. Nagel. 1977. Humeral hypertrophy in response to exercise. *Journal of Bone and Joint Surgery* 59A: 204-208.

37. Jones, D.A., D.J. Newham, J.M. Round, T.S.E. Jolfree. 1986. Experimental human muscle damage: morphological changes in relation to other indices of damage. *Journal of Physiology* 375: 435-448.

38. Jones, H.B., J.M. Harris, T.N. Vinh, C. Rubin. 1989. Exercise-induced stress fractures and stress reactions of bone: epidemiology, etiology, and classification. *Exercise and Sport Science Reviews* 17: 379-422.

41. Kibler, W.B., C. McQueen, T. Uhl. 1988. Fitness evaluations and fitness findings in competitive junior tennis players. *Clinics in Sports Medicine* 7: 403-416.

42. Kibler, W.B., T.J. Chandler, T. Uhl, R.E. Maddux. 1989. A musculoskeletal approach to the preparticipation physical examination: preventing injury and improving performance. *American Journal of Sports Medicine* 17: 525-531.

43. Kibler, W.B. 1990. Clinical aspects of muscle injury. *Medicine and Science in Sports and Exercise* 22: 450-452.

44. Kibler, W.B., C. Goldberg, T.J. Chandler. 1991. Functional biomechanical deficits in running athletes with plantar fasciitis. *American Journal of Sports Medicine* 19: 66-71.

45. Kibler, W. B. 1994. Clinical biomechanics of the elbow in tennis: implications for evaluation and diagnosis. *Medicine and Science in Sports and Exercise* 26: 1203-1206.

46. Kibler, W.B., T.J. Chandler, B.P. Livingston, E.P. Roetert. 1996. Shoulder range of motion in elite tennis players: the effect of age and years of tournament play. *American Journal of Sports Medicine* 24: 1-7.

47. Kibler, W.B. 1997. Diagnosis, treatment, and rehabilitation principles in complete tendon ruptures in sport. *Scandinavian Journal of Medicine in Sports* 7:119-129.

48. Knapik, J.J., C.L. Bauman, B.H. Jones, J.M. Harris, L. Vaughn. 1991. Preseason strength and flexibility imbalances associated with athletic injuries in female collegiate athletes. *American Journal of Sports Medicine* 19: 76-81.

49. Kovanen, V., H. Suominen, E. Heikkinen. 1980. Connective tissue of fast and slow skeletal muscle in rats: effects of endurance training. *Acta Physiologica Scandinavica* 108: 173-180.

50. Kovanen, V., H. Suominen, E. Heikkinen. 1984. Collagen of slow and fast twitch muscle fibers in different types of rat skeletal muscle. *European Journal of Applied Physiology* 52: 235-243.

51. Krolner, B., E. Tondevold, B. Toft, B. Berthelsen, S. Pors Nielsen. 1982. Bone mass of the axial and appendicular skeleton in women with Colles' fracture: its relation to physical activity. *Clinical Physiology* 2: 147-157.

52. Kuland, D.N., F.C. McCue, D.A. Rockwell, J A. Gieck. 1979. Tennis injuries: prevention and treatment. *American Journal of Sports Medicine* 7: 249-253.

53. Kuipers, H., H.A. Keizer. 1988. Overtraining in elite athletes. *Sports Medicine* 6: 79-92.

54. Kuipers, H. 1994. Exercise-induced muscle damage. *International Journal of Sports Medicine* 15: 132-135.

55. Kujala, U.M., M. Kvist, K. Osterman, O. Friberg, T. Aalto. 1986. Factors predisposing army conscripts to knee exertion injuries incurred in a physical

training program. *Clinical Orthopaedics and Related Research* 210: 203-212.

56. Laurent, G.J., M.P. Sparrow, P.C. Bates, D.J. Milward. 1978. Collagen content and turnover in cardiac and skeletal muscles of the adult fowl and the changes during stretch-induced growth. *Biochemistry Journal* 176: 419-427.

57. Leiber, R.L., J. Friden. 1988. Selective damage of fast glycolytic muscle fibers with eccentric contraction of the rabbit tibialis anterior. *Acta Physiologica Scandinavica* 133: 587-588.

58. Leiber, R.L., J. Friden. 1993. Muscle damage is not a function of muscle force but active muscle strain. *Journal of Applied Physiology* 74: 520-526.

59. McBride, A.M. 1985. Stress fractures in runners. *Clinics in Sports Medicine* 4: 737-752.

60. McMaster, W.C., S.C. Long, V.J. Caiozzo. 1991. Isokinetic torque imbalances in the rotator cuff of the elite water polo player. *American Journal of Sports Medicine* 19: 72-75.

61. Maffulli, N. 1990. Intensive training in young athletes. *Sports Medicine* 9: 229-243.

62. Meuwisse, W.H. 1994. Assessing causation in sport injury: a multifactoral model. *Clinical Journal of Sports Medicine* 4: 166-170.

63. Mellerowicz, H., D.K. Barron. 1971. Overtraining. In *Encyclopedia of sport sciences and medicine*, eds. Larson, Leonard, 1310-1312. New York: Macmillan.

64. Micheli, L.J. 1982. Upper extremity injuries: overuse injuries in the recreational adult. In *The exercising adult*, ed. R.C. Cantu, 121-128. Lexington, MA: Collamore.

65. Montgomery, L.C., F.R.T. Nelson, J.P. Norton, P.A. Deuster. 1989. Orthopedic history and examination in the etiology of overuse injuries. *Medicine and Science in Sport and Exercise* 21: 237-243.

66. Montoye, H.J., E.L. Smith, F.D. Fardon, E.T. Howley. 1980. Bone mineral in senior tennis players. *Scandinavian Journal of Sports Science* 2: 26-32.

67. Nilsson, B.E., N.E Westlin. 1971. Bone density in athletes. *Clinical Orthopaedics and Related Research* 77: 179-182.

68. Nikolaou, P.K., B.L. Macdonald, R.R. Glisson, A.V. Seaber, W.E. Garrett. 1987. Biomechanical and histological evaluation of muscle after controlled strain injury. *American Journal of Sports Medicine* 5: 914.

69. Nirschl, R.P. 1979. Tennis elbow. *Journal of Bone and Joint Surgery* 61A: 832-839.

70. Ogilvie, R.W., H. Hoppeler, R.B. Armstrong. 1985. Decreased muscle function following eccentric exercise in the rat. *Medicine and Science in Sports Exercise* 17: 195.

71. Ogilvie, R.W., R.B. Armstrong, K.E. Baird, C.L Bottoms. 1988. Lesions in the rat soleus muscle following eccentrically-biased exercise. *American Journal of Anatomy* 182: 335-346.

72. Pappas, A.M., R.M. Zawacki, C.F. McCarthy. 1985. Rehabilitation of the pitching shoulder. *American Journal of Sports Medicine* 13: 223-235.

73. Parker, M.G., R.O. Ruhling, D. Holt, E. Bauman, M. Drayna. 1983. Descriptive analysis of quadriceps and hamstrings muscle torque in high school football players. *Journal of Orthopedic and Sports Physical Therapy* 5: 26.

74. Priest, J.D., D.A. Nagle. 1976. Tennis shoulder. *American Journal of Sports Medicine* 4: 28-42.

75. Renstrom, P., R.J. Johnson. 1985. Overuse injuries in sports. *Sports Medicine* 2: 316-333.

76. Rettig, A.C., D. Shelbourne, K. McCarroll, Bisesi, J. Watts. 1988. The natural history and treatment of delayed union stress fractures of the anterior cortex of the tibia. *American Journal of Sports Medicine* 16: 250-255.

77. Ryan, A. J., R.L. Brown, E.C. Fredrick, H.L. Falsetti, R.E. Burke. 1983. Overtraining of athletes: a round table. *Physician and Sportsmedicine* 11: 93-110.

78. Southmayd, M., M. Hoffman, 1981. *Sports health: the complete book of athletic injuries.* New York: Quick Fox.

79. Stauber, W.T. 1989. Eccentric action of muscles: physiology, injury, and adaptation. In *Exercise and sport science review*, ed. K.B. Pandolf. Baltimore: Williams and Wilkins.

80. Stone, M.H. 1990. Muscle conditioning and muscle injuries. *Medicine and Science in Sports and Exercise* 22: 457-462.

81. Taimela, S., U.M. Kujala, K. Osterman. 1989. Individual characteristics are related to musculoskeletal injuries. *Paavo Nurmi Congress*, Turku, Finland, August 28 - September 1.

82. Taimela, S., U.M. Kujala, K. Osterman. 1990. Intrinsic risk factors and athletic injuries. *Sports Medicine* 9: 205-215.

83. Talag, T.S. 1973. Residual muscular soreness as influenced by concentric, eccentric, and static contractions. *Research Quarterly* 44: 458-469.

84. Taunton, J.E., D.C. McKenzie, D.B. Clement. 1988. The role of biomechanics in the epidemiology of injuries. *Sports Medicine* 6: 107-120.

85. Teague, B.N., J.A. Schwane. 1995. Effect of intermittent eccentric contractions on symptoms of muscle microinjury. *Medicine and Science in Sport and Exercise* 27: 1378-1384.

86. Tipton, C.M., S.L. James, W. Mergner, T. Tcheng. 1970. Influence of exercise on strength of medial collateral knee ligaments of dogs. *American Journal of Physiology* 218: 894-902.

87. Tipton, C.M., R.D. Matthes, J.A. Maynard, R.A. Carey. 1975. The influence of physical activity on ligaments and tendons. *Medicine and Science in Sports* 7: 165-175.

88. Torg, J.S., H. Pavlof, L.H. Cooley, M.H. Bryant, S.P. Arnoczky, J. Bergfeld, L.Y. Hunter. 1982. Stress fractures of the tarsal navicular. *Journal of Bone and Joint Surgery* 64A: 700-712.

89. Turto, H., S. Lindy, J. Halme. 1974. Protocollagen proline hydroxylase activity in work-induced hypertrophy of rat muscle. *American Journal of Physiology* 226: 63-65.

90. Valliant, P.M. 1981. Personality and injury in competitive runners. *Perceptual and Motor Skills* 53: 251-253.

91. Vailas, A.C., W.P. Morgan, J.C. Vailas. 1989. Physiologic and cellular basis for overtraining. In *Sports induced inflammation: workshop*, eds. W.B. Leadbetter, J.A. Buckwalter, S.L. Gordon, 677-686. Bethesda, MD: American Academy of Orthopedic Surgeons.

92. Vogel, J.M., M.W. Whittel. 1976. Bone mineral content changes in the Skylab astronauts. *American Journal of Roentgenology* 126: 1296.

93. Watson, R.C. 1974. Bone growth and physical activity in young males. *International Conference on Bone Mineral Measurements*, U.S. Department of Health, Education, and Welfare, publication number NIH 75683: 380-385.

94. Watson, A.W.S. 1981. Factors predisposing to sports injury in school boy rugby players. *Journal of Sports Medicine and Physical Fitness* 21: 417-422.

Immunological Considerations of Overreaching and Overtraining

CHAPTER 10

Effects of Athletic Endurance Training on Infection Rates and Immunity

David C. Nieman, DrPH

Introduction

As will be reviewed later in this chapter, epidemiological data suggest that endurance athletes are at increased risk for upper respiratory tract infection (URTI) during periods of heavy training and the 1-2 week period following prolonged and intensive aerobic exercise. Despite expanding and fervid investigation to discover an immune link, there still is no clear indication that either acute or chronic alterations in immune function explain the increased risk. Very little data exist on the effects of resistance training on infection rates or immunity in power and strength athletes, so the emphasis in this chapter will be on endurance athletes.

Data reviewed in this chapter will show that following acute bouts of prolonged heavy endurance exercise, several components of the immune system demonstrate suppressed function for several hours. This has led to the concept of the "open window," described as the three- to 24-hour time period following prolonged endurance exercise when host defense is decreased and risk of URTI is elevated.

As will be explained in the next section, chronic exercise training has mixed effects on the resting immune system. For example, while several researchers have reported a diminished neutrophil function in athletes during periods of intense and heavy training, others have shown an enhanced function of the natural killer cells.

It will be emphasized in this chapter that most exercise immunologists feel that further research is needed to provide a better understanding of exercise-induced changes in immunity before definitive clinical applications can be formulated. At present, researchers disagree on the mechanisms underlying the suppression of immune function, and their importance to host protection. Nonetheless, there is sufficient evidence at this time to caution athletes to practice various hygienic measures to lower their risk of URTI, and to avoid heavy exertion during systemic illness to circumvent potential medical complications. These recommendations will be reviewed in the last section of this chapter.

In this chapter, terms associated with the immune system will be defined so that the reader who has only a limited understanding will be able to grasp basic concepts. The immune system is comprised of two functional divisions: the innate, that acts as a first line of defense against infectious agents, and the adaptive, that when activated, produces a specific reaction and immunologic memory to each infectious agent. The innate immune system is comprised of cells (natural killer cells and phagocytes including neutrophils, eosinophils, basophils, monocytes, and macrophages) and soluble factors (acute phase proteins, complement, lysozyme, and interferons). The adaptive immune system is also comprised of cells (B and T lymphocytes) and soluble factors (immunoglobulins). Significant populations of cells found in these two systems and of interest in exercise immunology are outlined in figure 10.1.

Resting Immunity in Endurance Athletes Versus Nonathletes

Relatively few researchers have published cross-sectional comparisons of immune function in aerobic endurance athletes and nonathletes. This type of study design is a logical starting point when considering the question of prolonged aerobic exercise and immunity. Table 10.1 summarizes the data currently available on cross-sectional comparisons of human endurance athletes and nonathletes for natural killer cell activity (NKCA) (4, 45, 46, 50, 60, 78), neutrophil function (2, 18-20, 32, 66, 73), and lymphocyte proliferative response (2, 45, 46, 50, 58, 59).

Natural Killer Cell Activity

Natural killer (NK) cells are large granular lymphocytes that can mediate cytolytic reactions against a variety of neoplastic and virally infected cells (33). NK cells also exhibit key noncytolytic functions, and can inhibit microbial colonization and growth of certain viruses, bacteria, fungi, and parasites. Natural killer cell activity (NKCA) is measured with a four-hour ^{51}Cr-release assay where certain types of ^{51}Cr-labeled cancer or virally infected cells are mixed with blood lymphocytes and monocytes. NK cells, which represent about 10-15% of blood lymphocytes, respond quickly and within four hours can lyse a significant proportion of the ^{51}Cr-labeled target cells. The released ^{51}Cr is collected into filters, and then measured with a gamma counter.

Of the six studies listed in table 10.1 for NKCA, four support the finding of enhanced NKCA in athletes when compared to nonathletes, in both younger and older groups (46, 50, 60, 78). It is of interest that the two studies not supporting this relationship used college students (4, 45). Perhaps the difference in aerobic fitness and physical activity levels between college athletes and nonathletes is too small for a significant difference in NKCA to emerge.

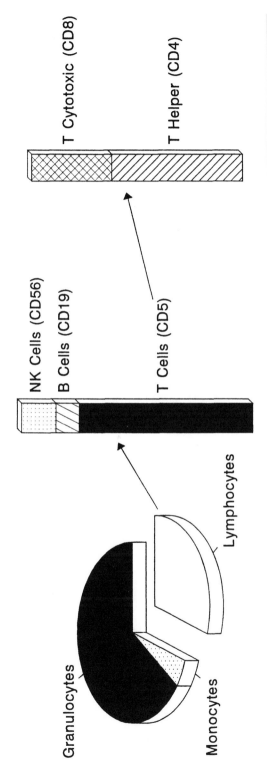

Figure 10.1   Leukocyte and lymphocyte subsets and normal proportions within the blood compartment.

Table 10.1 Resting Immune Function in Human Endurance Athletes and Nonathletes

Author/year	Subjects	Major results
Natural killer cell activity		
Brahmi et al. (4) (1985)	5 trained male athletes 10 untrained males and females	↔ No significant difference
Pedersen et al. (60) (1989)	27 elite male cyclists 15 untrained males	↑ 26% higher in athletes
Tvede et al. (78) (1991)	29 elite male cyclists (low training) 15 untrained males	↑ 27% higher in athletes
	14 elite male cyclists (high training) 10 untrained males	↑ 64% higher in athletes
Nieman et al. (50) (1993)	12 elderly female athletes 32 sedentary elderly females	↑ 55% higher in athletes
Nieman et al. (45) (1995)	18 male endurance athletes 11 untrained males	↔ No significant difference
Nieman et al. (46) (1995)	22 male marathon runners 18 untrained males	↑ 57% higher in athletes
Neutrophil function		
Green et al. (18) (1981)	20 male marathon runners Pool of untrained males	↔ No significant difference
Lewicki et al. (32) (1987)	20 elite male cyclists 19 untrained males	↓ 90% lower in athletes
Smith et al. (73) (1990)	11 elite male cyclists 9 untrained males	↓ 50% lower in athletes
Hack et al. (19) (1992)	20 elite male runners/ triathletes 10 untrained males	↔ No significant difference during low training period
Pyne (66) (1994)	12 elite male and female swimmers 11 untrained females and males	↓ 50% lower in athletes before and after 16 weeks of intensive training
Hack et al. (20) (1994)	7 elite male runners 10 untrained males	↔ No significant difference during low training period; ↓ lower in athletes during high training period

Table 10.1 *continued*

Author/year	Subjects	Major results
Baj et al. (2) (1994)	15 elite male cyclists 16 untrained males	↔ No significant difference during low training period; ↓ 33% lower in athletes during high training period (PMA stimulated)

Lymphocyte proliferative response

Oshida et al. (58) (1988)	6 elite male runners 5 untrained males	↔ No significant difference
Papa et al. (59) (1989)	7 elite water polo players 7 untrained males	↓ PHA 28% lower in athletes; PWM 60% lower in athletes
Tvede et al. (78) (1991)	29 elite male cyclists (low training) 15 untrained males	↔ No significant difference
	14 elite male cyclists (high training) 10 untrained males	↔ No significant difference
Nieman et al. (50) (1993)	12 elderly female athletes 32 sedentary elderly females	↑ PHA 56% higher in athletes
Baj et al. (2) (1994)	15 elite male cyclists 16 untrained males	↔ No significant difference during low training period ↑ PHA 35%, anti-CD3 mAb 50% higher in athletes during high training period (no effect on Con A, PWM)
Nieman et al. (45) (1995)	18 male endurance athletes 11 untrained males	↔ No significant difference
Nieman et al. (46) (1995)	22 male marathon runners 18 untrained males	↔ No significant difference

Figure 10.2 summarizes the data from a study comparing 22 marathoners and 18 untrained males (46). The average marathoner in the study was very experienced, having run a mean of 24 marathons over a 12-year period. Although the two groups were of similar age (38.7 ± 1.5 and 43.9 ± 2.2 years, respectively), the marathon runners were significantly leaner and possessed a VO_2max 60% higher

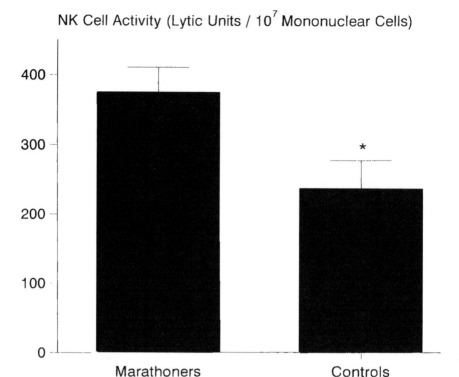

NK Cell Activity (Lytic Units / 10^7 Mononuclear Cells)

Figure 10.2   Natural killer cell activity (total lytic units) in 22 experienced marathon runners versus 18 sedentary controls. *p < .05.
Based on data from Nieman et al. 1995.

than that of the controls. The NKCA was 57% higher in the marathoners, and correlated negatively with percent body fat (r = -.48, p = .002).

The data of Tvede et al. (78) support a higher NKCA in elite cyclists during the summer (intensive training period) compared to the winter (low training period). Several prospective studies utilizing moderate endurance training regimens for eight to 15 weeks have reported no significant elevation in NKCA relative to sedentary controls (47, 50). Together, these data imply that endurance exercise may have to be intensive and prolonged (i.e., at athletic levels) before NKCA is chronically elevated.

Researchers disagree on whether the higher NKCA is due to a greater concentration of circulating NK cells or an enhanced activity of each NK cell (46, 50, 60, 62, 78). Data by Tvede et al. (78) and Nieman et al. (46, 50) suggest that the answer to this question may vary according to the time of the year, training intensity, age, and degree of body fatness.

Neutrophil Function

Neutrophils are an important component of the innate immune system, aiding in the phagocytosis of many bacterial and viral pathogens, and in the release of immunomodulatory cytokines (66). Neutrophils are considered to be the body's best phagocytes, and are critical in the early control of invading infectious agents. Neutrophil function can be expressed as a measure of an ability to engulf pathogens (phagocytic ability), and a facility to kill the pathogens once engulfed (the oxidative burst).

The cross-sectional NKCA data are in contrast to those for neutrophil function. As summarized in table 10.1, no researcher has reported an elevation in neutrophil function (either phagocytic ability or oxidative burst) among athletes when compared to nonathletes. Instead, during periods of high intensity training, neutrophil function has been reported to be suppressed in athletes. This is especially apparent in the studies by Hack et al. (20) and Baj et al. (2) where neutrophil function in athletes was similar to controls during periods of low training workloads, but significantly suppressed during the summer months of intensive training. Pyne (66) reported that elite swimmers undertaking intensive training have a significantly lower neutrophil oxidative activity at rest than do age- and sex-matched sedentary individuals, and that function is further suppressed during periods of strenuous training prior to national-level competition.

The decrease in neutrophil function with intensive athletic training is an important finding and warrants further research. The increased URTI risk that athletes experience following heavy exertion may well be related to the chronic suppression of these important phagocytes, even despite the fact that NK cells are functioning at an increased level.

Lymphocyte Proliferative Response

Determination of the proliferative response of human lymphocytes upon stimulation with various mitogens in vitro is a well-established test to evaluate the functional capacity of T and B lymphocytes. Mitogen stimulation of lymphocytes in vitro using optimal and suboptimal doses is believed to mimic events that occur after antigen stimulation of lymphocytes in vivo. When lymphocytes are exposed to a foreign pathogen, their ability to "divide and conquer" is an important component of the adaptive immune system. In the laboratory, researchers expose lymphocytes to various types of mitogens for three days, and then add a radioactively labeled compound, thymidine (methyl)-^{3}H, to the lymphocytes during the last four hours before harvesting. The thymidine (methyl)-^{3}H is incorporated into the DNA of the dividing lymphocytes, and the incorporated amount is then measured using a liquid scintillation beta counter.

Data on the lymphocyte proliferative response to athletic endeavor are less clear than for NK cells and neutrophils, but generally support no significant difference between athletes and nonathletes (see table 10.1). Baj et al. (2) reported no difference between elite cyclists and nonathletes during low training periods (March), but increased levels in the athletes for the mitogens PHA and anti-CD3 mAb (but not the mitogens Con A or PWM) during intensive training. Interleukin-2 generation (a cytokine released by activated lymphocytes), however, was suppressed in the athletes versus controls during intensive training. These data contrast with that of Tvede et al. (78) who found no difference between athletes and nonathletes during either low- or high-training periods.

Among highly conditioned elderly women, PHA-induced lymphocyte proliferative response was reported to be 56% higher than among sedentary controls (50) (see figure 10.3). These data are interesting because T cell function tends to diminish with age. However, moderate exercise training for 12 weeks failed to alter T cell function in elderly women, indicating that an unusual commitment to vigorous exercise may be necessary before an effect on T cell function can be measured in the elderly population. Nonetheless, large volumes of exercise training may have a greater effect on T cell function among older compared to younger individuals.

Other Measures of Immunity

Other components of immunity have been less well studied among human athletes and nonathletes. Tomasi et al. (75) reported that resting salivary IgA levels were lower in elite cross-country skiers than in age-matched controls, but this was not confirmed in a follow-up study of elite cyclists (36). As reviewed by Mackinnon and Hooper (37), the secretory immune system of the mucosal tissues of the upper respiratory tract is considered the first barrier to colonization by pathogens, with IgA the major effector of host defense. Secretory IgA inhibits attachment and replication of pathogens, preventing their entry into the body. While several studies have shown that salivary IgA concentration decreases after a single bout of intense

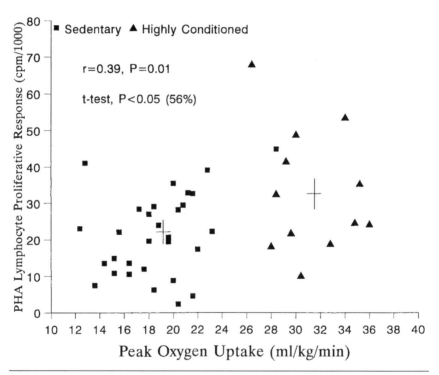

Figure 10.3 Correlation between PHA-induced lymphocyte proliferation and VO$_2$max in 12 highly conditioned and 30 sedentary elderly women. Crosses represent mean ±SE for each group, with the proliferative response 56% higher in highly conditioned versus sedentary subjects, and VO$_2$max 67% higher.
Based on data from Nieman et al. 1993.

endurance exercise, further research is needed to determine the overall chronic effect (36-38).

Complement but not serum immunoglobulins (especially when adjusted for the higher plasma volumes of athletes) has been reported to be lower in marathon runners versus sedentary controls (55, 57). Production of tumor necrosis factor in lipopolysaccharide-stimulated blood has been described as significantly depressed in elite cross-country skiers relative to controls (31). Most studies have failed to demonstrate any important effects of regular exercise training on concentrations of circulating total leukocytes or lymphocytes or their various subpopulations (18, 45, 46, 50).

Together these data support the concept that the innate immune system responds differentially to the chronic stress of intensive exercise, with NKCA tending to be enhanced while neutrophil function is suppressed. The adaptive immune system in general seems to be largely unaffected, although the research data at present are mixed. Further research is needed with larger groups of athletes and nonathletes to allow a more definitive comparison.

The Acute Immuno Response to Prolonged Aerobic Exercise

In the next section, epidemiological studies will be reviewed showing that marathon and ultramarathon race events are associated with a significant increase in risk of URTI during the one- or two-week recovery period (52, 63-65). In light of the mixed results regarding the effect of chronic, intensive training on resting immune function, several authors have posited that prolonged cardiorespiratory endurance exercise (defined in this chapter as >90 min) leads to transient but significant perturbations in immunity and host defense, providing a physiological rationale for the epidemiological data (37, 62).

For example, NKCA (3, 35, 44, 71), mitogen-induced lymphocyte proliferation (12, 17, 56, 58, 71), in vivo cell-mediated immunity (5), upper airway neutrophil function (41), and salivary IgA concentration (36-38, 75), have all been reported to be suppressed for at least several hours during recovery from prolonged, intense endurance exercise. During this "open window" of decreased host protection, the theory goes, viruses and bacteria may gain a foothold, increasing the risk of subclinical and clinical infection. This may be especially apparent when the athlete goes through repeated cycles of heavy exertion (66). Although this is an attractive hypothesis, no one has yet demonstrated conclusively that athletes showing the most extreme immunosuppression are those that contract an infection.

Natural Killer Cell Activity

The acute response of NKCA to prolonged endurance exercise by 50 marathon runners is depicted in table 10.2. These data, based on several studies conducted in the author's laboratory, show that NKCA is decreased 45-62% for at least six hours following 2.5-3 h of high intensity running (3, 44). Although some reason that the drop in NKCA can be ascribed to numerical shifts in NK cells (13, 44, 54, 72), others report that prostaglandins from activated monocytes and neutrophils (61,

62, 77) or elevated stress hormone levels (3) suppress the ability of NK cells to function appropriately. In the next section of this chapter, data will be presented demonstrating that indomethacin (which inhibits prostaglandin production) had no effect on the decrease in NKCA following 2.5 h of running (44). Notice from table 10.2 that the decrease in NKCA paralleled the drop in NK cells.

Influence of Cortisol

Serum cortisol concentration was significantly elevated above the preexercise and control levels for several hours following the 2.5-3 h run in this group of 50 marathon runners (3, 44) (see figure 10.4). Notice from this figure that following 2.5-3 h of running, serum cortisol is elevated for a prolonged period of time relative to sitting controls or other forms of exercise conducted for shorter periods of time. (These data are based on studies conducted in the author's laboratory.) (3, 44, 53).

Cortisol has been related to many of the immunosuppressive changes experienced during recovery (10, 16, 22, 24, 76). Glucocorticoids administered in vivo have been reported to cause neutrophilia, eosinopenia, lymphocytopenia, and a suppression of both NK and T cell function, all of which occur during recovery from prolonged, high-intensity cardiorespiratory exercise (see table 10.2). Notice

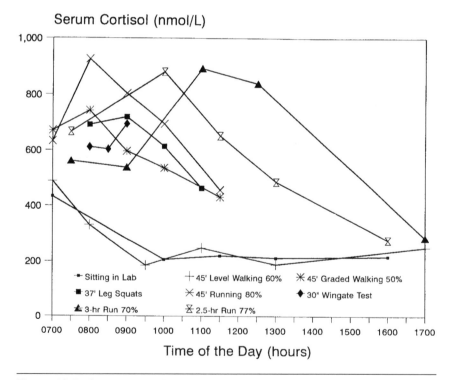

Figure 10.4 Serum cortisol response to different forms of exercise. Notice the large, sustained elevation in serum cortisol relative to sitting in the lab or 45 minutes of level walking at 60% VO₂max.

Based on data from references 3, 44, and unpublished data from author's laboratory.

Table 10.2 Acute Response of Immune System in Marathon Runners to 2.5-3 h Treadmill Running at 75.9 ± 0.9% VO_2max

Immune variable	Time	Preexercise 0715	Postexercise 1000	1.5-h Post-ex 1130	3-h Post-ex 1300	6-h Post-ex 1600
Neutrophils (10^9/L)		2.83 ± 0.14	7.86 ± 0.48***	9.49 ± 0.49***	10.1 ± 0.5***	7.43 ± 0.36***
Lymphocytes (10^9/L)		2.05 ± 0.14	1.93 ± 0.08	1.17 ± 0.06***	1.35 ± 0.07***	1.63 ± 0.07**
Neutrophil/lymphocyte ratio		1.60 ± 0.12	4.35 ± 0.30***	9.44 ± 0.77***	8.97 ± 0.80***	5.26 ± 0.48***
Monocytes (10^9/L)		0.46 ± 0.03	0.66 ± 0.05***	0.72 ± 0.06***	0.72 ± 0.05***	0.62 ± 0.03***
Eosinophils (10^9/L)		0.19 ± 0.02	0.10 ± 0.01***	0.06 ± 0.01***	0.06 ± 0.01***	0.11 ± 0.01***
NK lymphocytes (10^9/L); N = 37		0.28 ± 0.03	0.46 ± 0.05***	0.11 ± 0.01***	0.16 ± 0.02***	0.18 ± 0.02***
NK cell activity (lytic units)		309 ± 33	320 ± 35	116 ± 13***	133 ± 13***	169 ± 21***
T lymphocytes (10^9/L); N = 37		1.40 ± 0.10	1.43 ± 0.09	0.94 ± 0.06***	1.18 ± 0.08**	1.37 ± 0.07
Con A-induced lymphocyte proliferation (cpm $\times$ 10^{-3}); N = 40		13.9 ± 1.7	10.9 ± 0.9*	10.5 ± 1.1**	10.8 ± 1.1*	14.7 ± 1.3
Granulocyte phagocytosis mean florescence channel phagocytosis, FITC; N = 30		1276 ± 68	1727 ± 67***	1761 ± 67***	1844 ± 65***	1774 ± 44***
Granulocyte oxidative burst mean florescence channel oxidative burst, DCF; N = 30		435 ± 18	461 ± 15	415 ± 19	424 ± 22	376 ± 18***
Monocyte phagocytosis mean florescence channel phagocytosis, FITC; N = 30		887 ± 37	1333 ± 65***	1510 ± 65***	1544 ± 75***	1386 ± 61***
Monocyte oxidative burst mean florescence channel oxidative burst, DCF; N = 30		32.2 ± 1.9	28.9 ± 1.7	38.3 ± 2.5	35.4 ± 1.7	29.2 ± 1.3

*p < .05, ** p < .01, *** p < .001, comparison with preexercise.
N = 50 except where noted; NK = natural killer; FITC = fluoresceinisothiocyanate; DCF = dichlorofluorescein.
Based on data from 44, 56, and unpublished data from author's laboratory.

from table 10.2 that the Con A-induced lymphocyte proliferative response was significantly decreased for more than three hours following 2.5 h of intensive running, an effect that has been related to high cortisol concentrations (12, 56).

Figure 10.5 demonstrates that a significant correlation exists between the change in serum cortisol and the change in the neutrophil/lymphocyte ratio following 2.5-3 h of running (56). The neutrophil/lymphocyte ratio has been proposed as an excellent index of the physiologic stress on the immune system following intensive exercise (56).

Neutrophil and Monocyte Function

Notice from table 10.2 that while some aspects of immunity are suppressed during recovery from intensive exercise, granulocyte and monocyte phagocytosis are increased. Granulocyte oxidative burst activity, however, shows a significant decrease by 6 h postexercise.

Following prolonged, high-intensity running, substances released from injured muscle cells initiate an inflammatory response (74). Monocytes and neutrophils invade the area, and phagocytose debris. The increase in granulocyte and monocyte phagocytosis may therefore represent a part of the inflammatory response to

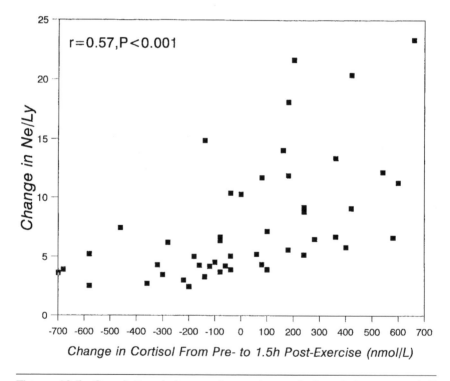

Figure 10.5 Correlation between change in cortisol and the neutrophil/ lymphocyte ratio from pre- to 1.5 h postexercise in 50 marathon runners who ran 2.5-3 h at $75.9 \pm 0.9\%$ VO_2max.

Based on data from references 3, 56, and unpublished data from author's laboratory.

acute muscle injury (79). The slight decrease in granulocyte oxidative burst may represent a tendency for reduced killing capacity by blood neutrophils (on a per cell basis) due to stress and overloading (15, 66).

Of interest are the data of Müns (41) depicted in figure 10.6. Using nasal lavage samples, the capacity of phagocytes to ingest *E. coli* was significantly suppressed for more than three days after a 20 km running race. Other data from Müns et al. (42) have also shown that IgA concentration in nasal secretions is decreased by nearly 70% for at least 18 h after racing 31 kilometers. Following a marathon race, nasal mucociliary clearance is significantly slower for nearly a week compared to control subjects (43). These data suggest that host protection in the upper airway passages is significantly suppressed for a prolonged time after endurance running races. These data may be the most important evidence to date linking risk of respiratory infection with athletic endeavor.

Taken together, these data suggest that the immune system is suppressed and stressed following prolonged endurance exercise, decreasing host protection against viruses and bacteria. There are no convincing data at this time, however, that exercise-induced changes in immune function explain the increased risk of URTI suggested by epidemiological data. Researchers disagree on the mechanistic interpretation of their findings, and none has provided follow-up data of large numbers of subjects to determine if various changes in immunity translate to altered host protection. Further research is needed to settle these issues, and to determine if the

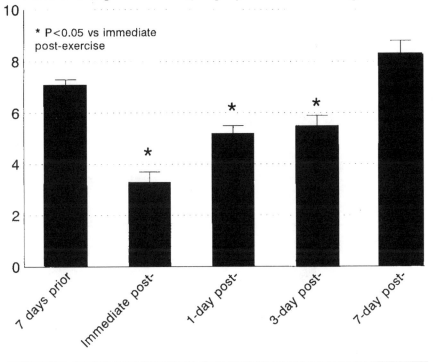

Figure 10.6 Number of ingested *E. coli* per phagocyte taken from nasal lavage samples from runners before and after a 20 km race.
Based on data from Müns 1993.

large but transient perturbations in leucocyte cell concentrations in both blood and peripheral lymphoid tissue (that often underlie reported in vitro functional alterations) are important from a clinical viewpoint.

Risk of URTI in Endurance Athletes

Among elite athletes and their coaches, a common perception is that heavy exertion lowers resistance and is a predisposing factor to upper respiratory tract infections (21, 28). During the Winter and Summer Olympic Games, it has been regularly reported by clinicians that "upper respiratory infections abound" (21) and that "the most irksome troubles with the athletes were infections" (28).

Several epidemiological reports suggest that athletes engaging in marathon-type events or very heavy training are at increased risk of URTI (23, 34, 52, 63-65). Reviews of this information are available (70) and in this chapter emphasis will be placed on studies of endurance athletes (see table 10.3).

Peters and Bateman (64) were the first to report an increased risk of URTI among athletes following heavy exertion (see figure 10.7). They studied the

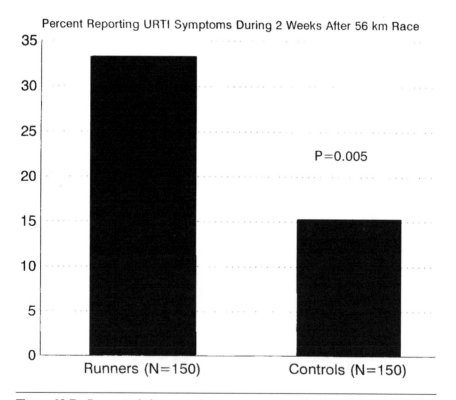

Figure 10.7   Percent of ultramarathon runners versus live-in controls reporting upper respiratory tract infection symptoms (URTI) during the two-week period following a 56 km race.
Based on data from Peters and Bateman 1983.

Table 10.3 Epidemiological Research on the Relationship Between Prolonged Endurance Exercise and Upper Respiratory Tract Infection (URTI)

Investigators	Subjects	Method of determining URTI	Major finding
Peters & Bateman (64) (1983)	141 South African marathon runners vs. 124 live-in controls	2 week recall of URTI incidence and duration after 56 km race	URTI incidence twice as high in runners after 56 km race vs. controls (33.3% vs. 15.3%)
Linde (34) (1987)	44 Danish elite orienteers vs. 44 matched nonathletes	URTI symptoms self-recorded in daily log for 1 year	Orienteers vs. controls had 2.5 vs. 1.7 URTIs during year
Nieman, Johanssen, & Lee (51) (1989)	294 California runners training for race	2 month recall of URTI incidence 1 week recall after March 5, 10 km, 21 km races	Training 42 vs. 12 km/wk associated with lower URTI; no effect of race participation on URTI
Peters (63) (1990)	108 South African marathon runners vs. 108 live-in controls	2 week recall of URTI incidence and duration after 56 km race	URTI incidence 28.7% in runners vs. 12.9% in controls after 56 km race
Nieman et al. (52) (1990)	2,311 Los Angeles marathon runners	2 month recall of URTI incidence during training for marathon; 1 week recall after March race	Runners training $\geq$ 97 vs. < 32 km/wk at higher URTI risk; odds ratio 5.9 for participants vs. nonparticipants 1 week after 42.2 km race
Heath et al. (23) (1991)	530 runners, South Carolina	1 year daily log using self-reported, precoded symptoms	Increase in running distance positively related to increased URTI risk
Peters et al. (65) (1993)	84 South African marathon runners vs. 73 nonrunner controls; runners using vitamin C vs. 53% of controls	2 week recall of URTI incidence and duration after 90 km race	URTI incidence 68% in runners vs. 45% in controls after 56 km race; 33% in runners using vitamin C vs 54% of controls
Nieman (1993) (unpublished data)	170 North Carolina marathon runners during week after summer race	1 week recall of URTI incidence after July marathon race	URTI reported by only 3% of marathoners

incidence of URTI in 150 randomly selected runners who took part in a 56 km Cape Town race in comparison to matched controls who did not run. Symptoms of URTI occurred in 33.3% of runners compared with 15.3% of controls during the two-week period following the race, and were most common in those who achieved the faster race times. Two subsequent studies from this group of researchers have confirmed this finding (63, 65).

Nieman et al. (52) researched the incidence of URTI in a group of 2,311 marathon runners who varied widely in running ability and training habits. Runners retrospectively self-reported demographic, training, and URTI episode and symptom data for the two-month period (January, February) prior to and the one-week period immediately following the 1987 Los Angeles Marathon race. During the week following the race, 12.9% of the marathoners reported an URTI compared to only 2.2% of control runners who did not participate (odds ratio, 5.9). Forty percent of the runners reported at least one URTI episode during the two-month winter period prior to the marathon race. Controlling for various confounders, it was determined that runners training more than 96 km/wk doubled their odds for sickness compared to those training less than 32 km/wk.

Other epidemiological data support these findings (34, 51) (see table 10.3). Together, these epidemiological studies imply that heavy acute or chronic exercise is associated with an increased risk of URTI. The risk appears to be especially high during the one- or two-week period following marathon-type race events. Among runners varying widely in training habits, the risk for URTI is slightly elevated for the highest distance runners, but only when several confounding factors are controlled.

Management of the Athlete During an Infectious Episode: Practical Applications

Endurance athletes are often uncertain of whether they should exercise or rest during an infectious episode. There are few data available in humans to provide definitive answers. Most clinical authorities in this area recommend that if the athlete has symptoms of a common cold with no constitutional involvement, then regular training may be safely resumed a few days after the resolution of symptoms (6, 67-69). Mild exercise during sickness with a common cold does not appear to be contraindicated but there is insufficient evidence at present to say one way or the other. However, if there are symptoms or signs of systemic involvement (fever, extreme tiredness, muscle aches, swollen lymph glands, etc.), then 2-4 weeks should probably be allowed before resumption of intensive training.

These recommendations are speculative, however, and are primarily based on animal studies and some case reports among humans who died following bouts of vigorous exercise during an acute viral illness. Depending on the pathogen (with some more affected by exercise than others), animal studies generally support the finding that one or two periods of exhaustive exercise following inoculation of the animal leads to a more frequent appearance of infection and a higher fatality rate (8).

It is well established that various measures of physical performance capability are reduced during an infectious episode (11, 14, 25, 67). Several case histories have been published demonstrating that a sudden and unexplained deterioration in athletic performance can in some individuals be traced to either recent URTI or subclinical viral infections that run a protracted course (67-69). In some athletes, a viral infection may lead to a severely debilitating state known as postviral fatigue syndrome (PVFS) (39). The symptoms of PVFS can persist for several months, and include lethargy, easy fatigability, and myalgia.

For elite athletes who may be undergoing heavy exercise stress in preparation for competition, several precautions may help them reduce their risk of URTI. Considerable evidence indicates that two other environmental factors, improper nutrition and psychological stress (9, 29), can compound the negative influence that heavy exertion has on the immune system. Based on current understanding, the athlete is urged to eat a well-balanced diet, keep other life stresses to a minimum, avoid overtraining and chronic fatigue, obtain adequate sleep, and space vigorous workouts and race events as far apart as possible. Immune system function appears to be suppressed during periods of low caloric intake and weight reduction (30), so when necessary, the athlete is advised to lose weight slowly during noncompetitive training phases. Cold viruses are spread by both personal contact and breathing the air near sick people (1, 26, 27). Therefore, if at all possible, athletes should avoid being around sick people before and after important events. If the athlete is competing during the winter months, a flu shot is recommended.

Can the athlete follow certain practices to attenuate the negative changes in immunity following heavy exertion? This is a matter of active study at present. Indomethacin, which inhibits prostaglandin production, has been administered to athletes prior to exercise, or used in vitro to determine whether the drop in NKCA

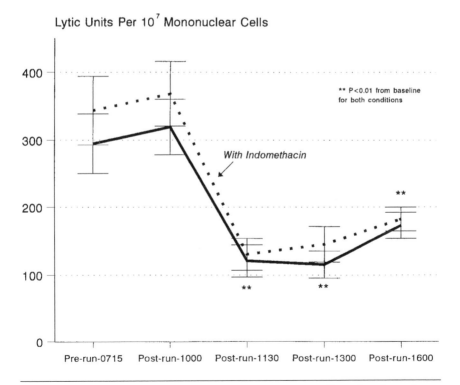

Lytic Units Per 10^7 Mononuclear Cells

** P<0.01 from baseline for both conditions

With Indomethacin

Pre-run-0715   Post-run-1000   Post-run-1130   Post-run-1300   Post-run-1600

Figure 10.8 In vitro incubation of blood mononuclear cells with indomethacin had no effect on the decrease in natural killer cell activity following 2.5 hours of intensive running by 12 marathon runners.

Based on data from Nieman et al. 1995.

can be countered. Figure 10.8 shows that following 2.5 hours of intensive running, indomethacin had no effect in countering the steep drop in NKCA (44). Other medications such as aspirin or ibuprofen are currently being studied for their effects on the immune system following heavy exertion.

A double-blind, placebo, randomized study investigated the effect of carbohydrate fluid (Gatorade, 6% carbohydrate) ingestion on the immune response to 2.5 hours of running (49). In prior research, carbohydrate versus water ingestion during prolonged endurance exercise had been associated with an attenuated cortisol and epinephrine response through its effect on the blood glucose (40). Could carbohydrate fluid ingestion during 2.5-3 h of high-intensity running ameliorate immune suppression during recovery through its influence on cortisol?

On the test day, following a blood sample at 0700 h in a 12 h fasted state, subjects drank 750 ml of either a carbohydrate or placebo beverage. At 0730 h, subjects began running at 75-80% VO_2max for 2.5 h, and ingested 250 ml carbohydrate or placebo beverage every 15 min. Immediately after the 2.5 h run (1000 h), another blood sample was taken, followed by 1.5 h, 3 h, and 6 h recovery samples. Subjects drank 500 ml/h of carbohydrate or placebo during the first 1.5 h of recovery, and then 250 ml/h during the last 4.5 h of recovery.

Figures 10.9 and 10.10 show that drinking the carbohydrate beverage before, during, and after 2.5 hours of running attenuated the rise in both cortisol and the

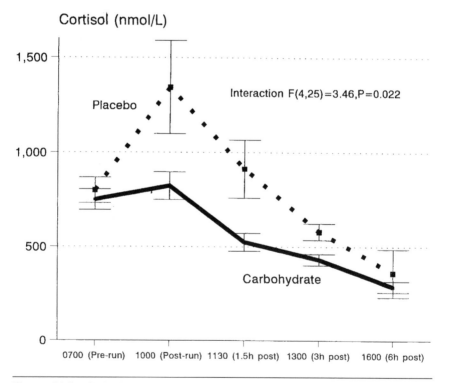

Figure 10.9 Carbohydrate ingestion (Gatorade) before, during, and after 2.5 hours of intensive running attenuated the cortisol response relative to a placebo drink.

Based on data from Nieman et al. in press.

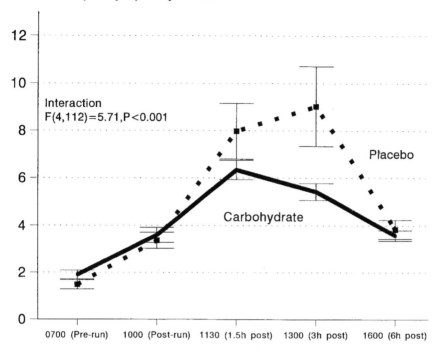

Neutrophil/lymphocyte ratio

Interaction
$F_{(4,112)} = 5.71, P < 0.001$

Placebo

Carbohydrate

0700 (Pre-run)   1000 (Post-run)   1130 (1.5h post)   1300 (3h post)   1600 (6h post)

Figure 10.10   Carbohydrate ingestion (Gatorade) before, during, and after 2.5 hours of intensive running attenuated the neutrophil/lymphocyte ratio response relative to a placebo drink.

Based on data from Nieman et al. in press.

neutrophil/lymphocyte ratio. The immediate post-run blood glucose level was significantly higher in the carbohydrate versus placebo group and was negatively correlated with cortisol ($r = -.67$, $p < .001$). Trafficking of most leukocyte and lymphocyte subsets was lessened in accordance with the lower cortisol levels in the carbohydrate subjects. These data suggest that carbohydrate ingestion before, during, and after prolonged endurance exercise may help to lessen the stress on the immune system.

What about nutrient supplements? Can they play a role in lowering risk of respiratory infection following heavy exertion? In a study by Peters et al. (65), 68% of runners reported the development of symptoms of URTI within two weeks after the 90 km Comrades Ultramarathon. The incidence of URTI was greatest among the runners who trained the hardest coming into the race (85% versus 45% of the low- or medium-training status runners). Using a double-blind, placebo research design, it was determined that only 33% of runners taking a 600 mg vitamin C supplement daily for three weeks prior to the race developed URTI symptoms. The authors suggested that because heavy exertion enhances the production of free oxygen radicals, vitamin C, which has antioxidant properties, may be required in increased quantities. This is an interesting finding, and further research will help to determine if this finding also applies to runners racing shorter distances (for example, a typical marathon of 42.2 km). It should be noted that one double-blind, placebo-controlled study was unable to establish that vitamin C supplementation (1,000 mg/d for 8 days) had any significant effect in altering the immune response to 2.5 h of intensive running (48).

There is some evidence that glutamine, a nonessential amino acid, is important for immune function, and that supplementation during heavy exertion may lessen the incidence of infection (7). These results are intriguing, and further investigation is warranted using careful research designs.

Table 10.4 summarizes the practical applications discussed in this section. Although they must be considered tentative, athletes are urged to adopt them until investigators prove otherwise.

Table 10.4 A Summary of the Practical Guidelines Athletes Can Follow to Lower Their Risk of Upper Respiratory Tract Infection Following Heavy and Prolonged Exertion

1. Keep other life stresses to a minimum. Mental stress in and of itself has been linked to an increased risk of upper respiratory tract infection.

2. Eat a well-balanced diet to keep vitamin and mineral pools in the body at optimal levels. Although there is insufficient evidence to recommend nutrient supplements, ultramarathon runners may benefit by taking vitamin C supplements before ultramarathon races.

3. Avoid overtraining and chronic fatigue.

4. Obtain adequate sleep on a regular schedule. Sleep disruption has been linked to suppressed immunity.

5. Avoid rapid weight loss (which has also been linked to negative immune changes).

6. Avoid putting hands to the eyes and nose (which is a primary route of introducing viruses into the body). Before important race events, avoid sick people and large crowds when possible.

7. For athletes competing during the winter months, flu shots are recommended.

8. Use carbohydrate beverages before, during, and after marathon-type race events or unusually heavy training bouts.

References

1. Ansari, S.A., V.S. Springthorpe, S.A. Sattar, S. Rivard, M. Rahman. 1991. Potential role of hands in the spread of respiratory viral infections: studies with human parainfluenza virus 3 and rhinovirus 14. *Journal of Clinical Microbiology* 29: 2115-2119.
2. Baj, Z., J. Kantorski, E. Majewska, K. Zeman, L. Pokoca, E. Fornalczyk, H. Tchórzewski, Z. Sulowka, R. Lewicki. 1994. Immunological status of competitive cyclists before and after the training season. *International Journal of Sports Medicine* 15: 319-324.

3. Berk, L.S., D.C. Nieman, W.S. Youngberg, K. Arabatzis, M. Simpson-Westerberg, J.W. Lee, S.A. Tan, W.C. Eby. 1990. The effect of long endurance running on natural killer cells in marathoners. *Medicine and Science in Sports and Exercise* 22: 207-212.

4. Brahmi, Z., J.E. Thomas, M. Park, M. Park, I.A.G. Dowdeswell. 1985. The effect of acute exercise on natural killer-cell activity of trained and sedentary human subjects. *Journal of Clinical Immunology* 5: 321-328.

5. Bruunsgaard, H., A. Hartkopp, T. Mohr. (1997). Decreased in vivo cell-mediated immunity, but normal vaccination response following intense, long-term exercise. *Medicine and Science in Sports and Exercise* 28: In press.

6. Burch, G.E. 1979. Viral diseases of the heart. *Acta Cardiologica* 34: 5-9.

7. Castell, L.M., J.R. Poortmans, E.A. Newsholme. 1996. Does glutamine have a role in reducing infections in athletes? *European Journal of Applied Physiology* 73: 488-451.

8. Chao, C.C., F. Strgar, M. Tsang, P.K. Peterson. 1992. Effects of swimming exercise on the pathogenesis of acute murine Toxoplasma gondii Me49 infection. *Clinical Immunology Immunopathology* 62: 220-226.

9. Cohen, S., D.A. Tyrrell, A.P. Smith. 1991. Psychological stress and susceptibility to the common cold. *New England Journal of Medicine* 325: 606-612.

10. Cupps, T.R., A.S. Fauci. 1982. Corticosteroid-mediated immunoregulation in man. *Immunological Reviews* 65: 133-155.

11. Daniels, W.L., D.S. Sharp, J.E. Wright, J.A. Vogel, G. Friman, W.R. Beisel, J.J. Knapik. 1985. Effects of virus infection on physical performance in man. *Military Medicine* 150: 8-14.

12. Eskola, J., O. Ruuskanen, E. Soppi, M.K. Viljanen, M. Järvinen, H. Toivonen, K. Kouvalainen. 1978. Effect of sport stress on lymphocyte transformation and antibody formation. *Clinical and Experimental Immunology* 32: 339-345.

13. Field, C.J., R. Gougeon, E.B. Marliss. 1991. Circulating mononuclear cell numbers and function during intense exercise and recovery. *Journal of Applied Physiology* 71: 1089-1097.

14. Friman, G., N.G. Ilbäck, D.J. Crawford, H.A. Neufeld. 1991. Metabolic responses to swimming exercise in Streptococcus pneumoniae infected rats. *Medicine and Science in Sports and Exercise* 23: 415-421.

15. Gabriel, H., H.J. Müller, A. Urhausen, W. Kinderman. 1994. Suppressed PMA-induced oxidative burst and unimpaired phagocytosis of circulating granulocytes one week after a long endurance exercise. *International Journal of Sports Medicine* 15: 441-445.

16. Gatti, G., R. Cavallo, M.L. Sartori, D. del Ponte, R. Masera, A. Salvadori, R. Carignola, A. Angeli. 1987. Inhibition by cortisol of human natural killer (NK) cell activity. *Journal of Steroid Biochemistry* 26: 49-58.

17. Gmünder, F.K., G. Lorenzi, B. Bechler, P. Joller, J. Müller, W.H. Ziegler, A. Cogoli. 1988. Effect of long-term physical exercise on lymphocyte reactivity: similarity to spaceflight reactions. *Aviation and Space Environmental Medicine* 59: 146-151.

18. Green, R.L., S.S. Kaplan, B.S. Rabin, C.L. Stanitski, U. Zdziarski. 1981. Immune function in marathon runners. *Annals of Allergy* 47: 73-75.

19. Hack, V., G. Strobel, J-P Rau, H. Weicker. 1992. The effect of maximal exercise on the activity of neutrophil granulocytes in highly trained athletes in a moderate training period. *European Journal of Applied Physiology* 65: 520-524.

20. Hack, V., G. Strobel, M. Weiss, H. Weicker. 1994. PMN cell counts and phagocytic activity of highly trained athletes depend on training period. *Journal of Applied Physiology* 77: 1731-1735.

21. Hanley, D.F. 1976. Medical care of the U.S. Olympic team. *Journal of the American Medical Association* 12:236: 147-148.

22. Haq, A., K. Al-Hussein, J. Lee, S. Al-Sedairy. 1993. Changes in peripheral blood lymphocyte subsets associated with marathon running. *Medicine and Science in Sports and Exercise* 25: 186-190.

23. Heath, G.W., E.S. Ford, T.E. Craven, C.A. Macera, K.L. Jackson, R.R. Pate. 1991. Exercise and the incidence of upper respiratory tract infections. *Medicine and Science in Sports and Exercise* 23:152-157.

24. Holbrook, N.J., W.I. Cox, H.C. Horner. 1983. Direct suppression of natural killer activity in human peripheral blood leukocyte cultures by glucocorticoids and its modulation by interferon. *Cancer Research* 43: 4019-4025.

25. Ilbäck, N.G., G. Friman, D.J. Crawford, H.A. Neufeld. 1991. Effects of training on metabolic responses and performance capacity in streptococcus pneumoniae infected rats. *Medicine and Science in Sports and Exercise* 23: 422-427.

26. Jackson, G.G., H.G. Dowling, T.O. Anderson, L. Riff, J. Saporta, M. Turck. 1960. Susceptibility and immunity to common upper respiratory viral infections—the common cold. *Annals of Internal Medicine* 53: 719-738.

27. Jennings, L.C., E.C. Dick. 1987. Transmission and control of rhinovirus colds. *European Journal of Epidemiology* 3: 327-335.

28. Jokl, E. 1974. The immunological status of athletes. *Journal of Sports Medicine* 14: 165-167.

29. Khansari, D.N., A.J. Murgo, R.E. Faith. 1990. Effects of stress on the immune system. *Immunology Today* 11: 170-175.

30. Kono, I., H. Kitao, M. Matsuda, S. Haga, H. Fukushima, H. Kashiwagi. 1988. Weight reduction in athletes may adversely affect the phagocytic function of monocytes. *Physician and Sportsmedicine* 16: 56-65.

31. Kvernmo, H., J.O. Olsen, B. Osterud. 1992. Changes in blood cell response following strenuous physical exercise. *European Journal of Applied Physiology* 64: 318-322.

32. Lewicki, R., H. Tchórzewski, A. Denys, M. Kowalska, A. Golinska. 1987. Effect of physical exercise on some parameters of immunity in conditioned sportsmen. *International Journal of Sports Medicine* 8: 309-314.

33. Lewis, C.E., J.O.D. McGee. 1992. *The natural killer cell*, 175-203. New York: Oxford University Press.

34. Linde, F. 1987. Running and upper respiratory tract infections. *Scandinavian Journal of Sports Science* 9: 21-23.

35. Mackinnon, L.T., T.W. Chick, A. Van As, T.B. Tomasi. 1988. Effects of prolonged intense exercise on natural killer cell number and function. *Exercise Physiology: Current Selected Research* 3: 77-89.

36. Mackinnon, L.T., T.W. Chick, A. Van As, T.B. Tomasi. 1987. The effect of exercise on secretory and natural immunity. *Advances in Experimental Medicine and Biology* 216A: 869-876.

37. Mackinnon, L.T., S. Hooper. 1994. Mucosal (secretory) immune system responses to exercise of varying intensity and during overtraining. *International Journal of Sports Medicine* 15: S179-S183.

38. Mackinnon, L.T., D.G. Jenkins. 1993. Decreased salivary immunoglobulins after intense interval exercise before and after training. *Medicine and Science in Sports and Exercise* 25: 678-683.

39. Maffulli, N., V. Testa, G. Capasso. 1993. Post-viral fatigue syndrome: a longitudinal assessment in varsity athletes. *Journal of Sports Medicine and Physical Fitness* 33: 392-399.

40. Mitchell, J.B., D.L. Costill, J.A. Houmard, M.G. Flynn, W.J. Fink, J.D. Beltz. 1990. Influence of carbohydrate ingestion on counterregulatory hormones during prolonged exercise. *International Journal of Sports Medicine* 11: 33-36.

41. Müns, G. 1993. Effect of long-distance running on polymorphonuclear neutrophil phagocytic function of the upper airways. *International Journal of Sports Medicine* 15: 96-99.

42. Müns, G., H. Liesen, H. Riedel, K.-Ch. Bergmann. 1989. Einfluß von langstreckenlauf auf den IgA-gehalt in nasensekret und speichel. *Deutsche Zeitschrift Für Sportmedizin* 40: 63-65.

43. Müns, G., P. Singer, F. Wolf, I. Rubinstein. 1995. Impaired nasal mucociliary clearance in long-distance runners. *International Journal of Sports Medicine* 16: 209-213.

44. Nieman, D.C., J.C. Ahle, D.A. Henson, B.J. Warren, J. Suttles, J.M. Davis, K.S. Buckley, S. Simandle, D.E. Butterworth, O.R. Fagoaga, S.L. Nehlsen-Cannarella. 1995. Indomethacin does not alter the natural killer cell response to 2.5 hours of running. *Journal of Applied Physiology* 79: 748-755.

45. Nieman, D.C., D. Brendle, D.A. Henson, J. Suttles, V.D. Cook, B.J. Warren, D.E. Butterworth, O.R. Fagoaga, S.L. Nehlsen-Cannarella. 1995. Immune function in athletes versus nonathletes. *International Journal of Sports Medicine* 16: 329-333.

46. Nieman, D.C., K.S. Buckley, D.A. Henson, B.J. Warren, J. Suttles, J.C. Ahle, S. Simandle, O.R. Fagoaga, S.L. Nehlsen-Cannarella. 1995. Immune function in marathon runners versus sedentary controls. *Medicine and Science in Sports and Exercise* 27: 986-992.

47. Nieman, D.C., V.D. Cook, D.A. Henson, J. Suttles, W.J. Rejeski, P.M. Ribisl, O.R. Fagoaga, S.L. Nehlsen-Cannarella. 1995. Moderate exercise training and natural killer cell cytotoxic activity in breast cancer patients. *International Journal of Sports Medicine* 16: 334-337.

48. Nieman, D.C., D.A. Henson, D.E. Butterworth, B.J. Warren, J.M. Davis, O.R. Fagoaga, S.L. Nehlsen-Cannarella. 1997. Vitamin C supplementation does not alter the immune response to 2.5 hours of running. *International Journal of Sports Nutrition,* in press.

49. Nieman, D.C., D.A. Henson, E.B. Garner, D.E. Butterworth, B.J. Warren, A. Utter, J.M. Davis, O.R. Fagoaga, S.L. Nehlsen-Cannarella. Carbohydrate affects natural killer cell redistribution but not activity after running. *Medicine and Science in Sports and Exercise,* in press.

50. Nieman, D.C., D.A. Henson, G. Gusewitch, B.J. Warren, R.C. Dotson, D.E. Butterworth, S.L. Nehlsen-Cannarella. 1993. Physical activity and immune function in elderly women. *Medicine and Science in Sports Exercise* 25: 823-831.

51. Nieman, D.C., L.M. Johanssen, J.W. Lee. 1989. Infectious episodes in runners before and after a road race. *Journal of Sports Medicine and Physical Fitness* 29: 289-296.

52. Nieman, D.C., L.M. Johanssen, J.W. Lee, J. Cermak, K. Arabatzis. 1990. Infectious episodes in runners before and after the Los Angeles Marathon. *Journal of Sports Medicine and Physical Fitness* 30: 316-328.

53. Nieman, D.C., A.R. Miller, D.A. Henson, B.J. Warren, G. Gusewitch, R.L. Johnson, J.M. Davis, D.E. Butterworth, J.L. Herring, S.L. Nehlsen-Cannarella. 1994. Effects of high- versus moderate-intensity exercise on circulating lymphocyte subpopulations and proliferative response. *International Journal of Sports Medicine* 15: 199-206.

54. Nieman, D.C., A.R. Miller, D.A. Henson, B.J. Warren, G. Gusewitch, R.L. Johnson, J.M. Davis, D.E. Butterworth, S.L. Nehlsen-Cannarella. 1993. The effects of high- versus moderate-intensity exercise on natural killer cell cytotoxic activity. *Medicine and Science in Sports and Exercise* 25: 1126-1134.

55. Nieman, D.C., S.L. Nehlsen-Cannarella. 1991. The effects of acute and chronic exercise on immunoglobulins. *Sports Medicine* 11:183-201.

56. Nieman, D.C., S. Simandle, D.A. Henson, B.J. Warren, J. Suttles, J.M. Davis, K.S. Buckley, J.C. Ahle, D.E. Butterworth, O.R. Fagoaga, S.L. Nehlsen-Cannarella. 1995. Lymphocyte proliferation response to 2.5 hours of running. *International Journal of Sports Medicine* 16: 406-410.

57. Nieman, D.C., S.A. Tan, J.W. Lee, L.S. Berk. 1989. Complement and immunoglobulin levels in athletes and sedentary controls. *International Journal of Sports Medicine* 10: 124-128.

58. Oshida, Y., K. Yamanouchi, S. Hayamizu, Y. Sato. 1988. Effect of acute physical exercise on lymphocyte subpopulations in trained and untrained subjects. *International Journal of Sports Medicine* 9: 137-140.

59. Papa, S., M. Vitale, G. Mazzotti, L.M. Neri, G. Monti, F.A. Manzoli. 1989. Impaired lymphocyte stimulation induced by long-term training. *Immunology Letters* 22: 29-33.

60. Pedersen, B.K., N. Tvede, L.D. Christensen, K. Klarlund, S. Kragbak, J. Halkjær-Kristensen. 1989. Natural killer cell activity in peripheral blood of highly trained and untrained persons. *International Journal of Sports Medicine* 10: 129-131.

61. Pedersen, B.K., N. Tvede, K. Klarlund, L.D. Christensen, F.R. Hansen, H. Galbo, A. Kharazmi, J. Halkjær-Kristensen. 1990. Indomethacin in vitro and in vivo abolishes post-exercise suppression of natural killer cell activity in peripheral blood. *International Journal of Sports Medicine* 11: 127-131.

62. Pedersen, B.K., H. Ullum. 1994. NK cell response to physical activity: possible mechanisms of action. *Medicine and Science in Sports and Exercise* 26: 140-146.

63. Peters, E.M. 1990. Altitude fails to increase susceptibility of ultramarathon runners to post-race upper respiratory tract infections. *South African Journal of Sports Medicine* 5: 4-8.

64. Peters, E.M., E.D. Bateman. 1983. Respiratory tract infections: an epidemiological survey. *South African Medical Journal* 64: 582-584.

65. Peters, E.M., J.M. Goetzsche, B. Grobbelaar, T.D. Noakes. 1993. Vitamin C supplementation reduces the incidence of postrace symptoms of upper-respiratory-tract infection in ultramarathon runners. *American Journal of Clinical Nutrition* 57: 170-174.

66. Pyne, D.B. 1994. Regulation of neutrophil function during exercise. *Sports Medicine* 17: 245-258.

67. Roberts, J.A. 1985. Loss of form in young athletes due to viral infection. *British Journal of Medicine* 290: 357-358.
68. Roberts, J.A. 1986. Viral illnesses and sports performance. *Sports Medicine* 3: 296-303.
69. Sharp, J.C.M. 1989. Viruses and the athlete. *British Journal of Sports Medicine* 23: 47-48.
70. Shephard, R.J., P.N. Shek. 1994. Infectious diseases in athletes: new interest for an old problem. *Journal of Sports Medicine and Physical Fitness* 34: 11-21.
71. Shinkai, S., Y. Kurokawa, S. Hino, M. Hirose, J. Torii, S. Watanabe, S. Watanabe, S. Shiraishi, K. Oka, T. Watanabe. 1993. Triathlon competition induced a transient immunosuppressive change in the peripheral blood of athletes. *Journal of Sports Medicine and Physical Fitness* 33: 70-78.
72. Shinkai, S., S. Shore, P.N. Shek, R.J. Shephard. 1992. Acute exercise and immune function: relationship between lymphocyte activity and changes in subset counts. *International Journal of Sports Medicine* 13: 452-461.
73. Smith, J.A., R.D. Telford, I.B. Mason, M.J. Weidemann. 1990. Exercise, training and neutrophil microbicidal activity. *International Journal of Sports Medicine* 11: 179-187.
74. Tidball, J.G. 1995. Inflammatory cell response to acute muscle injury. *Medicine and Science in Sports and Exercise* 27: 1022-1032.
75. Tomasi, T.B., F.B. Trudeau, D. Czerwinski, S. Erredge. 1982. Immune parameters in athletes before and after strenuous exercise. *Journal of Clinical Immunology* 2: 173-178.
76. Tonnesen, E., N.J. Christensen, M.M. Brinklov. 1987. Natural killer cell activity during cortisol and adrenaline infusion in healthy volunteers. *European Journal of Clinical Investigation* 17: 497-503.
77. Tvede, N., M. Kappel, J. Halkjær-Kristensen, H. Galbo, B.K. Pedersen. 1993. The effect of light, moderate, and severe bicycle exercise on lymphocyte subsets, natural and lymphokine activated killer cells, lymphocyte proliferative response and interleukin 2 production. *International Journal of Sports Medicine* 14: 275-282.
78. Tvede, N., J. Steensberg, B. Baslund, Baslund, J.H. Kristensen, B.K. Pedersen. 1991. Cellular immunity in highly-trained elite racing cyclists and controls during periods of training with high and low intensity. *Scandinavian Journal of Sports Medicine* 1: 163-166.
79. Weight, L.M., D. Alexander, P. Jacobs. 1991. Strenuous exercise: analogous to the acute-phase response? *Clinical Science* 81: 677-683.

CHAPTER 11

Effects of Overreaching and Overtraining on Immune Function

Laurel Traeger Mackinnon, PhD

Introduction

Overtraining has long been associated with increased susceptibility to infectious illness, particularly upper respiratory tract infection (URTI). As shown in the previous chapter, acute exercise and exercise training induce profound changes in various aspects of immune function. Intensive exercise is generally associated with immune suppression and with increased risk of URTI. Thus, it is not unexpected that overtraining resulting from prolonged periods of high-intensity exercise training may lead to changes in immune function in athletes.

Models to Study Overtraining and Immune Function

There are two general models to study overtraining and immune function. In one model, athletes are followed over the course of a normal training season, usually of three to eight months duration. Immune variables are sampled at several time points corresponding with periods of low-and high-intensity training. Immune function may then be compared across the season within the same athletes, or between athletes showing symptoms of overtraining with those considered well trained (i.e., not overtrained). This model provides information about the immune response during prolonged periods of intensive training in the athlete's natural environment, that is, during normal training and competition. However, not all studies include appropriate controls to account for seasonal variability. In addition, it is not always possible to control for other potentially confounding factors such as psychological stress, competition, travel, diet, and changes in training programs. In the second model, training is intensified over a specific time period, usually of one to four weeks duration; for ethical reasons, four weeks is considered the maximum time that athletes can tolerate intensified training (8, 26, 27, 38, 64). Immune parameters are then compared from before to after intensified training, or less commonly, between overtrained and well-trained athletes. While it may be possible to control the volume and intensity of training, not all athletes respond similarly to increased training. Moreover, the increases in training volume and intensity often used (e.g., doubling of volume within a few weeks) do not reflect normal training regimes with their more gradual increases in volume and intensity, as performed by elite athletes. Despite these limitations, by combining data from both models it is possible to gain an understanding of the immune response to overtraining in athletes.

Because few papers on exercise and immune function specifically document symptoms of overtraining or directly compare immune function in overtrained with nonovertrained athletes, a broader view will be taken in this chapter to more fully describe the relationship among overtraining, intensive training, and immune function.

Illness Rates in Overtrained Athletes

Despite the widely held belief among exercise scientists, coaches, and athletes (7, 10, 19, 28, 43, 49, 54, 66) that frequent illness is a symptom or outcome of overtraining, there are surprisingly few studies that have directly addressed this issue (15, 20, 38, 50, 67). Viral URTI is by far the most common infection affecting elite athletes (67). As shown in the previous chapter, the incidence of URTI is higher among endurance athletes compared with nonathletes. Moreover, the risk of URTI among athletes increases after competition and with training volume and intensity. However, despite this association between exercise training and risk of URTI, few studies have compared illness rates in overtrained with nonovertrained athletes.

In a recent study on URTI and intensified training, illness rates were compared among swimmers showing symptoms of overreaching and those considered well trained (38). Training was intensified over four weeks in 24 elite swimmers (16 female, 8 male) by increasing both swim and dryland (resistance) training volume progressively by 10% on each of the four weeks. Overreaching was identified in eight (33%) of the swimmers (6 female, 2 male) based on performance decrements in a criterion swim, persistent high fatigue ratings (1-7 scale) recorded in daily log books, and supporting comments from the swimmers in the log books indicating poor adaptation to the training loads (21, 22, 38). Of the 24 swimmers, 10 or 42% exhibited URTI during the four weeks. Surprisingly, the incidence of URTI over the four weeks was far higher in the well-trained swimmers (nine of 16, 56%) compared with overreached swimmers (one of eight, 12.5%). It was suggested that increased risk of URTI may not necessarily be associated with overreaching (i.e., presence of symptoms), but may occur as a consequence of intensive training in all athletes. At present, this paper appears to be the only study to specifically assess the incidence of URTI among athletes showing symptoms of overreaching or overtraining. Obviously, larger prospective studies are needed to follow different types of athletes over extended periods to determine whether increased risk of URTI is in actuality a consequence of overtraining.

Immune Parameters in Overtrained Athletes

Immune Cells

Immune cells are the major effectors of host resistance to disease, mediating the immune response via both direct and indirect actions. Immune cells can directly kill foreign organisms or infected cells; immune cells also produce soluble mediators that activate various immune cells and induce killing of pathogenic microorganisms. The responses of immune cells and soluble factors to overtraining and intensive training are summarized in tables 11.1 and 11.2.

Leukocyte Number.   Low resting leukocyte counts have been reported in endurance athletes such as distance runners (17, 25, 27). Although values were reported to be within the clinically normal range of 4-11 $\times$ 10^9/L for most athletes, mean levels were at the low end of the range and counts in several athletes were below the lower limit. These low leukocyte counts may suggest long-term suppression of circulating leukocyte number in endurance athletes. However, many other studies have shown resting leukocyte counts to be well within the normal range (5, 10, 11, 15, 22, 46) in various well-trained athletes.

Lehmann et al. (27) reported a progressive decrease in leukocyte counts over four weeks of intensified training in male distance runners. Training distance was increased by 33% each week until volume was twice baseline levels (e.g., from mean of 85.9 to 174.6 km/wk). Subjects also rated complaints such as severe muscular stiffness or fatigue, on a 1-4 scale, daily throughout the study. By the end of the four weeks, subjects showed clear signs of overreaching, such as a significant increase in the complaints index after day 20, decreased maximal exercise heart rate, decreased running speed at the 4 mmol/L lactate threshold and maximal blood lactate levels, and lower urinary catecholamine excretion. Leukocyte counts declined significantly from a mean 5.4 $\times$ 10^9/L before the study to 4.9 $\times$ 10^9/L after two weeks and 4.2 $\times$ 10^9/L after four weeks of intensified training. For ethical reasons, the study was not extended past four weeks, and it is unclear whether leukocyte numbers would have continued to decline, possibly below clinically normal levels, with continued intensive training.

Keen et al. (25) reported resting leukocyte numbers in well-trained cyclists that were in the clinically low end of the normal range; leukocyte counts were 5.04 and 7.5 $\times$ 10^9/L in cyclists and nonathletes, respectively. However, when measured over 12 days of intensive exercise during competition (a multistage cycling race over 117-185 km/d), neither early morning preexercise nor evening (5-7 h postrace) leukocyte counts varied significantly in the cyclists.

Decreased resting leukocyte counts were also reported in male competitive cyclists after, compared with before, a five month training cycle (5). However, mean counts were well within the normal range (7.6 and 6.8 $\times$ 10^9/L before and after training, respectively) and there were no significant differences between posttraining values in athletes and those obtained at the same time from control subjects. Taken together, data on leukocyte counts presented thus far suggest that daily intensive exercise over shorter periods (less than two weeks) may not significantly alter circulating leukocyte number and that cell counts change only after prolonged periods (i.e., weeks to months) of intensive, high-volume exercise in well-trained athletes.

Table 11.1 Changes in Immune Cells During Overtraining and Intensified Training

Cell	Athlete	Condition	Findings	Reference
Leukocytes	M distance runners	4 wk ↑ training	progressive ↓ in cell number	27
	elite M, F swimmers	6 mo training season	↑ cell number in OT vs. WT at taper only	22
	M cyclists	12 day race	no change in cell number	25
Neutrophils	elite M, F swimmers	12 wk ↑ training	↓ oxidative activity	50
	elite M, F swimmers	6 mo training season	↑ cell number in OT vs. WT at taper only	22
	M distance runners	moderate vs. intensive training	↓ resting and post-ex phagocytic activity during intensive training	18
Lymphocytes	endurance trained M soldiers	10 d ↑ training + 5 d recovery	↑ number activated (CD25+, HLA-DR+) cells	11,12
	M distance runners	3 wk ↑ training	↑ proliferation at rest but ↓ after submax ex; ↓ CD4, ↑ CD8 number, ↓ CD4/CD8 ratio	64,65
	elite M, F swimmers	7 mo training season	no change in cell number or % except for NK cells (see below)	15
	M, F various	3 mo intensive training	↑ CD4 & CD8 cell numbers but no change in ratio	14
	M distance runners	10 d ↑ training volume and intensity	no change resting CD3, CD4, CD8 numbers or ratio, or activated (HLA-DR+) cell number	23
NK cells	elite M, F swimmers	7 mo training season	↓ CD56+ cell number	15
	endurance trained M soldiers	10 d ↑ training + 5 d recovery	↓ CD56+ cell number	11,12
	M, F various	3 mo intensive training	↓ CD56+ cell number	14

Abbreviations: ↑ = increase; ↓ = decrease; M = male; F = female; OT = overtrained; WT = well trained; submax ex = submaximal exercise; post-ex = postexercise.

Table 11.2 Changes in Soluble Factors During Overtraining and Intensified Training

Variable	Athlete	Condition	Findings	Reference
Serum Ig	elite M, F swimmers	7 mo training season	no change in serum IgA, IgG, IgM over season	15
	M, F various	3 mo intensive training	↓ IgM, IgG, IgG$_1$, IgG$_2$	14
	M distance runners	10 d ↑ training intensity and volume	no change IgA, IgG, IgM	23
Secretory IgA	M collegiate swimmers	4 mo training season plus taper	↓ resting & postexercise IgA; partial recovery during taper	59
	elite M kayakers	1 wk ↑ training	↓ resting & postexercise IgA	34
	elite M, F hockey athletes	10 d training camp	↓ postexercise in those who developed URTI within 2 d	35
	elite M, F swimmers	6 mo training season	lower resting values in OT vs. WT	36
	elite M, F swimmers	7 mo training season	↓ resting & postexercise IgA	15
	M distance runners	intensive endurance exercise 3 d	↓ post-ex IgA concentration days 2 & 3 vs. day 1	36
	elite F hockey	5 d major competition	↓ pre- & post-match IgA concentration over 5 d	32
Cytokines	endurance trained M soldiers	10 d ↑ training plus 5 d recovery	↑ serum IL-2 & in vitro IL-2 production	12
Plasma glutamine	various elite	6 mo training season	lower in OT vs. WT	47
	endurance trained M soldiers	10 d ↑ training plus 5 d recovery	↓ during intensified training and returned to baseline during recovery	24
	elite M, F swimmers	4 wk ↑ training	↑ in WT but not OT; lower in OT vs. WT; glutamine levels not related to URTI	38

Abbreviations: ↑ = increase; ↓ = decrease; M = male; F = female; OT = overtrained; WT = well trained; post-ex = postexercise.

In contrast to the low leukocyte counts described previously, normal leukocyte counts were reported in elite swimmers sampled at five times during a six month season (early-, mid-, and late-season, during taper and postcompetition) (22). Gleeson et al. (15) and Tvede et al. (62) also reported no differences in resting leukocyte counts between low- and high-intensity training periods in elite swimmers and cyclists, respectively.

Although leukocyte count may increase acutely after exercise (see previous chapter), very long exercise (e.g., hours) may cause a persistent suppression of circulating cell number. For example, compared with resting levels, leukocyte counts were lower for up to 40 h after a 24 h, 120 km march in endurance trained males (13). Elite endurance athletes often train several hours each day, and it is possible that low resting leukocyte counts may reflect a long-lasting effect of prior exercise sessions. Alternatively, low leukocyte counts may also arise from migration of cells out of the circulation to tissue sites (see following discussion).

Neutrophil Number and Function.   Low resting neutrophil counts have been noted in some athletes compared with nonathlete controls or clinical norms (25, 50). Morphological evidence of immature neutrophils (e.g., less lobulated and indented nuclei) was observed in cyclists exhibiting low neutrophil counts (25), suggesting increased turnover of cells, which may account for the lower neutrophil number. In contrast, several other studies have reported normal neutrophil counts in athletes (11, 22, 62), and one recent paper reported a significant increase in neutrophil count during 12 wk intensive training in elite swimmers (50). In general, neutrophil counts do not appear to change during periods of intensified training ranging from 10 days to seven months (11, 15, 50). However, one paper reported lower counts in male distance runners during intensive compared with moderate training periods and with control subjects (18). In another study on elite swimmers followed over a six-month season, no differences in neutrophil counts were observed between overtrained and nonovertrained swimmers, except in overtrained swimmers during the taper before major competition (22). During the taper, neutrophil counts were significantly elevated by 80% in swimmers diagnosed as overtrained (persistent fatigue, performance decrements, subjective comments from swimmers in daily log books) (21, 22). The high neutrophil counts in overtrained swimmers during the taper period were attributed to a long-lasting effect of exercise prior to blood sampling and to maintenance of high-volume, high-intensity work by these swimmers, who, in fact, did not reduce training (i.e., they did not taper).

Although neutrophil number may remain relatively constant, neutrophil function is markedly affected by intensive exercise training. Smith et al. (56) noted lower resting and postexercise neutrophil microbicidal activity in elite male cyclists compared with untrained controls; exercise consisted of 60 min cycling at 60% $\dot{V}O_2$max. In another study on male distance runners, resting and postexercise phagocytic activity were lower during moderate training compared with intensive training and with nonathlete control subjects (18); low-intensity training involved 89 km/wk endurance training, and high-intensity training included 102 km/wk with intensive daily interval work. Although phagocytic activity was lower at all times before and up to 24 h after exercise during intensive compared with moderate training, the greatest deficit was observed in blood sampled 24 h after the exercise test (graded test of aerobic power). These data suggest that daily intensive exercise training exerts a prolonged suppressive effect on neutrophil function lasting at least 24 h; since athletes are usually rested for 24-36 h prior to laboratory testing, this prolonged effect may account for the lower resting neutrophil activity observed in athletes (18, 56). It has been suggested (18, 56) that decreased neutrophil function in athletes may reflect partial suppression of the inflammatory response to chronic low-level tissue damage resulting from intensive daily exercise.

Neutrophil oxidative activity on a per cell basis was significantly lower in elite male and female swimmers compared with nonathlete controls at the start of a 12-week intensive training cycle (50). In swimmers, neutrophil oxidative activity decreased significantly during the 12-week intensive training period, with lowest values observed during the peak training phase; activity recovered partially during a rest phase at the end of the intensive training period (see figure 11.1). The authors suggested two possible mechanisms to explain why fewer neutrophils reacted positively to stimulation in the in vitro assay system: release into the circulation of immature neutrophils with intrinsically lower activity or chronic reduction in the ability of neutrophils to respond to pathogens. However, despite the marked decrease in neutrophil oxidative activity, there was no correlation with the appearance of URTI during the 12-week intensive training, suggesting that such changes may not always compromise immune function in athletes.

Lymphocyte Number, Activation, and Proliferation.   Low resting lymphocyte counts have been reported in some studies on endurance athletes. For example, Green et al. (17) noted low lymphocyte counts ($< 1.5 \times 10^9$/L) in 10 of 20 male

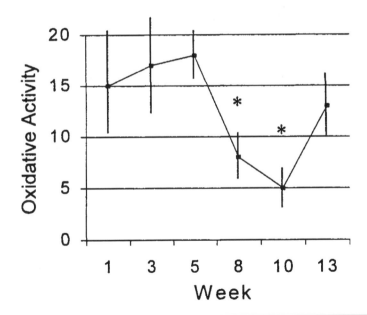

Figure 11.1   Neutrophil oxidative activity in sedentary control subjects and elite swimmers measured at three times, at start (wk 0), after 3 weeks (wk 3), and one week after (wk 13) 12 weeks of intensive swim training. Data are expressed as mean channel number of positively responding cells to PMA stimulation ± SD. *p < .01 compared with controls.
Based on data from Pyne et al. 1995.

distance runners, five of whom were considered elite and training intensively at the time. However, blood samples were obtained from several runners within hours of completing long training runs, and a long-lasting acute effect of prior exercise could not be discounted. Keen et al. (25) also noted low resting lymphocyte counts (e.g., mean 1.78, range 1.1-2.5 $\times$ 10^9/L) in male cyclists.

Overtraining and overreaching do not appear to significantly alter circulating lymphocyte number, the relative proportion of T and B cells, or the ratio of T cell subsets; in contrast, NK cell number may decrease during overtraining (see following discussion). For example, no differences were observed over a six-month season in lymphocyte counts or T or B cell numbers between elite swimmers diagnosed as overtrained (performance decrements, persistent fatigue, and subjective comments by the athletes) compared with well-trained swimmers (21). Similarly, Gleeson et al. (15) reported no changes over a seven-month season in elite swimmers in a variety of cell numbers (e.g., total lymphocyte count and percentage, CD19 (B cells), CD3 (T cells), CD4 (T helper cells), CD8 (T suppressor cells), and the CD4/CD8 ratio). Moreover, cell counts did not differ from nonathlete control subjects measured at the same times. Several other studies have also reported no changes in lymphocyte counts or subpopulations after short-term intensified training of up to three weeks (5, 11, 23) or between periods of low- and high-intensity training (62).

Lymphocyte Activation. Intensified training via large increases in training volume may induce activation of lymphocytes. For example, increases in the number of cells displaying nonspecific activation markers CD25 and HLA-DR were noted within the first six days of a 10-day intensive interval training in endurance-trained soldiers (12). The CD25+ cell number increased 2.4 and 2.9 times after six and 10 days of training, respectively; HLA-DR+ cell number increased 1.3 and 1.8 times during the same period. The number of cells expressing these activation markers remained elevated throughout five days of recovery. An increase in serum IL-2 concentration and in vitro IL-2 production in the same subjects also suggests activation of lymphocytes (see following discussion). Evidence for activation of lymphocytes by endurance training also comes from a recent paper showing increased expression of the high affinity IL-2 receptor (IL-2R) in distance runners compared with nonathletes (51). R.W. Fry et al. (12) suggested that nonspecific activation of the immune system may be related to symptoms of overreaching experienced by these subjects, such as sleep disturbances, poor appetite, fatigue, and irritability. For example, it was noted that nonspecific immune activation also occurs during chronic fatigue syndrome, the symptoms of which resemble many of the features of overtraining syndrome (49).

In contrast to the increased number of lymphocytes displaying activation markers after short-term intensified training (12), no changes in HLA-DR+ cell number were observed during a seven-month training season in elite swimmers (15), despite other evidence of immune suppression as the season progressed (e.g., decreasing secretory IgA concentration, low serum immunoglobulin levels, and declining NK cell number, each to be discussed). Kajiura et al. (23) also found no changes in HLA-DR+ cell numbers in runners after 10 days of intensified training using various protocols including increased volume, increased intensity, and a combination of the two. The reasons for the differences between studies are unclear at present, and may relate to the different protocols for intensifying training or to duration of the studies.

Lymphocyte Proliferation. The ability of lymphocytes to proliferate in response to an antigenic challenge provides an indirect measure of lymphocyte activation and function. In male distance runners, stimulation of proliferation by the T cell mitogen concanavalin A (Con A) increased 32% in resting blood samples obtained after, compared with before, three weeks of intensified training involving a 38% increase in training volume (64, 65). Lymphocyte proliferative responses remained elevated in resting blood samples obtained after a subsequent three-week normal training period. However, after the intensified training period, proliferation was significantly depressed by 18% in postexercise compared with preexercise blood samples; exercise consisted of 30 min running at 80% VO_2max, relatively moderate exercise in these runners. Although six of ten subjects showed some symptoms, such as fatigue and mood-state changes, these runners were not deemed to be overtrained because of a lack of changes in other parameters such as sleep disturbances or muscle soreness. These data showing increased resting lymphocyte proliferation are consistent with an increased activation of lymphocytes after short-term intensified training (previously discussed). The decreased proliferative response postexercise after three weeks of intensified training is more difficult to explain, and cannot be accounted for by changes in T cell number or subset (CD4, CD8) percentages. It is possible that fatigue induced by heavy training altered the acute response to submaximal exercise; however, ratings of perceived exertion (RPE) did not differ after exercise at the different time points.

NK Cells.   Higher circulating NK cell number and NK cytotoxic activity (NKCA) have been observed in cross-sectional comparisons of nonelite endurance trained athletes with nonathlete control subjects (44, 48, 62; see chapter 10). However, despite this apparent stimulation of NKCA as a result of endurance training, recent reports suggest that NK cell number may decline during prolonged periods of intensive training. For example, Gleeson et al. (15) noted decreasing NK cell (CD56[+]) number and percentage over a seven-month intensive training period in elite male and female swimmers. NK cell number declined by 43% on an absolute basis and as percentage of lymphocytes by 32% from the start to the end of the seven months. In contrast, numbers and percentages of other cell types (CD3, CD4, CD8, CD19, and CD4/CD8 ratio) did not change over the training period, suggesting a specific effect on NK cells. The low NK cell number persisted even after a taper at the end of the season, suggesting long-lasting effects of prolonged periods of intensive training on NK cell number.

R.W. Fry et al. (12) reported a similar decline in NK cell number (CD56[+]) after 10 days intensive twice-daily interval training in endurance-trained male soldiers. NK cell number decreased by 40% after 10 days training and a further 10% after five days recovery (light endurance training) (see figure 11.2). These subjects showed clear signs of overreaching, including poor performance, mood-state changes, and high fatigue ratings. The observation that NK cell numbers continued to decrease during five days recovery suggests that NK cell number is not rapidly restored after even relatively short periods of very intensive exercise training. It was suggested that the decline in circulating NK cell number may reflect redistribution of cells within the body. Presumably, NK cells would migrate from the circulation to sites of inflammation within damaged skeletal muscle fibers; all subjects reported extensive muscle soreness, suggestive of cellular damage, throughout the ten-day intensive training period. Changes in NKCA may occur independently of changes in NK cell number (30, 44), and it is currently unknown whether NKCA also decreases during prolonged periods of intensive exercise training.

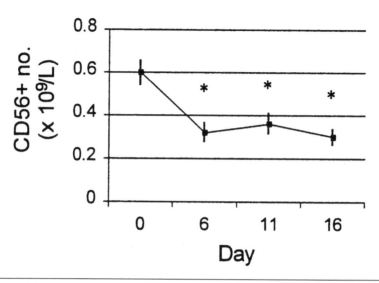

Figure 11.2   NK cell number in peripheral blood from male endurance-trained soldiers before, during, and after 10 days intensive sprint training on a motorized treadmill. Day 0 = start of intensified training, day 6 = after six days training, day 11 = 1 day recovery, day 16 = 5 days recovery. Data are expressed as mean number of CD56+ cells ± SEM. *p < .01 compared with day 0.
Based on data from Fry et al. 1994.

Soluble Factors

Besides various cells, the immune response also involves many soluble factors, including: immunoglobulin produced by activated B cells, cytokines produced by various immune and nonimmune cells, components of the complement system, and substrates such as the amino acid glutamine. Soluble factors act as chemical messengers between cells and as regulatory agents (cytokines); they are also involved in activation of immune cells (cytokines, immunoglobulin), in killing or neutralizing microorganisms, tumor cells, or infected cells (cytokines, immunoglobulin, complement), and providing substrates for cell division and metabolism (glutamine). Relatively little is known about the response of these factors to overtraining and overreaching (see table 11.2).

Serum Immunoglobulin (Ig).   Clinically normal resting serum Ig concentrations have been reported in various athletes, and acute exercise does not alter serum Ig concentration when adjusted for changes in plasma volume (37). However, recent reports indicate low serum Ig levels during prolonged periods of intensive training among elite athletes (14-16). For example, a recent prospective study of elite swimmers suggested possible long-term suppression of serum Ig (15). Serum levels of IgA, IgG, IgM and the IgG_2 subclass were significantly lower throughout the seven-month season in swimmers compared with age-matched nonathlete control subjects. Moreover, compared with clinical norms, values for the swimmers were in the lowest 10th percentile for their population. There were no differences between swimmers and controls in relative or absolute numbers of B, T, or activated lymphocytes (HLA-DR+), nor were there changes in these cells over the season in swimmers, indicating that the lower Ig levels were not due to fewer antibody-producing cells or to changes in the relative proportion of T to B cells. These data suggest that elite athletes who train intensively for many years may exhibit compromised serum Ig production.

Another recent report on 60 male and female athletes from various sports (swimming, track and field, cycling, football (soccer), basketball, tennis, and triathlon) showed declining serum Ig levels during three months of intensive training (14). Total Ig, as well as IgM, IgG, and subclasses IgG_1 and IgG_2 concentrations, declined progressively from the start to the end of the three months; training included sessions of approximately 130-140 min, five to seven days per week. Concentrations of total Ig, IgG, and IgM decreased approximately 7- 20%, with much larger decreases observed for the two IgG subclasses (30-45% and 10-30% decreases for IgG_1 and IgG_2, respectively). In contrast, shorter periods of intensive training (e.g., 10 days) have not been associated with changes in serum Ig (23).

The low levels of serum IgG_2 observed in these studies may relate to an increase in endotoxin known to occur during endurance exercise such as distance running, presumably due to microtrauma to the gut endothelium resulting from repeated impact (1); it is not clear whether endotoxemia also occurs after nonimpact activities such as swimming. IgG_2 interacts specifically with endotoxin, and its levels decrease in response to increasing levels of endotoxin in the blood. Since IgG_2 deficiency is associated with recurrent upper and lower respiratory tract infection (68), chronically low levels of this subclass may contribute to the high incidence of URTI among athletes. However, although in the study by Garagiola et al. (14) 24% of the athletes exhibited at least one episode of infectious illness during the three months (mainly upper and lower respiratory tract infection), changes in Ig levels were not related to the appearance of illness, and it cannot be discerned whether such changes compromised immunity to infection in these athletes.

Secretory Ig.   The body's external surfaces provide a large surface area that can be colonized by pathogenic microorganisms. The secretory immune system is a major effector of host resistance on external surfaces of the eyes, nose, upper and lower respiratory tracts, gastrointestinal tract, and genitourinary tract. The humoral immune response of mucosal surfaces is mediated mainly by antibodies of the IgA class. Secretory IgA has been shown to inhibit attachment and replication of certain viruses and bacteria thus preventing their entry into the body, to neutralize toxins and some viruses, and to mediate antibody-dependent cytotoxicity (ADCC), another anti-viral defense mechanism (61). Secretory IgA is important to host defense against certain viruses which are not carried in the blood, especially those causing upper respiratory tract infections. The level of secretory IgA contained in mucosal fluids correlates more closely than do serum antibodies with resistance to certain infections caused by viruses, such as upper respiratory tract infection (37, 61). Since secretory IgA is a major effector of resistance to URTI, several groups have focused on the IgA response to exercise to identify possible mechanisms related to this high incidence of illness among athletes. IgA concentration and secretion rate have been measured in saliva, which is easily obtained and can be used as a marker of mucosal immune status. Several studies have documented an acute suppression of salivary IgA after both brief and prolonged intensive exercise (15, 31-34, 59, 60). Moreover, intensive daily exercise appears to exert a cumulative suppressive effect on secretory IgA levels (32, 36).

The first study to report changes in IgA during intensive exercise found that elite male and female Nordic skiers (U.S. national team) exhibited significantly lower, by 50%, resting salivary IgA concentration compared with age-matched nonathletes (60). It was suggested that the low resting IgA levels may reflect chronic suppression due to daily intensive training and possibly to psychological stress leading up to major competition.

Intensive exercise training has been associated with decreased salivary IgA levels. In a study on male collegiate swimmers followed over a four-month season, IgA concentration declined progressively with increasing training intensity (59). Both resting and postexercise IgA concentrations decreased significantly by about 25% from early to late in the season as training intensity increased from light to heavy. IgA concentration partially recovered during reduced training (taper) leading up to major competition, but still remained significantly lower than early in the season. These data are consistent with the suggestion by Tomasi et al. (60) that cumulative suppression of IgA may occur in athletes who train intensively over prolonged periods. A more recent study of elite male and female swimmers also reported progressively decreasing resting and postexercise salivary IgA concentrations over a seven-month season (15) (see figure 11.3). In contrast to the data of Tharp and Barnes (59) described earlier, however, IgA concentration continued to decrease during the taper period, suggesting that the taper (of unspecified length, but usually two to three weeks) may have been insufficient to overcome the long-term suppression of mucosal immunity resulting from several months of intensive training. Alternatively, immunosuppressive effects of psychological stress associated with the upcoming major competition may have obscured or prevented recovery of IgA levels with easing of the physical training stress. It was further noted in this study that salivary IgM was most often detected in samples obtained at the end of the season just before major competition, and it was suggested that increasing IgM may partially compensate for decreasing IgA concentration.

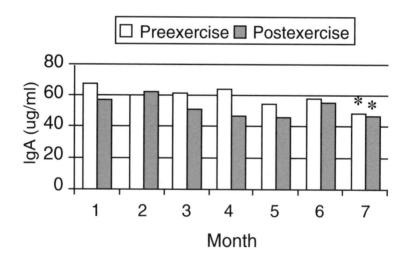

Figure 11.3 Mean salivary IgA concentration in elite swimmers before and after normal swim training sessions measured monthly during a seven-month season. *p < .05 compared with month 1.

Reprinted from Gleeson et al. 1995.

In another study on elite male and female swimmers, salivary IgA concentration was significantly lower, by 18-32%, throughout a six-month competition season in athletes showing symptoms of the overtraining syndrome (OT, e.g., prolonged high fatigue levels, poor performance, comments in daily log books consistent with the syndrome) compared with those considered well trained (WT)) (36) (see figure 11.4). In contrast to data on other groups of swimmers followed over a competitive season as previously discussed (15, 59), however, IgA levels did not change significantly in either group over the six months. The differences between studies may relate to the particular athletes involved and possibly to their training regimes. In the latter study (36), swimmers began the season with high training volumes and intensity, and may not have experienced the more gradual increase in training load as described in the other studies. These swimmers may also have been under considerable psychological stress from early in the season as they prepared for the upcoming world titles at the end of that season.

A cumulative suppressive effect of daily intensive endurance exercise on salivary IgA levels has also been observed over the short term in runners (36) and in hockey athletes (32). In the study on runners (36), male distance runners ran on a treadmill for 90 min at 75% VO_2max at the same time of day on three consecutive days; saliva was sampled before and after exercise on each day and at rest on day 4. IgA secretion rate declined significantly, by approximately 30-40%, after exercise on days 2 and 3, but not on day 1. Moreover, postexercise IgA secretion rate was significantly lower, by 27-37%, on days 2 and 3 compared with day 1, suggesting a cumulative effect of daily exercise. In the study on elite hockey athletes (32), both pre- and postexercise IgA concentrations declined progressively, by

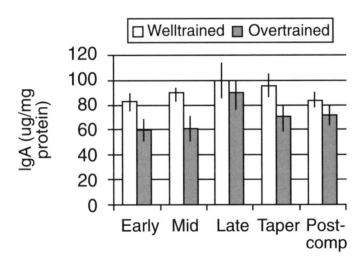

Figure 11.4 Resting salivary IgA concentration in elite swimmers measured at five times during a six-month season. Swimmers were diagnosed as overtrained based on performance decrements, persistent fatigue (rated in daily log books), and statements in daily log books by the swimmers indicating poor adaptation to training loads (21, 22); those not considered overtrained were deemed well trained. Data are expressed as mean ± SEM. p < .02 main effect of group, ANOVA.
Based on data from Mackinnon and Hooper 1994.

approximately 20%, during five days of major competition (national hockey tournament).

Given that secretory IgA is a major effector of host resistance to URTI, and IgA levels are related to resistance to infection, suppression of IgA levels during prolonged periods of intensive training and in overtrained athletes suggests that such suppression of IgA output may be one mechanism related to the high incidence of URTI among endurance athletes and may partially explain the perceived elevation of risk of infection during periods of intensive training (7, 15, 33-35, 60).

Few studies have addressed the question of whether changes in secretory immunity are directly associated with the increase in the incidence of infectious illness among athletes. In a prospective study of elite hockey and squash athletes (35), exercise-induced decreases in salivary IgA concentration were temporally associated with subsequent appearance of URTI. Saliva samples were obtained before and after normal training sessions at the same time each day during a nine-day intensive training camp in 19 male and female national team hockey athletes, and before and after normal training sessions at the same time each day on the same day of the week over a ten-week period of intensive training in 14 national team squash athletes. Athletes completed daily log books detailing the presence, nature, duration, and severity of symptoms of all illnesses; team physicians also documented the severity, symptoms, duration, and cause of each episode of illness. Of the episodes of viral URTI, six of seven episodes in squash athletes, and all five episodes in hockey athletes were preceded, within two days, by a decrease in IgA concentration, by an average of 20-25%, during training sessions. In contrast, IgA concentration tended to either increase or to decrease only slightly (< 10%) in athletes who did not develop URTI within two days of sampling. Although obtained on a relatively small sample of only two types of elite athletes, these data provide suggestive evidence of a link between exercise-induced decreases in salivary IgA and subsequent URTI. Whether such association occurs in all endurance athletes and is directly related to susceptibility to URTI awaits further, larger studies on athletes from various sports.

Plasma Glutamine.   Lymphocytes rely on glutamine as an energy source as well as for nucleotide synthesis. Low plasma glutamine levels have been associated with immune suppression after trauma such as burns or surgery (41). Marginally lower plasma glutamine concentration has been reported in athletes showing symptoms of overtraining syndrome compared with well-trained athletes (47). Plasma glutamine concentration decreases acutely after exercise (24, 47) and declines markedly during intensive training (24). Because of the apparent requirement of lymphocytes for glutamine, it has been suggested that low plasma glutamine levels associated with intensive training and overtraining may compromise lymphocyte function and possibly contribute to an increased incidence of infectious illness in competitive athletes (24, 41, 47).

Parry-Billings et al. (47) reported a 9% lower plasma glutamine concentration in 40 athletes from various endurance sports who were clinically diagnosed as overtrained (e.g., poor performance, fatigue lasting longer than three weeks, mood-state disturbances) compared with similar nonovertrained athletes. However, despite lower plasma glutamine concentration, lymphocytes from the overtrained athletes exhibited normal proliferative responses to mitogenic challenge. Thus, the lower resting plasma glutamine concentration in overtrained athletes was not reflected in impaired lymphocyte function measured in an in vitro system, although the possibility of impaired in vivo function could not be discounted.

In endurance-trained male soldiers, plasma glutamine concentration decreased by 48% during 10 days of intensified interval training (24). Glutamine levels remained depressed during four days of recovery, and returned to baseline only after five days of recovery (reduced training). By the end of the 10 days of intensified training, subjects showed clear signs of overreaching, including decrements in running performance and time to exhaustion, mood-state disturbances (POMS), and self-reported symptoms such as fatigue, sleep disturbances, difficulty concentrating on specific tasks, and loss of appetite. However, despite the decreasing plasma glutamine concentration during intensified training and recovery, there was little evidence of immunosuppression. As previously described, lymphocyte activation (number of CD25+ and HLA-DR+ cells) and IL-2 production appeared to increase, and the only evidence of suppressed immune function was a decline in NK cell numbers.

Although plasma glutamine concentration may decrease in response to prolonged periods of intensive exercise training, it is still unclear whether such changes are related to immunosuppression in athletes. In a recent study, 24 elite male and female swimmers underwent four weeks of intensified training (increased swim volume and resistance work) (38). Eight or 33% of the swimmers (six female, two male) exhibited symptoms of overreaching, including performance decrements and high fatigue ratings in daily log books along with accompanying comments consistent with the presence of the syndrome. Plasma glutamine concentration increased significantly, by approximately 20%, from before to after the four weeks of increased training only in the well-trained swimmers, remaining unchanged in the overreached swimmers. Glutamine level was significantly lower in the overreached compared with well-trained swimmers after two weeks. Although 10 of the 24 swimmers, or 42%, exhibited symptoms of URTI during the four weeks, plasma glutamine concentration did not differ between those exhibiting URTI and those who did not. Taken together, these data suggest that, although the concentration may differ between overtrained or overreached and well-trained athletes, plasma glutamine level may not necessarily decrease during extended periods of intensified training, and that the appearance of upper respiratory tract infection is not related to changes in plasma glutamine concentration during intensified training in elite athletes.

Cytokines.   Cytokine production appears to increase acutely in response to exercise (45), as shown by elevation in urinary excretion of substances such as IL-1β, IL-6, TNF-α, and IFN-γ after prolonged exercise such as 20 km running (58). The appearance of pro-inflammatory cytokines such as IL-1, IL-6 and TNF-α after prolonged weight-bearing exercise such as running is possibly due to local inflammatory processes within muscle fibers damaged by intensive exercise. For example, IL-1β has been shown to localize to damaged skeletal muscle fibers after eccentric exercise (6). Overtraining is associated with persistent muscle soreness (28), suggestive of muscle fiber damage and possibly local inflammation (55). Thus, elevated production of the pro-inflammatory cytokines may be expected during overtraining. Alternatively, there is evidence of down-regulation of the general inflammatory response in athletes. For example, lower serum concentration of complement components C3 and C4 (42) have been reported in athletes compared with nonathletes. As noted earlier, neutrophil activity also appears to be suppressed as a result of intensive exercise training (18, 50, 56). Moreover, training attenuates the exercise-induced release of acute phase reactants, glycoproteins associated with inflammation (4, 29).

Few studies have examined the response of cytokines to prolonged periods of intensive training or overtraining. In a recent study on endurance-trained soldiers, serum IL-2 concentration increased by threefold after 10 days intensive interval running training (twice-daily sessions of maximal sprints), and further increased sixfold during a five-day recovery period of light exercise (12). In vitro IL-2 production by mitogen-stimulated lymphocytes also increased ninefold during the 10 days of intensified training, remaining at that level throughout the five-day recovery period. All subjects showed clear symptoms of overreaching, including high fatigue, decrements in exercise performance and mood-state disturbances.

In a recent paper, expression of the high affinity IL-2 receptor α-chain was higher in lymphocytes from endurance-trained compared with sedentary men (51); the number of cells expressing the lower affinity IL-2 receptor β-chain did not differ between groups. The number of NK cells (CD56$^+$) with the high affinity receptor was also higher in trained compared with untrained subjects. Aerobic power ($\dot{V}O_2$max) was highly correlated (r = .91) with β-chain expression. These data suggest that IL-2 receptor expression is enhanced by endurance training. However, the significance, if any, of increased IL-2 production and receptor expression is not known, and it is unclear whether secretion of, or receptors for, other cytokines change during overtraining.

Does Illness Contribute to Overtraining?

Frequent illness, especially viral URTI, is considered a common symptom or outcome of overtraining, and much attention has focused on understanding immunomodulation accompanying overtraining. However, less frequently asked is the converse: Does illness due to intensive training cause or contribute to overtraining? There is some similarity between symptoms of infectious illness and overtraining, such as persistent fatigue, poor performance, lethargy, and increased sleep (10, 28, 49, 52, 53).

Roberts (52) presented several case studies of athletes performing poorly for no obvious reasons. In these athletes there was evidence of prior viral infection such as high antiviral, antibody titers, despite the absence of clinical symptoms of illness. Febrile illness causes decrements in submaximal exercise capacity (3) and muscular strength (3, 9). For example, muscular strength declines by 10-30% during febrile illness, returning to normal levels upon recovery (3). Infection may also disrupt cellular structures or energy metabolism in cardiac and skeletal muscle (2). Moreover, IL-1 released during infection induces sleep and stimulates muscle proteolysis, which may contribute to feelings of fatigue, lethargy, and muscle soreness; stimulation of proteolysis may also limit the capacity of skeletal muscle to adapt to exercise training.

Thus, the presence of subclinical illness may mimic or cause some symptoms of overtraining, in particular the inability to train effectively, poor performance, persistent fatigue, lethargy, and muscle soreness. The presence or absence of infectious illness should be discounted or documented when diagnosing overtraining among athletes.

Mechanisms to Explain Immune Suppression in Overtraining and Intensive Training

There are several possible mechanisms to explain the apparent immunosuppression observed during overtraining and prolonged periods of intensive training; it is unlikely that a single mechanism is responsible, given the complexity of the immune system and its response to exercise. It is likely that neuroendocrine changes accompanying overtraining are very much related to alterations in immune function.

Low resting leukocyte and lymphocyte counts may reflect increased turnover and a shorter life span of these cells. Increased turnover observed in red blood cells (57) has been attributed to mechanical, oxidative, and osmotic stress during intensive exercise. It is unknown whether leukocyte turnover also increases in response to these stresses. In reporting low leukocyte, neutrophil, and lymphocyte counts in cyclists compared with clinical norms, Keen et al. (25) noted neutrophil morphology indicative of a younger cell population, consistent with increased turnover. Alternatively, intensive exercise causes transient decreases in lymphocyte proliferation (40). It is possible that low cell counts result from frequent exercise-induced suppression of cell proliferation, since most competitive athletes train intensively at least once per day.

In human studies, immune cells are sampled from peripheral blood that contains only a small percentage (less than 2%) of the body's immune cells at any given time. During and after exercise, leukocytes and lymphocyte subsets are redistributed among the circulation and various lymphoid and other tissues; for example, during exercise leukocytes are released from marginated pools in the lungs and spleen in response to increased catecholamine levels. After exercise, some leukocytes, especially neutrophils, monocytes, and NK cells, may migrate to sites of injury or inflammation in skeletal muscle, possibly in response to inflammatory mediators (e.g., IL-1) released locally (6). It is possible that frequent intensive exercise enhances movement of cells to tissue sites, effectively lowering their concentration in the blood. Such mechanism may explain the progressive decline in peripheral blood NK cell number observed during periods of intensive training (discussed previously).

Regulation of neutrophil function is complex. A variety of hormones and cytokines are known to influence neutrophil activity, including catecholamines, corticosteroids, β-endorphin, IL-1, TNF-α, and colony stimulating factor (CSF) (see 18, 50, 56); some of these factors change during overtraining and intensive training (63; also discussed in chapters 10 and 12). Decreased neutrophil function during periods of intensive training may relate to changes in hormones or cytokines (18, 56) or to the presence of newly matured cells of lower functional capacity in the circulation (50, 56). For example, epinephrine inhibits several aspects of neutrophil function. In cyclists, the higher epinephrine concentration observed during intensive compared with moderate training, and significant negative correlation between epinephrine and neutrophil oxidative activity are consistent with a suppressive effect of this hormone (18) on neutrophil function. Neutrophils are important cells in the inflammatory process, appearing quite early in sites of tissue injury and inflammation, including skeletal muscle after eccentric exercise (6). It has been suggested (56) that partial suppression of neutrophil activity resulting from intensive training may be a positive adaptation in one sense, by limiting neutrophil involvement in inflammatory processes within tissue, particularly skeletal muscle.

The mechanisms responsible for low secretory and serum Ig concentrations observed during overtraining and periods of intensive training are likely to involve complex interaction of several factors (37). Regulation of Ig synthesis involves input from the neuroendocrine system. Ig and antibody synthesis can be influenced by the neuroendocrine system both directly, via B cells, and indirectly, via regulatory T cells or via control of blood flow and distribution of immune cells throughout the body. Antigen-producing B (and other immune) cells express β-adrenergic receptors, and norepinephrine enhances antibody synthesis in response to antigen exposure. Exercise training is associated with down-regulation of β-adrenergic receptors, and catecholamine depletion has been observed during intensive training and overtraining (27, 63; see also chapters 10 and 12). It is possible that, during overtraining, catecholamine depletion coupled with down-regulation of β-adrenergic receptors may result in impairment of Ig and antibody synthesis.

The mechanisms responsible for decreases in secretory IgA during overtraining and intensive training are likely to be equally complex and to also involve neuroendocrine factors. Intensive exercise affects only secretory Ig (i.e., IgA and IgM, but not IgG in mucosal secretions), suggesting a specific effect on the local production of secretory Ig. It is possible that intensive exercise alters the process of local IgA secretion, perhaps via changes in soluble mediators (e.g., hormones or cytokines) or by mechanical or structural changes to the mucosal epithelium resulting from high ventilation over prolonged periods. Sympathetic output may influence flow of saliva, the vehicle for IgA secretion, and thus the amount and concentration of IgA appearing on oral and nasal mucosal surfaces exposed to pathogens causing URTI. Sympathetic output may also affect migration of IgA-secreting B cells to the oral submucosal areas via vasoconstriction of blood flow to the region.

Overtraining also has a significant psychological component, and is characterized by mood-state changes (see chapter 16). Psychological stress influences secretory IgA levels; for example, in university students salivary IgA concentration decreases with increasing perceived stress levels during the academic year and is lower in students with poor coping strategies (39). Susceptibility to URTI is influenced by both psychological and physical stress, and it is likely that susceptibility to URTI among highly trained athletes reflects influence of both physical and psychological factors.

Other endocrine changes occurring during intensive training and overtraining may include increased corticosteroid, and decreased androgen (in males) and growth hormone levels (chapters 10 and 12; see also 63). These hormones may have profound effects, the first generally suppressing and the latter two stimulating immune function. Lymphocytes express receptors for several hormones including catecholamines, steroids, endorphins, and enkaphalins, all of which are released acutely during exercise. Moreover, most lymphoid tissues are innervated by sympathetic neurons, with lymphocytes in close proximity to nerve endings.

Summary

Intensive exercise training is associated with an increased risk of viral illness, in particular URTI. It is unclear at present whether the high incidence of URTI in athletes is a consequence of overtraining per se or simply results from the high volume, high intensity training undertaken by most high-performance athletes.

Large prospective studies are needed to determine the contribution of overtraining to the risk of infectious illness, and conversely, whether frequent illness contributes to the symptoms of overtraining.

Few studies have documented both overtraining and immune function in athletes during periods of intensive training. Secretory IgA and plasma glutamine concentrations have been shown to be lower in overtrained compared with well-trained athletes. In the past few years, evidence has begun to suggest suppression of several immune parameters during long periods of intensive training. Intensive training has been associated with low resting leukocyte and lymphocyte counts compared with clinical norms, decreasing NK cell number, low serum and secretory Ig concentrations, and suppression of neutrophil antimicrobial activity. To date, the only exercise-induced change in an immune parameter that has been directly associated with the appearance of URTI in athletes is secretory IgA (35); however, few studies have attempted to correlate changes in specific immune parameters with the incidence of URTI. Although many of these alterations are relatively small, and athletes are not immune deficient in a clinical sense, it is possible that immune function is compromised by the additive and interactive effects of small changes in several parameters important to host defense.

The mechanisms for such changes are unknown at present and are likely to be complex. Overtraining involves both physical and psychological causes and consequences. It is probable that the hormonal changes accompanying overtraining and overreaching play a significant role in immunomodulation during overtraining.

References

1. Bosenberg, A.T., J.G. Brock-Utne, S.L. Gaffin, M.T.B. Wells, G.T.W. Blake. 1988. Strenuous exercise causes systemic endotoxemia. *Journal of Applied Physiology* 65: 106-108.
2. Cabinian, A.E., R.J. Kiel, F. Smith, K.L. Ho, R. Khatib, M.P. Reyes. 1990. Modification of exercise-aggravated coxsackie virus B3 murine myocarditis by T lymphocyte suppression in an inbred model. *Journal of Laboratory and Clinical Medicine* 115: 454-462.
3. Daniels, W.L., D. S. Sharp, J.E. Wright, J.A. Vogel, G. Friman, W.R. Beisel, J.J. Knapik. 1985. Effects of virus infection on physical performance in man. *Military Medicine* 150: 1-8.
4. Dufaux, B., U. Order, H. Geyer, W. Hollmann. 1984. C-reactive protein serum concentration in well-trained athletes. *International Journal of Sports Medicine* 5: 102-106.
5. Ferry, A., F. Picard, A. Duvallet, B. Weill, M. Rieu. 1990. Changes in blood leucocyte populations induced by acute maximal and chronic submaximal exercise. *European Journal of Applied Physiology* 59: 435-442.
6. Fielding, R.A., T.J. Manfredi, W. Ding, M.A. Fiatarone, W.J. Evans, J.G. Cannon. 1993. Acute phase response in exercise III. Neutrophil and IL-1 (accumulation in skeletal muscle). *American Journal of Physiology* 265: R166-R172.
7. Fitzgerald, L. 1991. Overtraining increases the susceptibility to infection. *International Journal of Sports Medicine* 12: S5-S8.
8. Flynn, M.G., F.X. Pizza, J.B. Boone Jr., F.F. Andres, T.A. Michaud, J.R. Rodriguez-Zayas. 1994. Indices of training stress during competitive running

and swimming seasons. *International Journal of Sports Medicine* 15: 21-16.

9. Friman, G. 1977. Effect of acute infectious disease on isometric muscle strength. *Scandinavian Journal of Clinical and Laboratory Investigation* 37: 303-308.

10. Fry, R.W., A.R. Morton, D. Keast. 1991. Overtraining in athletes: an update. *Sports Medicine* 12: 32-65.

11. Fry, R.W., A.R. Morton, G.P.M. Crawford, D. Keast. 1992. Cell numbers and in vitro responses of leucocytes and lymphocyte subpopulations following maximal exercise and interval training sessions of different intensities. *European Journal of Applied Physiology* 64: 218-227.

12. Fry, R.W., J.R. Grove, A.R. Morton, P.M. Zeroni, S. Gauderi, D. Keast. 1994. Psychological and immunological correlates of acute overtraining. *British Journal of Sports Medicine* 28: 241-246.

13. Galun, E., R. Burstein, E. Assia, I. Tur-Kaspa, J. Rosenblum, Y. Epstein. 1987. Changes of white blood cell count during prolonged exercise. *International Journal of Sports Medicine* 8: 252-255.

14. Garagiola, U., M. Buzzetti, E. Cardella, F. Confalonieri, E. Giani, V. Polini, P. Ferrante, R. Mancuso, M. Montanari, E. Grossi, A. Pecori. 1995. Immunological patterns during regular intensive training in athletes: quantification and evaluation of a preventive pharmacological approach. *Journal of International Medical Research* 23: 85-95.

15. Gleeson, M., W.A. McDonald, A.W. Cripps, D.B. Pyne, R.L. Clancy, P.A. Fricker. 1995. The effect on immunity of long term intensive training in elite swimmers. *Clinical and Experimental Immunology* 102: 210-216.

16. Gmunder, F.K., P.W. Joller, H.I. Joller-Jemelka, B. Bechler, M. Cogoli, W.H. Ziegler, J. Muller, R.E. Aeppli, A. Cogoli. 1990. Effect of a herbal yeast food supplements and long-distance running on immunological parameters. *British Journal of Sports Medicine* 24: 103-112.

17. Green, R.L., S.S. Kaplan, B.S. Rabin, L. Stanitski, U. Zdziarski. 1981. Immune function in marathon runners. *Annals of Allergy* 47: 73-75.

18. Hack, V., G. Strobel, M. Weiss, H. Weicker. 1994. PMN cell counts and phagocytic activity of highly trained athletes depend on training period. *Journal of Applied Physiology* 77: 1731-1735.

19. Hackney, A.C., S.N. Pearman, J.M. Nowacki. 1990. Physiological profiles of overtrained and stale athletes: a review. *Applied Sport Psychology* 2: 21-33.

20. Heath, G.W., C.A. Macera, D.C. Nieman. 1992. Exercise and upper respiratory tract infections: is there a relationship? *Sports Medicine* 14: 353-365.

21. Hooper, S., L.T. Mackinnon, R.D. Gordon, A.W. Bachmann. 1993. Hormonal responses of elite swimmers to overtraining. *Medicine and Science in Sports and Exercise* 25: 741-747.

22. Hooper, S., L.T. Mackinnon, A. Howard, R.D. Gordon, A.W. Bachmann. 1995. Markers for monitoring overtraining and recovery in elite swimmers. *Medicine and Science in Sports and Exercise* 27: 106-112.

23. Kajiura, J.S., J.D. MacDougall, P.B. Ernest, E.V. Younglai. 1995. Immune response to changes in training intensity and volume in runners. *Medicine and Science in Sports and Exercise* 27: 1111-1117.

24. Keast, D., D. Arstein, W. Harper, R.W. Fry, A.R. Morton. 1995. Depression of plasma glutamine concentration after exercise stress and its possible influence on the immune system. *Medical Journal of Australia* 162: 15-18.

25. Keen, P., D.A. McCarthy, L. Passfield, H.A.A. Shaker, A.J. Wade. 1995. Leucocyte and erythrocyte counts during a multi-stage cycling race ('The Milk Race'). *British Journal of Sports Medicine* 29: 61-65.

26. Lehmann, M., H.H. Dickhuth, G. Gendrisch, W. Lazar, M. Thum, R. Kaminski, J.F. Aramendi, E. Peterke, W. Wieland, J. Keul. 1991. Training-overtraining: a prospective, experimental study with experienced middle- and long-distance runners. *International Journal of Sports Medicine* 12: 444-452.

27. Lehmann, M., P. Baumgartl, C. Wiesenack, A. Seidel, J. Baumann, S. Fischer, U. Spori, G. Genmdrisch, R. Kaminski, J. Keul. 1992. Training-overtraining: influence of a defined increase in training volume vs training intensity on performance, catecholamines and some metabolic parameters in experienced middle- and long-distance runners. *European Journal of Applied Physiology* 64: 169-177.

28. Lehmann, M., C. Foster, J. Keul. 1993. Overtraining in endurance athletes: a brief review. *Medicine and Science in Sports and Exercise* 25: 854-862, 1993.

29. Liesen, H., B. Dufaux, W. Hollmann. 1977. Modifications of serum glycoproteins the days following a prolonged physical exercise and the influence of physical training. *European Journal of Applied Physiology* 37: 243-254.

30. Mackinnon, L.T. 1989. Exercise and natural killer cells: what is the relationship? *Sports Medicine* 7: 141-149.

31. Mackinnon, L.T., T.W. Chick, A. van As, T.B. Tomasi. 1989. Decreased secretory immunoglobulins following intensive prolonged exercise. *Sports Training, Medicine, and Rehabilitation* 1: 1-10.

32. Mackinnon, L.T., E. Ginn, G. Seymour. 1991. Effects of exercise during sports training and competition on salivary IgA levels. In *Behaviour and immunity*, ed. A.J. Husband, 169-177. Boca Raton, FL: CRC Press.

33. Mackinnon, L.T., D.G Jenkins. 1993. Decreased salivary IgA after intensive interval exercise before and after training. *Medicine and Science in Sports and Exercise* 25: 678-683.

34. Mackinnon, L.T., E. Ginn, G.J. Seymour. 1993. Decreased salivary immunoglobulin A secretion rate after intensive interval training in elite kayakers. *European Journal of Applied Physiology* 67: 180-184.

35. Mackinnon, L.T., E. Ginn, G.J. Seymour. 1993. Temporal relationship between exercise-induced decreases in salivary IgA and subsequent appearance of upper respiratory tract infection in elite athletes. *Australian Journal of Science and Medicine in Sport* 25: 94-99.

36. Mackinnon, L.T., S. Hooper. 1994. Mucosal (secretory) immune system responses to exercise of varying intensity and during overtraining. *International Journal of Sports Medicine* 15: S179-S183.

37. Mackinnon, L.T. 1996. Exercise, immunoglobulin and antibody. *Exercise Immunology Review* 2: 1-35.

38. Mackinnon, L.T., S.L. Hooper. 1996. Plasma glutamine and upper respiratory tract infection during intensified training in swimmers. *Medicine and Science in Sports and Exercise* 28: 285-290.

39. McLelland, D.C., E. Floor, R.J. Davidson, C. Saron. 1980. Stressed power motivation, sympathetic activation, immune function and illness. *Journal of Human Stress* 6: 11-19.

40. MacNeil, B., L. Hoffman-Goetz, A. Kendall, M. Houston, Y. Arumugam. 1991. Lymphocyte proliferation responses after exercise in men: fitness, intensity and duration effects. *Journal of Applied Physiology* 70: 179-185.

41. Newsholme, E.A. 1993. Biochemical mechanisms to explain immunosuppression in well-trained and overtrained athletes. *International Journal of Sports Medicine* 15: S142-S147.

42. Nieman, D.C., S.A. Tan, J.W. Lee, L.S. Berk. 1989. Complement and immunoglobulin levels in athletes and sedentary controls. *International Journal of Sports Medicine* 10: 124-128.

43. Nieman, D.C. 1994. Exercise, upper respiratory tract infection, and the immune system. *Medicine and Science in Sports and Exercise* 26: 128-139.

44. Nieman, D.C., K.S. Buckley,. D.A Henson, B.J. Warren, J. Suttles, J.C. Ahle, S. Simandle, O.R. Fagoaga, S.L. Nehlsen-Cannarella. 1995 Immune function in marathon runners versus sedentary controls. *Medicine and Science in Sports and Exercise* 27: 986-992.

45. Northoff, J., C. Weinstock, A. Berg. 1994. The cytokine response to strenuous exercise. *International Journal of Sports Medicine* 15: S167-S171.

46. Osterud, B., J.O. Olsen, L. Wilsgard. 1989. Effect of strenuous exercise on blood monocytes and their relation to coagulation. *Medicine and Science in Sports and Exercise* 21: 374-378.

47. Parry-Billings, M., R. Budgett, Y. Koutedakis, E. Blomstrand, S. Brooks, C. Williams, P.C. Calder, S. Pilling, R. Baigrie, E.A. Newsholme. 1992. Plasma amino acid concentrations in the overtraining syndrome: possible effects on the immune system. *Medicine and Science in Sports and Exercise* 24: 1353-1358.

48. Pedersen, B.K., N. Tvede, L.D. Christensen, K. Klarlund, S. Kragbak, J. Halkjær-Kristensen, 1989. Natural killer cell activity in peripheral blood of highly trained and untrained persons. *International Journal of Sports Medicine*, 10: 129-131.

49. Puffer, J.C., J.M. McShane. 1992. Depression and chronic fatigue in athletes. *Clinics in Sports Medicine* 11: 327-338.

50. Pyne, D.B., M.S. Baker, P.A. Fricker, W.A. McDonald, R.D. Telford, M.J. Weidemann. 1995. Effects of an intensive 12-wk training program by elite swimmers on neutrophil oxidative activity. *Medicine and Science in Sports and Exercise* 27: 536-542.

51. Rhind, S.G., P.N. Shek, S. Shinkai, R.J. Shephard. 1994. Differential expression of interleukin-2 receptor alpha and beta chains in relation to natural killer cell subsets and aerobic fitness. *International Journal of Sports Medicine* 15: 911-918.

52. Roberts, J.A. 1985. Loss of form in young athletes due to viral infection. *British Medical Journal* 290: 357-358.

53. Roberts, J.A., J.A. Wilson, G.B. Clements. 1988. Virus infection and sports performance: a prospective study. *British Journal of Sports Medicine* 22: 161-162.

54. Shephard, R.J., P.N. Shek. 1994. Infectious diseases in athletes: new interest for an old problem. *Journal of Sports Medicine and Physical Fitness* 34: 11-22.

55. Smith, L.L. 1991. Acute inflammation: the underlying mechanism in delayed onset muscle soreness? *Medicine and Science in Sports and Exercise* 23: 542-551.

56. Smith, J.A., R.D. Telford, I.B. Mason, M.J. Weidemann. 1990. Exercise, training and neutrophil microbicidal activity. *International Journal of Sports Medicine* 11: 179-187.

57. Smith, J.A. 1995. Exercise, training and red blood cell turnover. *Sports Medicine* 19: 9-31.
58. Sprenger, H., C. Jacobs, M. Nain, A.M. Gressner, H. Prinz, W. Wesemann, D. Gemsa. 1992. Enhanced release of cytokines, interleukin-2 receptors, and neopterin after long-distance running. *Clinical Immunology and Immunopathology* 63: 188-195.
59. Tharp, G.D., M.W. Barnes. 1990. Reduction of saliva immunoglobulin levels by swim training. *European Journal of Applied Physiology* 60: 61-64.
60. Tomasi, T.B., F.B. Trudeau, D. Czerwinski, S. Erredge. 1982. Immune parameters in athletes before and after strenuous exercise. *Journal of Clinical Immunology* 2: 173-178.
61. Tomasi, T.B., A.G. Plaut. 1985. Humoral aspects of mucosal immunity. In *Advances in host defense mechanisms*, vol. 4, eds. J.I. Gallin, A.S. Fauci, 31-61. New York: Raven Press.
62. Tvede, N., J. Steensberg, B. Baslund, J. Halkjær-Kristensen, B.K. Pedersen. 1991. Cellular immunity in highly trained elite racing cyclists during periods of training with high and low intensity. *Scandinavian Journal of Medicine and Science in Sports* 1: 163-166.
63. Urhausen, A., H. Gabriel, W. Kindermann. 1995. Blood hormones as markers of training stress and overtraining. *Sports Medicine* 20: 251-276.
64. Verde, T., S. Thomas, R.J. Shephard. 1992. Potential markers of heavy training in highly trained endurance runners. *British Journal of Sports Medicine*, 26: 167-175.
65. Verde, T.J., S.G. Thomas, P.N. Shek, and R.J. Shephard. 1993. The effects of heavy training on two in vitro assessments of cell-mediated immunity in conditioned athletes. *Clinical Journal of Sports Medicine* 3: 211-216.
66. van Borselen, F., N.H. Vos, A.C. Fry, W.J. Kraemer. 1992. The role of anaerobic exercise in overtraining. *National Strength and Conditioning Association Journal* 14: 74-79.
67. Weidner, T.G. 1994. Literature review: upper respiratory illness and sport and exercise. *International Journal of Sports Medicine* 15: 1-9.
68. Welliver, R.C., P.L. Ogra. 1988. Immunology of respiratory viral infections. *Annual Review of Medicine* 39: 147-162.

CHAPTER 12

Potential Interventions to Prevent Immunosuppression During Training

Elena P. Gotovtseva, PhD, Ida D. Surkina, PhD, and Peter N. Uchakin, PhD

Introduction

This chapter will review the incidence of respiratory infection and overreaching or overtraining in elite athletes during different stages of a year-round training cycle and during a four-year Olympic cycle. Data will indicate that athletes experience changes in immune function over different periods of training and racing, as well as demonstrate the association of the immune alterations with the high risk of respiratory infection and the onset of overreaching. The next section of this chapter will present data on the potential use of immunomodulators and antiviral medicines in athletes who are immunocompromized or highly prone to respiratory infection. As will be discussed, the immunopotentiating medicines produce differential effects on the immune function predominantly through modulating cell-mediated immunity.

The close relationship between the immune and neuroendocrine systems in adaptation to physiological and psychological stress in athletes will also be discussed. Based on the mutual role of these systems in preserving homeostasis, the neuroendocrine and immune profiles of competitive athletes during different training situations will be analyzed. Pilot data on the improvement of adaptation to athletic stress by the means of immunomodulation will be presented. Finally, the important role of trace elements, vitamins, and glutamine on the athlete's immune function will be reviewed.

The Use of Immunomodulators, Interferon Preparations, and Inducers in Prevention of Immunosuppression and Respiratory Infections in Athletes

Immune Suppression and Incidence of Viral Respiratory Infection in Elite Athletes

In recent years it has become clear that extreme athletic stress may alter the immune function in elite athletes. In this regard, elite athletes have been reported to be highly susceptible to respiratory infections during periods of heavy training and major competitions (28, 95). For example, medical service during the 1980 Winter Olympic Games reported (15) that nearly two-thirds of athletes' visits were for upper respiratory tract infections (URTI). Moreover, results of a longitudinal epidemiological study involving more than 500 elite Soviet athletes during the four years of training preceding the 1988 Olympic Games showed that the incidence of URTI was almost twice as high during the last two pre-Olympic years compared to the first two years (32). Respiratory infections were represented by acute viral infection and influenza (49.1%), laryngotracheitis (26.0%), bronchitis and tracheobronchitis (5%), and rhinitis (11.3%). Recurrent and protracted respiratory infections were reported in 15% of all cases. Attempts to find a relationship between the incidence of respiratory infection and immune alteration showed that the status of an athlete's immune system changed over the year-round training cycle depending on the stage of training, the athlete's fitness, and the initial state of immune function (4, 82).

However, other studies indicate that well-trained elite athletes possess a high level of adaptive mechanisms to physiological stress in that heavy endurance training did not produce significant immune changes. No alterations in immune function, as measured by the mitogen phytohemagglutinin (PHA)-induced lymphocyte proliferative activity, PHA-induced production of interleukin-2 (IL-2) and interferon-γK (IFN-γ), and Newcastle disease virus (NDV)-induced interferon-α (IFN-α) in vitro, were found in Soviet national- and regional-level cross-country skiers and speed skaters following several weeks of high-volume base training (34, 90).

Similar observations have been reported by Wolfarth et al. (97) and Tvede et al. (89). Neither changes in URTI rate nor immune alterations were registered in German national-and regional-level cross-country skiers and road cyclists, or in elite Danish racing cyclists during an eight-week and a four-week endurance training program, respectively. These data provide evidence that an elite athlete's immune system may be able to adequately adapt to heavy endurance training.

The high immunologic tolerance to stress is mostly observed at the beginning of a year-round training cycle, when the athlete's immune system is not yet exhausted. However, the further increase in training loads in the preparatory period, or the combined effect of physiological and psychological stress of competitions may result either in transient or sustained immunosuppression (53, 55). Monitoring of lymphokine production in U.S. elite and collegiate female middle- and long-distance runners revealed a progressive reduction of PHA-induced IFN-γ and IL-2 production in vitro during the periods of heavy training and competitions as compared to easy training (35) (see figure 12.1).

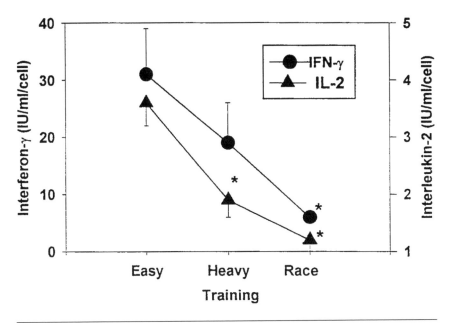

Figure 12.1 Whole blood PHA-induced interferon-γ and interleukin-2 production, adjusted to lymphocyte number, in 34 U.S. female long-distance runners over the periods of easy training, heavy training (alternative high mileage and intensive interval running), and racing/training. *p < .05.
Based on data from Gotovtseva et al. 1996.

Surprisingly, the incidence of colds among runners did not correlate with immune changes; 40% of the URTI were registered during easy training in comparison with 25% and 28% URTI reported during heavy training and racing/training periods, respectively. Although immune suppression may predispose athletes to viral infection, it seems that actual infections are due to a combination of immune status and environmental factors relating to exposure to virus.

Irrational training, exceeding the athlete's adaptive limits, may result in the accumulation of fatigue with the onset of overreaching and overtraining. One-third of 500 elite Soviet athletes were overreached and overtrained during the four-year training cycle prior to the 1988 Olympic Games (32). In the 1988 Olympic year, the incidence of overreaching and overtraining was seven times higher, than in the first year of a four-year training cycle. A high incidence of overreaching has been observed in rowers of academic style (39%) and in swimmers (27%). The incidence increased to 44% in swimmers during the 1988 Olympic year. The same high incidence of overreaching in swimmers (33%) has been reported by Mackinnon and Hooper (55).

Immunosuppression has been suggested as one of the markers of overreaching and overtraining (28). In our own studies with Russian regional-level cross-country skiers, we found that overreaching in athletes was accompanied by immune alterations (25, 34, 90). Following four weeks of intensified training with 60-80 km/d of skiing, athletes developed the whole spectrum of overreaching symptoms, including sustained fatigue, poor sleep, a loss of desire for training, gastrointestinal problems, dizziness, increased blood pressure, nose bleeds, and heart pain. Both cell-mediated and humoral immunity were suppressed in these athletes at the end of the training cycle as compared to the baseline (see figure 12.2).

The immune dysfunction and clinical signs of overreaching (sustained fatigue, poor sleep, and no desire for training) were less pronounced in another group of male cross-country skiers, who were followed during three weeks of training with an average of 40 km/d of skiing (90). Despite the similarity in the impaired cytokine production (IFN-γ, IFN-α, and IL-2), the PHA-induced lymphocyte proliferative response was significantly increased in these athletes. This points to disregulation in immune function in overreached athletes. Since all skiers were of the same fitness, the difference in the severity of overreached symptoms and in some immunologic variables was likely due to the diverse volume of training. During that time, URTIs were registered in most of the athletes (i.e., in 100% of the first team and in 75% of the second team).

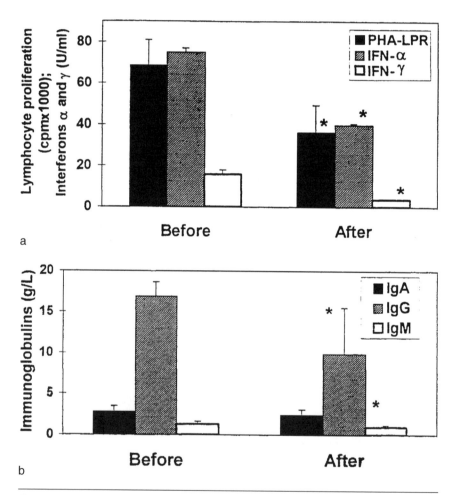

Figure 12.2   Cell-mediated (a) and humoral immune (b) response to a 4-week intensified training period in six regional-level male cross-country skiers. PHA-LPR = PHA-induced lymphocyte proliferative response. *p < .05.
Based on data from references 25 and 34.

The competitive season is yet another period of the year-round training cycle when athletes are highly prone to respiratory infection. During this time a whole complex of factors may affect the athlete's immune function, including a lack of rest due to the tight schedule of races, combined physiological and psychoemotional stress of competition, as well as exposure to an unfamiliar environment.

No recovery in altered IL-2, IFN-γ, and IFN-α production in vitro was found in Russian regional-level cross-country skiers eight days following a series of four races (90), or in U.S. elite and collegiate female long-distance runners 5-8 days following two consecutive weekends with races (35). A high rate of respiratory infection was registered in cross-country skiers (57%, winter) and runners (28%, spring or fall) during those competitive periods. It appeared that more than one week of recovery should be allowed in order to recuperate the athletes' immune function.

In another study with elite Soviet female speed skaters, the tendency to recovery in PHA-induced lymphocyte proliferative response, and the complete restoration of stimulated production of IFN-γ and IFN-α in vitro were found following two weeks of decreased training after a major competition (24, see figure 12.3). The duration of recuperative processes may be even longer in the overreached and overtrained athletes. In regard to other symptoms of overreaching, complete rest for 3-5 weeks in underperforming British elite athletes improved their aerobic capacity, but only slightly reduced negative states of tension, depression, fatigue, anger, and confusion (47).

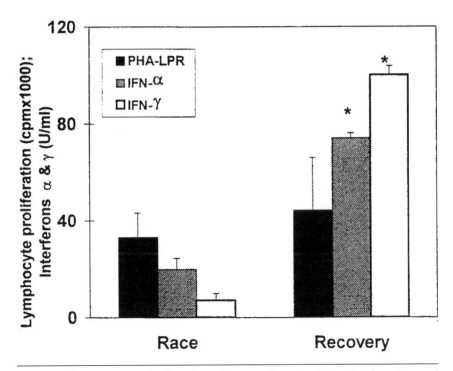

Figure 12.3   Recovery of cell-mediated immune function in ten elite female speed skaters following a 2-week reduced training after major competition. *p < .05.
Based on data from Ershov and Gotovtseva 1988.

Potential Immunomodulators for Practical Use in Athletes

Stress-related immune dysfunction and its clinical significance became the primary impetus for developing effective countermeasures in such life sciences as sports medicine, space, and military medicine. The preventive methods and tactics of an athlete's behavior during the infection episode have been discussed in the previous chapter. The following discussion emphasizes the potential pharmacological and nutritional interventions to prevent sustained immunosuppression and to reduce the high rate of respiratory infection reported among competitive athletes.

Although adequate rest, well-balanced diet, hydrotherapy, and physiotherapy are the first remedies to be considered for practical use in athletes, these measures may not be sufficient to provide a speedy and complete recovery. Vaccination, as well as the use of antiviral and immunomodulatory medicines have been proposed as prophylactics and to aid the immune function recovery in immunocompromised athletes. An annual flu shot has been consistently recommended, although to our knowledge, there is no scientific data to prove its efficacy in athletes. Research in this area would be valuable, since there is evidence that psychoemotional stress can modulate the immune response to a viral vaccine (45). Accordingly, prophylactic measures, other than flu shots may be more effective for athletes.

Extensive efforts have been devoted to the search for chemical and biological agents capable of enhancing immunity and possessing antiviral efficacy. Nowadays a number of such medicines are available for clinical use. They represent a wide spectrum of natural and recombinant medicines, as well as chemical compounds, including thymic hormones, interferons, and well-known drugs with immunopotentiating activity: Levamisole, Diuciphon, and Dipyridamole.

Thymic Hormones.   It has been known for years that thymic extracts have immunomodulatory activity. Several preparations have been obtained from partially purified bovine thymic extracts or synthesized in the laboratory. These natural and synthetic polypeptides, including thymostimulin, thymosin fraction 5, thymosin-α_1, thymulin, thymalin, and thymogen have been fully characterized and studied in clinics all over the world (33, 61). The biological properties of thymic factors include the induction of T cell differentiation, maturation, and proliferation. These polypeptides can also enhance interferon production, stimulate natural killer (NK) cell activity and activate phagocytosis (33). Moreover, thymic hormones also display neuroendocrine activity by increasing the production of corticotropic-releasing hormone, ACTH, and cortisol (33, 56). Clinical trials with thymic preparations have shown their efficacy in restoring immune responsiveness and augmenting specific lymphocyte activities in patients with primary and secondary T cell deficiencies (33).

Interferon Preparations.   Interferons (IFNs) are glycoproteins, classified according to physicochemical, antigenic, and biological properties into three types: IFN-α, IFN-β, and IFN-γ. The principal antiviral IFNs, IFN-α and IFN-β, are produced by a variety of cells in response to virus challenge, whereas IFN-γ is mainly an immunoregulatory cytokine produced by antigen- or mitogen-activated T lymphocytes. A wide spectrum of IFN activities includes antiviral, antiproliferative, immunomodulatory, and neuroendocrine effects (73). The antiviral effect of IFNs is based on the release of IFN-induced enzymes, that inhibit viral replication. Immunomodulatory effects of IFNs include the up-regulation of class I and class II antigens of major histocompatibility complex, activation of antigen-presenting cells, enhancement of cytotoxic T lymphocytes, augmentation of NK cell activity, and modulation of antibody production (73).

A number of IFN preparations have been manufactured for clinical application. Human interferon-α (HuIFN-α) is a crude or partially purified preparation derived from the virus-treated leukocytes of healthy donors (14). An alternative way of receiving IFN preparations is through recombinant DNA technology (85). Now several natural and recombinant human IFN (rhIFN) preparations are on the world market and their availability depends on the country where they may be licensed for use with certain indications, including treatment of viral diseases, neoplastic diseases, and AIDS (79).

Clinical, antiviral, and prophylactic effects of both intranasal HuIFN-α and rhIFN-α preparations have been demonstrated against experimental rhinovirus infection in volunteers and against natural respiratory viral infection in field trials (75). Unfortunately, the long-term use of large doses of intranasal interferon was associated with a high incidence of local inflammatory reaction and that was the major consideration for the cessation of further investigations in this area. However, since short-term IFN prophylaxis was shown to reduce the risk of side effects and still had 40% efficacy against respiratory infections among family members, the prospect of using IFN preparations still exists (37).

Levamisole. Levamisole hydrochloride (levamisole), available under the trade names Ergamisol (Janssen pharmaceutical, USA) and Decaris (Gedeon Richter, Hungary), is a synthetic compound, chemically known as (-)-(S)-2,3,5,6-tetrahydro-6-phenylimidazo [2,1-b] thiazole monohydrochloride. Levamisole was originally developed as an anthelmintic drug for human use: later its immunopotentiating activities were found and examined intensively in both experimental and clinical studies (1, 62).

The immunomodulative activities of levamisole include the enhancement of both mitogen- and antigen-induced T cell responses, the enhancement of delayed-type hypersensitivity reaction, the potentiation of monocyte and macrophage function, and the stimulation of antibody formation (96). The most convincing results of levamisole treatment were shown in patients with rheumatoid arthritis (1, 13). However, long-term treatment with levamisole may induce agranulocytosis, and the appropriate hematological monitoring is recommended routinely during the therapy.

Diuciphon. Diuciphon (available in Russia) is a synthetic compound, chemically known as para-(2,4-dioxo-6-methylpyrimidinil-5-sulfonamino)-diphenylsulphonate. Diuciphon, originally known as an antilepric drug, was later shown to induce T lymphocyte maturation (50). Clinical and immunologic efficacy of Diuciphon treatment (in combination with the basic therapy or in its monotherapy) has been demonstrated in patients with protracted acute pneumonia (50).

Dipyridamole. Dipyridamole, available under the trade names Persantine (Boehringer Ingelheim, USA) and Curantyl (Veb Berlin-Chemie, Germany), is a vasodilator and a platelet inhibitor, chemically described as 2,6-bis-(diethanolamino)-4,8-dipiperidino-pyrimido-(5,4-d) pyrimidine. This medicine has been in Western and Eastern pharmacopoeia for many decades and it has no serious side effects.

The first information on the antiviral effects of dipyridamole in vitro was published in the late '70s (87). Later its antiviral efficacy against influenza A virus (86) and its interferon-inducing capacity were demonstrated in rodents (29). In humans, production of IFN-α has been reported after a single oral administration of dipyridamole (30), and its prophylactic efficacy has been shown in a placebo-controlled field trial with 500 volunteers during an outbreak of natural respiratory infections in wintertime (48).

Similar mechanisms of action have been proposed for a wide spectrum of dipyridamole effects, including its antiviral, interferon-inducing, immunomodulating (12, 20), neuroendocrine (6), and analgetic (59) effects. The mechanisms have been attributed to adenosine-dependent and adenosine-independent pathways, involving the secondary messenger, cyclic AMP (11, 20).

The Use of Immunomodulators in Athletes

The availability of immunomodulators and antiviral preparations for clinical use in the former Soviet Union made it possible to test their prophylactic and immunocorrecting effects in athletes. The application of immunomodulators in athletes began in the early '80s, when the pilot study of a short-term levamisole treatment was conducted on immunocompromised walkers (81). Following medication of subnormal numbers of late and active E-RFC and Ea-RFC rosette-forming cells (cells that form rosettes with sheep erythrocytes, predominantly T lymphocytes), T lymphocytes were significantly increased and reached normal levels. No effect was observed in regard to B cell concentration. These results were comparable with those obtained in clinical populations. Although the underlying mechanism of this effect was not clear, levamisole has been postulated to decrease the cAMP level in lymphocytes, and thereby increase lymphocyte receptor activity (98).

Another immunomodulator, Diuciphon, has been used in a placebo-controlled study with 14 elite swimmers, the members of the USSR national team (15-19 years of age, 5-12 years of training background) (80). The indication for the immunotherapy was recurrent infections in the athletes, that dramatically affected their training. Five out of 14 swimmers had frequent respiratory infections and chronic tonsillitis, three athletes suffered from skin purulent diseases and one athlete had chronic ear infection.

Athletes were randomized into two groups; the experimental group was given Diuciphon (0.1 g × 3 times per day, six days a week) for three weeks. The control group received a placebo. The pretreatment assessment of immune function occurred during the preparation for the national championship. That period was characterized by training of high intensity and volume (2-3 sessions per day, 65 km of swimming per week with 40% of work performed at high intensity). The posttreatment assessment was done 10 days following the nationals during the reduced training and taper period (two sessions per day, 50 km of swimming per week at light to moderate intensities).

Figure 12.4 presents the effect of Diuciphon treatment on PHA-induced lymphocyte proliferation. The baseline PHA-stimulated lymphocyte proliferate response in all athletes was significantly lower than the normal values of nontrained healthy young men. Reduced training and Diuciphon treatment resulted in an 85% increase of PHA-induced lymphocyte proliferative response in the experimental group versus a 30% increase in the placebo group. These findings were observed only in regard to PHA stimulation (but not to the pokeweed mitogen, that stimulates B lymphocytes). No changes were observed in initially normal concentration of E-RFC and plasma level of immunoglobulins IgA, IgG, and IgM.

The incidence of URTI was markedly reduced following the medication. Only three cases of respiratory infections and one case of ear infection were registered in the experimental group during a 6-month follow-up period, while incidence of URTI was still high in the placebo group. The individual analysis of medical histories revealed that the athlete who suffered from chronic otitis had a recurrence of infection only once during a 6-month follow-up period. Another swimmer, who suffered from chronic tonsillitis, experienced only two attacks of tonsillitis during the follow-up. The third athlete, who had three respiratory infections during the five months preceding Diuciphon treatment, did not have any infection during the follow-up period. Seven out of eight swimmers (85%) of the experimental group improved their individual best results during the competition. In contrast, only two out of six athletes (33%) of the placebo group showed better performance. Thus, immunocorrection resulted in a better immunologic recovery from the major competition, reduced incidence of infection among athletes, and thus provided better performance.

Thymic hormone preparation has been applied to highly-trained male speed skaters, the members of the Russian Federation team. Athletes were followed during a 20-day heavy training and during participation in the prestigious national competition, Kiseleff's prize (unpublished personal data). Nine athletes of the experimental group were given a 10-day course of thymalin (20 mg × 3 times a day), while nine other speed skaters received a placebo. Due to the different race schedules in athletes of the two groups and because of the difficulties in conducting this study, the posttreatment blood was drawn in diverse competitive situations. In the experimental group blood was drawn the morning following the race, while blood

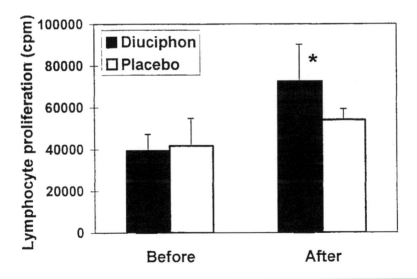

Figure 12.4   Effect of Diuciphon treatment versus placebo on the recovery of PHA-induced lymphocyte proliferative response in elite swimmers after Nationals. *p < .05.

Based on data from Surkina et al. 1983.

samples from the placebo group were collected the morning following the rest day, on the eve of the race. Assuming this discrepancy in the study protocol, the athletic conditions for the immune assessment may be considered easier for the athletes of the placebo group.

The baseline immune profile was similar in all athletes and characterized by normal production of virus-induced IFN-α, subnormal synthesis of IFN-γ, and low lymphocyte proliferative response to PHA stimulation (see figure 12.5). Thymalin treatment resulted in an 85% increase of PHA-induced lymphocyte proliferation compared to the unchanged low proliferative response of lymphocytes in the placebo group. Virus-induced IFN-α production slightly decreased in athletes of the thymalin group, while a 53% decrease of IFN-α synthesis was registered in the placebo group. No effect of thymalin treatment was observed in regard to PHA-induced IFN-γ production, that significantly decreased in all athletes. Taking into consideration the data on the immunosuppressive effect of heavy training in combination with the competitive stress (82), the positive changes in the lymphocyte proliferative response and the stable production of IFN-α in speed skaters of the thymalin group may be attributed to the immunomodulative effect of the medicine.

A three-day intranasal application of another thymic hormone preparation, Thymogen (available in Russia), in 89 young athletes (10-19 years of age, 47 boys and 42 girls) during base training in wintertime has been shown to reduce incidence of respiratory infection (27.7% vs. 63.3% in controls), to enhance mitogen-induced lymphocyte proliferative activity, to increase blood and salivary IgA concentrations, and to activate phagocytosis (51).

The availability of interferon preparations and interferon inducers for human use in the former USSR made it possible to apply them to athletes. It was proposed that the use of low concentrations of intranasal interferon may have a prophylactic effect and reduce incidence of URTI in athletes. Eighteen regional-level cross-country skiers (8 women and 10 men, 17-23 years of age) were followed during their first four-week intensified training session of the season in snow (25). In order to train in the snow, the athletes had to move to the northern part of Russia. In addition to changing training modes (from cycling and roller skiing to cross-country skiing) and dealing with jet lag, ambient temperatures changed from 3-5°C to –20-30°C. According to our personal observations and published data of other authors (38), cross-country skiers are usually highly prone to respiratory infections during this time.

In an attempt to reduce the incidence of URTI of these athletes, they were treated with intranasal interferon preparations twice a day during the first nine days of training. According to a double-blind, placebo-controlled protocol, athletes were randomized into three groups, where the first group of skiers was administered with HuIFN-α in a dose of 64 IU/0.5 ml, the second group was given rhIFN-α$_2$ (available under the trade name Reaferon) in a dose of 4×10^4 IU/0.5 ml, and the third group received a placebo (saline solution). Athletes were instructed to report any symptoms to the team's physician, who diagnosed the illness. No side effects have been registered during a short-term IFN treatment.

As a result of treatment no respiratory infections were diagnosed in skiers of the second group (Reaferon), and only 44% of the athletes of the first group (HuIFN-α) got colds with mild symptoms, such as runny nose and scratchy throat. These symptoms lasted not more than two days and did not cause an interruption

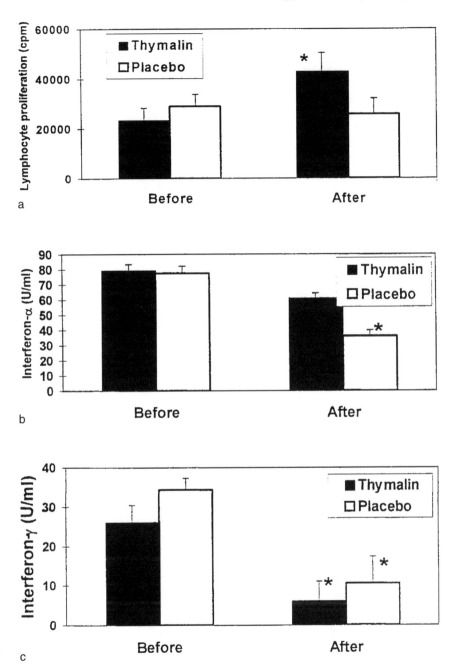

Figure 12.5   Effect of (a) thymalin treatment versus placebo on PHA-induced lymphocyte proliferative response, (b) virus-induced IFN-α production, and (c) PHA-induced IFN-γ production in 18 highly-trained male speed skaters. Personal unpublished data. *p < .05.

in training. To the contrary, all athletes of the placebo group (100%) had respiratory infection with more severe symptoms, including cough, hoarseness, nasal discharge, and in some cases fever. Due to systemic symptoms of illness training of the placebo group was interrupted for up to three days.

Figure 12.6 shows changes in the athletes' immune function at the end of a four-week training cycle and 20 days following interferon treatment. The immunomodulative effect of IFN preparations was registered in regard to PHA-induced IFN-γ production, which was initially below normal values. At the end of the training cycle its synthesis significantly increased in athletes who received medication in comparison with the reduced IFN-γ production in the placebo group. Also, changes in PHA-induced lymphocyte proliferation were less pronounced in IFN-treated athletes as compared with the placebo group. There was no effect of IFN preparations on virus-induced IFN-α production in vitro. Four weeks of heavy training resulted in significant decrease of its synthesis in all athletes.

A high prophylactic efficacy of interferon preparations in athletes triggered the further screening of compounds with an IFN-inducing activity and an easier means of administration. Dipyridamole was one of the medicines that satisfied the above-mentioned requirements. Dipyridamole was used in two separate placebo-controlled studies with endurance-trained athletes (cross-country skiers and rowers) according to the scheme proposed by Galabov and Mastikova (30) of 100 mg taken in two doses with a two-hour interval between doses, once a week.

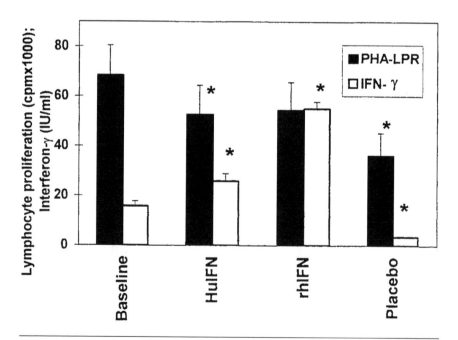

Figure 12.6 Effect of interferon preparations treatment versus placebo on PHA-induced lymphocyte proliferative response and IFN-γ production in 18 regional-level cross-country skiers. *p < .05.
Based on data from Ershov, Gotovtseva, and Surkina 1988.

A two-week course of dipyridamole, administered to six highly-trained male cross-country skiers during base training, resulted in a 182% increase of initially low IFN-γ production, and in stabilization of IL-2 production. On the contrary, both PHA-induced production of IFN-γ and IL-2 were reduced in six athletes of the placebo group (see figure 12.7). No effect of dipyridamole treatment was registered in regard to initially subnormal PHA-induced lymphocyte proliferation. The synthesis of IFN-α significantly increased in cross-country skiers of the placebo group, while its production was not changed in athletes of the dipyridamole group (83). The index of illness was quantified by the incidence of respiratory infection, its severity, and duration. In the dipyridamole-treated athletes, the index of illness was 1.8-times lower than in the placebo group during the study and a six-week follow-up period. A markedly increased production of IFN-α in the control group was likely a response of the IFN system to viral infection in the athletes.

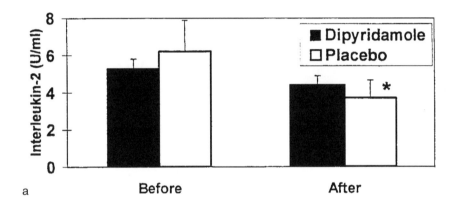

a   Before   After

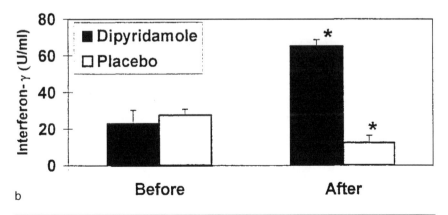

b   Before   After

Figure 12.7 Effect of dipyridamole administration versus placebo on PHA-induced production of IL-2 (a) and IFN-γ (b) in 12 highly-trained cross-country skiers. *p < .05.

Based on data from Surkina et al. 1993.

In another group of eight rowers a three-week course of dipyridamole administration during training at moderate altitude (1600 m) produced a 23% and 70% increase of IFN-α and IFN-γ production, respectively. No changes in the production of these IFNs were observed in seven athletes of the placebo group (90). The index of illness was 2.5-times lower in athletes of the dipyridamole group in comparison to the placebo group during a two-month follow-up period.

Neuro-Immunomodulations During Physiological and Psychological Stress in Elite Athletes

The Relationship Between the Neuroendocrine and Immune Systems

The underlying mechanisms of exercise-induced immunologic changes should be considered in view of the current research in the field of neuroendocrinimmunology. This research clearly demonstrates that the effect of stress on immunity is associated with the bidirectional interactions among the immune, nervous, and endocrine systems (44). All three systems participate in close relationship in an adaptive response to reestablish stress-disturbed homeostasis.

Regardless of their nature, stressors initiate central catecholamine liberation and the release of corticotrophin-releasing hormone with the subsequent activation of the autonomic nervous system and the pituitary-adrenal axis, leading, respectively, to catecholamine and glucocorticoid secretion. In addition to these two principal effectors of stress, many other neuroendocrine systems, including the opioid and reproductive systems, are involved in generalized stress response (36, 71).

Neuroendocrine molecules released in response to stress act on receptors within the immune system and thus modulate immune function (44). In this case, the status of the immune system will depend on the hormone concentration, target cell, and the specific immune function. In turn, immune molecules (cytokines) can modulate neuroendocrine activity through feedback mechanisms and thus complete the regulatory loop (39). In addition to soluble mediators of the neuroendocrine-immune interactions, lymphoid tissue is richly innervated, which provides neurologic control over the immune system (26).

Aspects of the exercise-induced neuroimmunomodulation have been studied mostly in respect to an acute bout of exercise performed in a laboratory setting. An exercise-induced rise of plasma catecholamines has been shown to initiate leukocytosis, to increase NK cells concentration, to reduce T helper (Th) cell numbers, and to inhibit a mitogen-induced lymphocyte proliferative response (39, 40, 57, 88). The transient postexercise leukocytosis and lymphopenia, and the reduced production of cytokines have been attributed to the immunosuppressive effects of cortisol (42, 96).

The opioid system plays a critical role in the reduction of pain, in the increase of stress tolerance, and in regulation of the immune function (63, 70). The modulation of NK cell activity by β-endorphin during exercise has been clearly demonstrated (27). Among a variety of other hormones involved in immune regulation are human growth hormone, prolactin (9, 43), and sex hormones (58). However, the role of these hormones in exercise-induced immune changes still remains to be elucidated.

The bidirectional interactions between the neuroendocrine and immune systems during exercise become more complicated when physical effort is accompanied by inflammation and a release of pro-inflammatory cytokines, that are now recognized as potent activators of the central stress response (10). Although the laboratory models of exercise-induced stress provide valuable information on neuroendocrine and immune modulation in humans, there is still a lack of knowledge regarding the role of regulative systems in the adaptive processes to extreme athletic stress. This information is particularly important since the extreme stress of exhaustive training and competition may produce disruption at any level of the neuroendocrine and immune network. This functional dissociation between the components of general adaptive response may result in a disease, that, in the case of elite sport, is an overtraining syndrome, or in other words, the disease of dysadaptation. Thus, understanding the adaptive mechanisms in elite athletes will help to optimize the training process and to diagnose early symptoms of dysadaptation, as well as to choose interventions to prevent overreaching and overtraining.

The Neuroendocrine and Immune Profile of Athletes at Rest, During Endurance Training, and Competitive Events

Monitoring immune and neuroendocrine functions in elite athletes during different periods of training allowed us to characterize the status of these systems in association with the particular athletic situation and with the initial state of regulative system functioning. Immune, opioid, and endocrine indices were evaluated in elite female and male speed skaters the morning following a rest day and during the base training period (82). At rest, neuroendocrine and immune parameters strongly correlated with each other in both groups of athletes. The positive correlation was observed despite the differences in the athletes' training programs (reduced volume of training in female athletes in contrast with a high volume of skating in male athletes), and in the state of immune function. In fact, increased number of Ea-RFC corresponded to increased opioid activity (plasma δ-ligand opioid receptor (δ-LOR) activity) in female athletes, while normal number of Ea-RFC and plasma opioid activity correlated with each other in male athletes. Serum cortisol, growth hormone, and insulin levels were within normal range in male speed skaters. At the same time PHA-induced lymphocyte proliferation, mitogen- or virus-induced production of IFN-γ and IFN-α, in these athletes was significantly below normal values. It is of interest, that low lymphocyte proliferation positively correlated with a normal level of serum cortisol ($r = .705$, $p < .05$).

The resting neuroendocrine-immune profile in overtrained athletes had other characteristics. Athletes who had completely ceased training for several months were admitted to the hospital with the symptoms of local myocardiodystrophy. Immune and endocrine assessment revealed a profound depression of lymphocyte proliferative activity and inverse diurnal catecholamine excretion (82). Adrenocortical dysfunction in response to ACTH stimulation was observed in two female athletes who experienced declined performance, frequent respiratory infections, and had pronounced immunosuppression. One athlete did not produce adrenocortical response, possibly due to exhaustion of the adrenal cortex. Initial adrenocortical activation in another athlete switched to a pronounced decrease of excretion of the corticosteroid metabolite, 17-ketosteroid, pointing to a reduced potential reserve of the pituitary corticotropic function (82, 92).

The fact that different athletic efforts may produce diverse endocrine responses has been reported by Weicker and Werle (94). Sympatho-adrenal and pituitary-adrenocortical responses measured immediately after different running distances (sprint, middle, and long) varied significantly. Thus, it was speculated that neuroendocrine regulation of immune function in athletes may also vary. Dissociation between neuroendocrine and immune functions has been revealed in the above-mentioned group of elite male speed skaters following a two-day training session of high intensity and volume (82). The signs of functional dysregulation were the further suppression of immune function, endocrine dysfunction, and no changes in plasma opioid activity (see table 12.1).

After a six-hour recovery from exercise, the serum cortisol level was significantly below normal baseline values (33.8 ng/ml, 42.5 ng/ml, 61.9 ng/ml versus 200.9 ng/ml at baseline) in three out of nine athletes (92). At the same time, the changes in serum cortisol level were reciprocally related to high blood growth hormone concentrations. Both of these hormones reached their normal values the morning following a second-day training session (18 h recovery), while the blood insulin level was very low or even undetectable at that period of recovery. These exercise-induced neuroendocrine and immune changes resulted in a loss of the correlative relations between regulative systems that had been previously observed at rest.

The same team of elite male speed skaters was followed during major competitions (82). Athletes raced in 500-m and 1000-m sprint distances during the pre-Olympic trials. Therefore intensive, but short-duration physical effort was accompanied by a very high emotional strain. In addition to the assessment of neuroendocrine and immune function, the psychological status of athletes was evaluated by questionnaires. The questionnaires included questions on the importance of the individual event, the actual result, and its conformity to the planned result, as well as a personal rating of the performance. All answers were coded and quantified into an index of emotional strain (IES), that could be positive (+) or negative (−).

The main immunologic characteristics, observed 6 h following the second race, were normal numbers of Ea-RFC and E-RFC, low mitogen-induced lymphocyte proliferative response, and a complete blockade of IFN-α and IFN-γ production by stimulated lymphocytes. At the same time point, plasma δ-LOR activity was subnormal (1.4-times lower than baseline) and it did not correlate to any immune parameter. The relationship between the index of emotional strain and plasma δ-LOR activity varied individually, with the general trend of the higher negative index corresponding to the lower opioid activity. However, the reverse relationship has also been found.

Endocrine indices were measured at rest, 30-40 min before the start, 10-12 min after the finish, and at 6 h following the race (see table 12.2). The prestart activation of the endocrine system was particularly pronounced with respect to the level of growth hormone, with a 633% increase in comparison with the baseline. The mean increase of blood cortisol level was mild (25%) and in half of the athletes it did not elevate at all. The peak of blood growth hormone and cortisol concentrations was observed immediately after the race.

The six-hour recovery period was characterized by a 40% decrease of blood growth hormone, a 56% decrease of cortisol and an 82% decrease of insulin levels, while individual analysis showed dramatically low or undetectable blood cortisol and insulin levels in three athletes (30%). These subnormal levels of hormones during recovery could be attributed to the exhaustion of the pituitary-adrenocortical system and pancreatic function (92).

Table 12.1 Endocrine, Immune, and Opioid Response to a 2-day Exhaustive Training in Elite Male Speed Skaters With the Signs of Immunodeficiency (N = 9; Mean ± SEM)

Time point	Cortisol (ng/ml) N = 6	GH (ng/ml) N = 5	Insulin (pmol/L)	LPR (cpm)	E-RFC (%)	IFN-α (U/ml)	δ-LOR (pmol-eqv DADLE/ml)
Rest	200.9 ± 19.9	1.5 ± 0.1	9.6 ± 0.9	14400 ± 3156	64.4 ± 2.6	44.3 ± 1.5	132.0 ± 8.0
Post 6 h	123.1 ± 28.7*	4.7 ± 2.9	9.15 ± 2.6	NM	NM	NM	NM
Post 18 h	210.4 ± 10.5	0.4 ± 0.2**	1.3 ± 0.5**	9331 ± 2178	47.0 ± 5.5*	14.5 ± 2.5*	127.0 ± 7.0

*p < .05, ** p < .01, comparison with rest. NM = non-measured; GH = growth hormone; LPR = mitogen-induced lymphocyte proliferative response; E-RFC = erythrocyte rosette-forming cells; IFN-α = interferon-α; δ-LOR = δ-type ligand of opioid receptors.
Based on data from Surkina and Gotovtseva 1991.

Table 12.2 Serum Cortisol, Growth Hormone (GH), and Insulin Response to 1000 m Race in Elite Male Speed Skaters With the Signs of Immunodeficiency (N = 9; Mean ± SEM)

	Cortisol ng/ml (N = 6)	GH ng/ml (N = 5)	Insulin pmol/L (N = 9)
Rest	170.3 ± 31.7	1.1 ± 0.4	69.8 ± 17.4
Prestart	23 ± 38.6	8.2 ± 3.7*	NM
Postrace, immediately	290.5 ± 37.0*	22.5 ± 8.3*	NM
Postrace, 6 hr	74.8 ± 14.2*	0.7 ± 0.2	12.9 ± 3.6*

*$p < .05$, comparison with rest; NM = non-measured.
Based on data from Surkina and Gotovtseva 1991.

Thus, these unfavorable changes of neuroendocrine function could affect the athlete's immunologic competence, energy metabolism, and performance. The insufficiency of glucocorticoid function in a prestart and race period is usually associated with poor sport results (92). In our study (82), the athlete who did not elicit cortisol response on competitive stress showed a significantly lower performance, compared to his previous and predicted results. Comparison of the immune status with the athletic results did not reveal any correlation which was likely due to similar profound immunosuppression in all athletes. However, in general, the performance in all speed skaters was worse, compared to the previously shown results.

Thus, monitoring neuroendocrine and immune parameters in the same elite athletes revealed the specific pattern of changes in response to different athletic situations. Similar results on the differential effects of physiological and psychological stress on the relationship between blood levels of pituitary hormones and cytokines have been reported by Schulz et al. (74).

Potential Intervention of Immunomodulative Compound in Athletes to Improve Adaptation to Endurance Training

Research on psychoneuroimmunology (PNI) and neuroendocrine immunology has demonstrated not only the close relationship between the neural, endocrine, and immune systems in maintaining homeostasis or their common role in the onset of diseases, but also has suggested new approaches in developing interventive measures. Such PNI interventions as hypnosis, relaxation, and moderate exercise have been shown to enhance stress- and disease-altered immune function (45, 46, 49).

In athletes this approach may be used for the improvement of adaptive and recuperative processes. Thus, it has been proposed that immunostimulating intervention in immunocompromised athletes could not only enhance immune function and reduce the rate of respiratory infection, but also modulate adaptive mechanisms to stress. To investigate the effect of immunomodulation on neuroendocrine and physiological responses to an acute bout of exercise in athletes, we conducted two studies (83, 84).

The effect of dipyridamole administration on adaptive mechanisms to an acute bout of exercise was investigated in a double-blind placebo-controlled study. Interferon, opioid, physiological, and metabolic responses to submaximal ergometer cycling test (60-90 min, 80% VO$_2$max until volitional exhaustion) were registered in 12 well-trained cross-country skiers (VO$_2$max 78.0 ± 0.9 ml/min/kg) before and after two weeks of endurance training and medication. Results showed that the baseline virus-induced IFN-α production in vitro was within the normal range in all athletes, although it was twice as high in the dipyridamole (experimental) group (see figure 12.8). The interferon response on the premedication exercise test was similar in both groups of skiers and it was characterized by the increased production of IFN-α in vitro during the late (18 h) recovery. At the end of the second week of training, the baseline production of IFN-α was enhanced in all athletes, while the pattern of interferon response on the postmedication exercise test was diverse. In contrast with the complete restoration of IFN-α production in the experimental group, a four-time decrease of IFN-α synthesis was registered in the control group during late (18 h) recovery.

Monitoring plasma opioid activity (plasma δ-LOR activity) during the premedication exercise test and the recovery period revealed two patterns of opioid responses: (1) adequate (N = 4), with an increase of opioid activity during the exercise and a peaking at the end of the test (136%) with a decrease to the baseline during the early (60 min) recovery period, and (2) inadequate (N = 8), with an unchanged opioid activity during the test and an increase or decrease of plasma δ-LOR activity during the early (5-15-60 min) and late (18 h) recovery period. This inadequate opioid response was considered to be the result of either a delayed

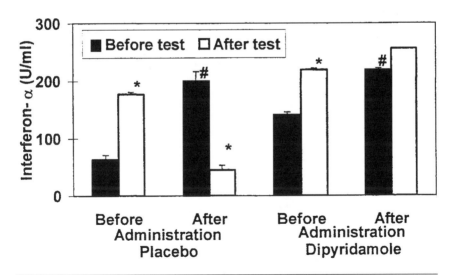

Figure 12.8   Response of IFN-α production to submaximal ergometer cycling test (60-90 min, 80% VO$_2$max) in 12 highly-trained cross-country skiers before and after dipyridamole administration versus placebo. *p < .05, comparison with pre- or posttest; #P < .05, comparison with baseline before administration.

Based on data from Surkina et al. 1993.

opioid reaction or a temporal exhaustion of opioid function. In four out of six athletes of the experimental group, the type of opioid response switched from inadequate to adequate response during the postmedication exercise test.

Furthermore, athletes of the experimental group displayed better physiological (oxygen uptake) and metabolic (blood lactate level) responses to the postmedication test than athletes of the control group (see figure 12.9). The results of this study

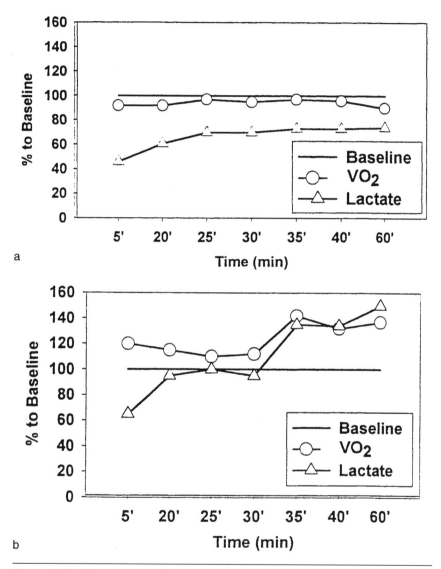

Figure 12.9 Physiological (oxygen uptake) and metabolic (blood lactate level) responses to submaximal ergometer cycling test (60-90 min, 80% VO_2max) in 12 highly-trained cross-country skiers before and after dipyridamole administration versus placebo. Data are presented as % of the results of the pre-administration response, taken as 100%. (a) = dipyridamole administration. (b) placebo administration. Based on data from Surkina et al. 1993.

suggest that the use of dipyridamole during endurance training can optimize the adaptation to an acute bout of exercise by modulating IFN-α and opioid response, and economizing energy uptake.

The effect of immunomodulation on endocrine response to an acute bout of exercise has been evaluated in another study (84). Seven male competitive rowers underwent two 7 min rowing ergometer tests (simulated 2,000-m race), separated by a three-week interval during which athletes started in 2-3 unimportant races and received dipyridamole. The immune response on the second rowing test, measured at late (18 h) recovery, was less pronounced. It was characterized by the complete recovery of PHA-induced IFN-γ production and a mild decrease (46%) of PHA-induced IL-2 synthesis in comparison with a declined production of IFN-γ and IL-2 (75% and 93%, respectively) after the first test. A marked 412% increase of blood cortisol level was registered immediately after the first test, although in three out of seven rowers it was peaking after 24 h recovery. Following a three-week course of medication the same exercise test produced a more adequate cortisol response. In all athletes the peak cortisol concentration was registered immediately after the test, although it was still elevated (89%) over baseline after 24 h recovery. Along with the less pronounced immunologic changes and more adequate endocrine response to the second rowing test, athletes showed better performance (a mean increase in rowing power of +0.9 ± 0.4 kgm/min/kg, and the mean improvement of time on calculated 2,000-m race was -2.5 ± 1.9 sec).

Unfortunately, this study was not placebo-controlled due to the difficulties of finding the properly matched controls. Thus, we may only speculate that the improvement in adaptation to exercise stress and better performance was more likely associated with the dipyridamole administration, than with the training itself. At the time of the study the athletes had finished a build-up training period and were close to their peak performances.

Micronutrients and Immune Function in Athletes

Role of Vitamins and Trace Elements in the Immune Function

The recognition of a modulative effect of nutrition on the interaction between exercise and immune function brought new aspects to the consideration of immunodeficiency in athletes (76). An inadequate, poorly balanced diet, or an unusual diet including megadoses of supplements, is not rare among competitive athletes who are attempting to control their weight or to improve their performance (7, 19).

Micronutrients may affect immune function because of either a deficiency or an excessive intake. Vitamins and trace elements play crucial roles in key metabolic processes and cell functions, including the functions of immunocompetent cells (16). A deficiency of vitamin A, B_6, C, or E results in impaired cell-mediated immunity, including decreased mitogen- and antigen-induced lymphocyte proliferation, altered delayed cutaneous hypersensitivity reaction, reduced T cell cytotoxicity, declined production of cytokines, decreased antibody synthesis, and altered phagocytosis (8, 21, 31). Zinc (Zn), magnesium (Mg), copper, and iron deficiencies have been associated with impaired immune functions, such as T cell mitogenic response,

NK cell activity, antibody production, and phagocytosis (16). At the same time, an excessive intake of zinc, vitamin A, or vitamin E has been shown to impair immune function (16).

Vitamin and Trace Element Profile in Athletes

Two common procedures—assessment of dietary intake and biochemical measurements—are widely used to determine vitamin and mineral status in athletes. A survey on nutritional habits in Dutch elite athletes showed low iron and vitamin A intake in female swimmers and gymnasts, respectively, and insufficient vitamin B_1 and B_6 intake in professional cyclists (91).

Low iron, zinc, magnesium, and calcium intake and their altered absorption are often observed in female runners and gymnasts who deliberately restrict food intake, consume an unbalanced diet with high carbohydrate and fiber contents, or eat vegetarian food for maintaining low body weights (22, 52, 60, 91). In another study with 1300 German athletes, low serum magnesium and iron levels have been found in 21.2% and 14.1%, respectively (93). Low plasma zinc concentrations have been measured in runners in comparison with the untrained population (23). Low ferritin level, as an indicator of iron depletion, was found in 20-35% of U.S. female cross-country skiers and long-distance runners (5, 18).

A deficiency of vitamins and minerals in athletes may also be the result of the increased loss of these micronutrients during heavy training. The increased urinary excretion of Zn and Mg has been found in athletes after strenuous exercise (2, 17). Gastrointestinal bleeding is the major cause of iron loss in long-distance runners (78). On the other hand, anti-oxidant vitamins C and E are the most widely used supplements, often in megadose quantities (41, 68). Thus, either a deficiency of micronutrients or their excessive intake may put athletes at a high risk of immune disorders and an increased susceptibility to infection. In the case of a micronutrient deficiency, it seems likely a nutrient supplement can improve immune function in athletes. Indeed, some positive immune changes have been found in a double-blind cross-over study with Zn supplementation in runners (77).

Potential Role of Glutamine in Exercise-Induced Immunosuppression and Overtraining

Recently, interest in the relationship between nutrition and the immune system has focused on the role of glutamine in the functioning of immunocompetent cells (69). Glutamine is utilized by macrophages and lymphocytes at a high rate for nucleotide synthesis and formation of DNA and RNA, which are required by cells for their proliferation and protein synthesis (3, 64, 65). Such immune functions as lymphocyte proliferation, IL-1 and IL-2 production, antibody synthesis, and macrophage phagocytosis have been shown to be glutamine-dependent (64), and thus, any decrease in glutamine concentration may result in impaired immune function.

Muscles are the major source of produced and released glutamine (66), and due to this close relationship between muscles and the immune system any muscular activity can modulate plasma glutamine levels and therefore immune function. A decrease of plasma glutamine level has been found immediately following a marathon race and in resting overtrained athletes (67, 72). However, the results of another study showed an unchanged plasma glutamine level in swimmers who became overreached following a four-week intensified training program. At the same time, plasma glutamine concentration increased significantly in those athletes who were not overreached (54). Furthermore, no correlation between plasma glutamine level and lymphocyte proliferative activity (69), lymphocyte subsets (72) or URTI rate among athletes (54) has been found. Thus, the clinical importance of plasma glutamine changes in athletes remains to be elucidated.

Summary

The presented data clearly demonstrate that competitive athletes may have reduced resting immune function and increased susceptibility to respiratory infections during heavy endurance training and competition. Training in extreme environmental conditions (e.g., cold or heat), exposure to altitude, and jet lag may aggravate the effect of athletic stress on immune function and thus predispose athletes to immune alterations and respiratory infection.

Along with the other classic symptoms of overreaching and overtraining, immune dysfunction and frequent colds have been found in overreached and overtrained athletes, and thus may be considered as markers of this athletic pathology. As discussed in the previous chapters, there is little doubt that preventive behavioral changes can reduce incidence of respiratory infection. However, the real life of competitive athletes dictates its own rules. In contrast with the general population, athletes live in close proximity in training camps, travel frequently to training and competitive sites, purposely train in cold or hot climates, and train at altitude in order to improve adaptation to extreme environmental conditions and performance. Exhaustive training and the tight schedules of competitions are not rare in athletic life. On the other hand, there is a lack of rehabilitative centers for competitive athletes. In this situation athletes may have insufficient recovery and develop chronic fatigue and overreaching.

All of the above-mentioned factors and the insistent requests for help from athletes, coaches, and team physicians made us consider the use of immunomodulators and antiviral medicines in athletes. Ten years of experience in the use of immunopotentiating medicines in athletes demonstrated their beneficial effects relative to the correction of an athlete's immune function, the reduction in the incidence of respiratory infection, and the improvement of performance. As it has been shown, the different compounds produced specific immunopotentiating effects on certain immunologic variables. These data are summarized in table 12.3. Diuciphon and thymic hormones were more effective in regard to lymphocyte proliferation, while IFN preparations and dipyridamole showed a stimulative effect predominantly on IFN-γ production. Thus, it is important to assess immune function and to diagnose the weak link in the immune system prior to the use of immunopotentiating medicines in athletes.

The underlying mechanisms of the immune alterations in athletes are closely associated with the exercise-induced neuroendocrine changes. At rest, profound endocrine and immune dysfunctions have been found only in overtrained athletes; whereas isolated immune alterations in nonovertrained athletes did not change correlative relations among the immune, endocrine, and opioid systems. At the same time, following exhaustive training and competitions these correlations were no longer observed, pointing to the potential disruption of the adaptive

Table 12.3 A Summary of the Effects of Immunomodulative Compounds on Immune Function in Athletes

Medicine	LPR	IFN-α	IFN-γ	IL-2
Diuciphon	+	NM	NM	NM
Thymalin	+	+	–	–
HuIFN-α	+	–	+	NM
rhIFN-α	+	–	+	NM
Dipyridamole	–	–	+	+

+ = positive effect; – = no effect; NM = non-measured; LPR = mitogen-stimulated lymphocyte proliferative response; IFN-α = virus-induced production of interferon-α; IFN-γ = mitogen-induced interferon-γ production; IL-2 = mitogen-induced interleukin-2 production.

mechanisms. The most pronounced changes in the regulative mechanisms have been found after major competitions.

The important role the immune system plays in preserving the balance of neuroendocrine-immune cooperation allows us to hypothesize that immuno-correction may improve adaptation to athletic stress. The presented pilot data showed that immunopharmacological intervention may optimize the neuroendocrine, physiological, and metabolic response to acute physical stress.

Another important factor for proper immune functioning is a sufficient, well-balanced intake of vitamins and microelements, since heavy exercise increases the demand for micronutrient intake. Recently, the role of glutamine in immune function, and the association of glutamine deficiency with an increased risk of respiratory infection and the onset of overreaching attracted wide attention. However this association is still unclear and remains to be elucidated.

References

1. Amery, W.K., C. Hörig. 1984. Levamisole. In *Immune modulation agents and their mechanisms*, eds. R.L. Fenichel, M.A. Chirigos, 383. New York: Marcel Dekker.
2. Anderson, R.A., M.M. Polansky, N.A. Bryden. 1983. Strenuous running: acute effects on chromium, copper, zinc, and selected clinical variables in urine and serum in male runners. *Biological Trace Element Research* 6: 327-335.
3. Ardawi, M.S.M., E.A. Newsholme. 1985. Metabolism in lymphocytes and its importance in the immune response. *Essays in Biochemistry* 21: 1-44.
4. Baj, Z., J. Kantorski, E. Majewska, K. Zeman, L. Pokoca, E. Fornalczyk, H. Tchorzewski, Z. Sulowska, R. Lewiski. 1994. Immunological status of competitive cyclists before and after the training season. *International Journal of Sports Medicine* 15: 319-324.

5. Balaban, E.P., J.V. Cox, P. Snell, R.H. Vaughan, E.P. Frenkel. 1989. The frequency of anemia and iron deficiency in the runner. *Medicine and Science in Sports and Exercise* 21: 643-648.

6. Balashov, A.M., I.D. Surkina, E.P. Gotovtseva, O.B. Petrichenko, L.F. Panchenko. 1990. Effect of dipyridamole on blood level of opioid peptides and alpha-interferon production. *Zurnal Experimentalnoi Biologii i Medicinyi* 5: 462-464.

7. van der Beek, E.J. 1985. Vitamins and endurance training: food for running or faddism claims. *Sports Medicine* 2: 175-197.

8. Bendich, A., R.K. Chandra. 1990. *Micronutrients and immune functions*. New York: New York Academy of Sciences.

9. Bernton, E.W., H.U. Bryant, J.W. Holaday. 1991. Prolactin and immune function. In *Psychoneuroimmunology*, eds. R. Ader, D.L. Felten, N. Cohen, 403-429. San Diego: Academic Press.

10. Blalock, J.E. 1988. Immunologically-mediated pituitary-adrenal activation. In *Mechanisms of physical and emotional stress*, eds. G.P. Chrousos, D.L. Loriaux, P.W. Gold, 217-224. New York: Plenum Press.

11. Blalock, J.E., E.M. Smith. 1986. Interferon and other hormones of the interferon system. In *Clinical application of interferons and their inducers*, ed. D.A. Stringfellow, 19-42. New York: Marcel Dekker.

12. Bruserud O. 1987. Dipyridamole inhibits activation of human T lymphocytes in vitro. *Clinical Immunology and Immunopathology* 42: 102-109.

13. Bunchuk, N.V., V.A. Nasonova, Y.A. Sigidin. 1980. Controlled (double-blind) trial of levamisole at rheumatoid arthritis. *Therapevticheskyi Archiv* 6: 93-98.

14. Cantell, K., S. Hirvonen. 1978. Large-scale production of human leukocyte interferon containing 10(8) units per ml. *Journal of General Virology* 39: 541-543.

15. Casey, T.M., E.C. Dick. 1990. Acute respiratory infections. In *Winter sports medicine*, eds. T.M. Casey, C. Foster, E.G. Hixon, 112-128. Philadelphia: Davis.

16. Chandra R.K. 1991. 1990 McCollum award lecture. Nutrition and immunity: lessons from the past and new insights into the future. *American Journal of Clinical Nutrition* 53: 1087-1101.

17. Clarkson, P.M., E.M. Haymes. 1995. Exercise and mineral status of athletes: calcium, magnesium, phosphorus, and iron. *Medicine and Science in Sports and Exercise* 27: 831-843.

18. Clement, D.B., D.R. Lloyd-Smith, J.G. MacIntyre, G.O. Matheson, R. Brock, M. Dupont. 1987. Iron status in Winter Olympic sports. *Journal of Sports Science* 5: 261-271.

19. Colgan M. Effects of multinutrient supplementation on athletic performance. 1986. In *Sports, health and nutrition*, ed. F.I. Katch, 21-51. Champaign, IL: Human Kinetics.

20. Colli, S., E. Tremoli. 1991. Multiple effects of dipyridamole on neutrophils and mononuclear leukocytes: adenosine-dependent and adenosine-independent mechanisms. *Journal of Laboratory and Clinical Medicine* 118: 136-145.

21. Cunningham-Rundles, S. 1993. *Nutrient modulation of the immune response*. New York: Marcel Dekker.

22. Deuster, P.A., S.B. Kyle, P.B. Moser, R.A. Vigersky, A. Singh, E.B. Schoomaker. 1986. Nutritional survey of highly trained women runners. *American Journal of Clinical Nutrition* 45: 954-962.

23. Dressendorfer, R.H., R. Sockolov. 1980. Hypozincemia in runners. *Physician and Sportsmedicine* 8: 97-100.
24. Ershov, F.I., E.P. Gotovtseva. 1989. Inteferon status under stresses. *Soviet Medicine Review of Experimental Virology* 3: 35-49.
25. Ershov, F.I., E.P. Gotovtseva, I.D. Surkina. 1988. The use of recombinant a-2 interferon in sportsmen. *Voprosy Virusologii* 6: 693-697.
26. Felten, S.Y., D.L. Felten. 1991. Innervation of lymphoid tissue. In *Psychoneuroimmunology,* eds. R. Ader, D.L. Felten, N. Cohen, 27-70. San Diego: Academic Press.
27. Fiatarone, M.A., J.E. Morley, E.T. Bloom, D. Benton, T. Makinodan, G.F. Solomon. 1988. Endogenous opioids and the exercise-induced augmentation of natural killer cell activity. *Journal of Laboratory and Clinical Medicine.* 112: 544-552.
28. Fitzgerald L. 1991. Overtraining increases the susceptibility to infection. *International Journal of Sports Medicine* 12: 5-8.
29. Galabov, A.S., M. Mastikova. 1983. Interferon-inducing activity of dipyridamole in mice. *Acta Virologica* 27: 356-358.
30. Galabov, A.S., M. Mastikova. 1984. Dipyridamole induces interferon in man. *Biomedicine and Pharmacotherapy* 38: 412-413.
31. Gershwin, M.E., R.S. Beach, L.S. Hurley. 1985. *Nutrition and immunity.* Orlando: Academic Press.
32. Geselevich V.A. 1990. Dynamics of incidence of injury and infection in athletes, participated in the summer 1988 Olympic games. *Nauchno-Sportivnyi Vestnik* 1: 27-30.
33. Goldstein A.L. 1983. Thymic hormones and lymphokines. *Basic chemistry and clinical applications.* New York: Plenum Press.
34. Gotovtseva, E.P. 1987. Immune and interferon status of high-qualified athletes. PhD diss., The D.I. Ivanovsky Institute of Virology, USSR Academy of Medical Sciences.
35. Gotovtseva E., P. Uchakin, R. Vaughan, J. Stray-Gundersen. 1996. Cytokine production and incidence of respiratory infection in female middle and long distance runners. *Medicine and Science in Sports and Exercise* 28: S91.
36. Grossman, A., J.R. Sutton. 1985. Endorphins: what are they? how are they measured? what is their role in exercise? *Medicine and Science in Sports and Exercise* 17: 74-81.
37. Hayden, F.G., J.K. Albrecht, D.L. Kaiser, J.M. Gwaltney Jr. 1986. Prevention of natural colds by contact prophylaxis with intranasal alpha2-interferon. *New England Journal of Medicine* 314: 71-75.
38. Heir, T., S. Larsen. 1995. The influence of training intensity, airway infections and environmental conditions on seasonal variations in bronchial responsiveness in cross-country skiers. *Scandinavian Journal of Medicine and Science in Sports* 5: 152-159.
39. Imura, H., J. Fukat, T. Mori. 1991. Cytokines and endocrine function: an interaction between the immune and neuroendocrine systems. *Clinical Endocrinology* 35: 107-115.
40. Kappel, M., N. Tvede, H. Galbo, P.M. Haahr, M. Kjær, M. Linstouw, K. Klarlund, B.K. Pedersen. 1991. Epinephrine can account for the effect of physical exercise on natural killer cell activity. *Journal of Applied Physiology* 70: 2530-2534.
41. Kavanagh, T., L.J. Lindley, R.J. Shephard, R. Campbell. 1988. Health competitor. *Annual of Sports Medicine* 4: 55-64.

42. Keast, D., K. Cameron, A.R. Morton. 1988. Exercise and immune response. *Sports Medicine* 5: 248-267.
43. Kelley, K.W. 1991. Growth hormone and immunobiology. In *Psychoneuroimmunology*, eds. R. Ader, D.L. Felten, N. Cohen, 377-403. San Diego: Academic Press.
44. Khansari, D.N., A.J. Murgo, R.E. Faith. 1990. Effects of stress on the immune system. *Immunology Today* 11:169-175.
45. Kiecolt-Glaser, J.K., R. Glaser, E. Strain, J. Stout, K. Tarr, J. Holliday, C.E. Speicher. 1986. Modulation of cellular immunity in medical students. *Journal of Behavioral Medicine* 9: 5-21.
46. Kiecolt-Glaser, J.K., R. Glaser. 1992. Psychoneuroimmunology: can psychological interventions modulate immunity? *Journal of Consulting and Clinical Psychology* 60: 569-575.
47. Koutedakis, Y., R. Budgett, L. Faulmann. 1990. Rest in underperforming elite competitors. *British Journal of Sports Medicine* 24: 248-252.
48. Kozucharova, M.S., 1986. Study of incidence of influenza and respiratory infection among Sofian factory workers and the prophylactic efficacy of some compounds. PhD diss., The D.I. Ivanovsky Institute of Virology, USSR Academy of Medical Sciences.
49. LaPerriere, A., M.A. Fletcher, M.H. Antoni, G. Ironson, N. Klimas, N. Schneiderman. 1991. Aerobic exercise training in an AIDS risk group. *International Journal of Sports Medicine* 12: S53-S57.
50. Leskov, V.P., L.E. Kostyuk, N.K. Gorlina, E.V. Ermakov, V.G. Novozhenov, N.M. Kolomoets. 1982. Some aspects of the activity of the new immunostimulator: Diuciphon. *Immunologiia* 5: 34-37.
51. Levin, M.Ya., V.Ch. Chavinson, V.Yu. Byazemsky, S.V. Seryi, B.S. Moldobaev. 1991. Thymogen prophylaxis of respiratory infections in young athletes. *Teoriya i Praktika Phyzicheskoyi Kulturyi* 8: 40-44.
52. Loosli, A.R., J. Benson, D.M. Gillien, K. Bourdet. 1986. Nutrition habits and knowledge in competitive adolescent female gymnasts. *Physician and Sportsmedicine* 14: 118-130.
53. Mackinnon, L.T., T.W. Chick, A. van As, T.B. Tomasi. 1989. Decreased secretory immunoglobulins following intense endurance exercise. *Sports Training and Medicine Rehabilitation* 1: 209-218.
54. Mackinnon, L.T., D.G. Jenkins. 1993. Decreased salivary immunoglobulins after intense interval exercise before and after training. *Medicine and Science in Sports and Exercise* 25: 678-683.
55. Mackinnon, L.T., S.L. Hooper. 1996. Plasma glutamine and upper respiratory tract infection during intensified training in swimmers. *Medicine and Science in Sports and Exercise* 28: 285-290.
56. Malaise, M.G., M.T. Hazee-Hagelstein, A.M. Reuter, Y.Vrinos-Gevaert, G. Goldstein, and P. Franchimont. 1987. Thymopoietin and thymopentin enhance the levels of ACTH, Beta-endorphin and Beta-lipotropin from rat pituitary cells in vitro. *Acta Endocrinologica* 115: 455-459.
57. McCarthy, D.A., M.M. Dale. 1988. The leucocytosis of exercise: a review and model. *Sports Medicine* 6: 333-363.
58. McCruden, A.B., W.H. Stimson. 1991. Sex hormones and immune function. In *Psychoneuroimmunology*, eds. R. Ader, D.L. Felten, N. Cohen, 475-495. San Diego: Academic Press.
59. Merskey, H., J.T. Hamilton. 1989. An open label trial of the possible analgesic effects of dipyridamole. *Journal of Pain and Symptom Management* 4: 34-37.

60. Moffatt, R.J. 1984. Dietary status of elite female high school gymnasts: inadequacy of vitamin and mineral intake. *Journal of American Dietetic Association* 84: 1361-1363.

61. Morozov, V.G., V.Ch. Chavinson. 1978. Characteristics of mechanisms of thymus factor (thymarin). *Doklady Akademii Nauk* 240: 1004-1007.

62. Mowat, A.G., T.L. Vischer. 1979. Levamisole: immunomodulation: a new approach to basic therapy of rheumatoid arthritis. Eular Bulletin, Monograph Series 5, Basel: Eular.

63. Murgo, A.J., R.E. Faith, N.P. Plotnikoff. 1986. In *Enkephalins and endorphins: stress-induced immunomodulation*, eds. N.P. Plotnikoff, R.E. Faith, A.J. Murgo, R.A. Good, 221-239. New York: Plenum Press.

64. Newsholme, E.A. 1994. Biochemical mechanisms to explain immunosuppression in well-trained and overtrained athletes. *International Journal of Sports Medicine* 15: S142-S147.

65. Newsholme, P., R. Curi, S. Gordon, E.A. Newsholme. 1986. Metabolism of glucose, glutamine, long-chain fatty acids and ketone bodies by murine macrophages. *Biochemistry Journal* 239: 121-125.

66. Newsholme, E.A., M. Parry-Billings. 1990. Properties of glutamine release from muscle and its importance for the immune system. *Journal of Parenteral and Enteral Nutrition* 14: 63-67.

67. Newsholme, E.A., M. Parry-Billings, N. McAndrew, R. Budgett. 1991. A biochemical mechanism to explain some characteristics of overtraining. *Medicine and Science in Sports and Exercise* 32: 79-93.

68. Parr, R.B., M.A. Porter, S.C. Hodgon. 1984. Nutrition knowledge and practice of coaches, trainers and athletes. *Physician and Sportsmedicine* 12: 127-138.

69. Parry-Billings, M., R. Budgett, Y. Koutedakis, E. Blomstrand, S. Brooks, C. Williams, P.C. Calder, S. Pilling, R. Baigrie, E.A. Newsholme. 1992. Plasma amino acid concentration in the overtraining syndrome: possible effects on the immune system. *Medicine and Science in Sports and Exercise* 24: 1353-1358.

70. Payan, D.G., J.P. McGillis, F.K. Renold, M. Mitsuhaschi, E.J. Goetzl. 1987. Neuropeptide modulation of leukocyte function. *Annals of New York Academy of Science* 496: 182-191.

71. Rabin, D., R.W. Gold, A.N. Margioris, G.P. Chrousos. 1988. Stress and reproduction: physiologic and pathophysiologic interactions between the stress and reproductive axes. In *Mechanisms of physical and emotional stress*, eds. G.P. Chrousos, D.L. Loriaux, P.W. Gold, 377-388. New York: Plenum Press.

72. Rowbottom, D.G., D. Keast, C. Goodman, A.R. Morton. 1995. The haematological, biochemical and immunological profile of athletes suffering from the overtraining syndrome. *European Journal of Applied Physiology* 70: 502-509.

73. Samuel, C.E. 1986. Molecular mechanisms of interferon action. In *Clinical application of interferons and their inducers*, ed. D.A. Stringfellow, 1-18. New York: Marcel Dekker.

74. Schulz, K.H., H. Schulz, K.M. Braumann, A. Flögel, V. Hentschel, T. Lauf, E. Taute, M. Puchner, D.K. Lüdecke, A. Raedler. 1994. Differential effects of physical and psychological stress on endocrinological and immunological parameters in athletes. *International Journal of Sports Medicine* 6: 360.

75. Scott, G.M. 1986. Clinical trials of interferons against viral diseases. In *Clinical application of interferons and their inducers*, ed. D.A. Stringfellow, 149-196. New York: Marcel Dekker.

76. Shephard, R.J., P.N. Shek. 1995. Heavy exercise, nutrition and immune function: is there a connection? *International Journal of Sports Medicine* 16: 491-497.

77. Singh, A., M.L. Failla, P.A. Deuster. 1994. Exercise-induced changes in immune function: effects of zinc supplementation. *Journal of Applied Physiology* 76: 2298-2303.

78. Stewart, J.G., D.A. Ahlquist, D.B. McGill, D.M. Ilstrup, S. Schwartz, R.A. Owen. 1984. Gastrointestinal blood loss and anemia in runners. *Annual Internal Medicine* 100: 843-845.

79. Stringfellow, D.A. 1986. *Clinical application of interferons and their inducers*. New York: Marcel Dekker.

80. Surkina, I.D., U.V. Borodin, L.N. Ovcharenko, G.S. Orlova, E.P. Schumayi. 1983. The experience of immunocorrection in the immunodeficient athletes (the preliminary data). *Teoriya i Praktika Phyzicheskoyi Kulturyi* 7: 18-20.

81. Surkina, I.D., U.V. Borodin, G.S. Orlova, N.A. Usakova. 1983. Immunity in swimmers under conditions of modern training. *Plavanie* 2: 22-24.

82. Surkina, I.D., E.P. Gotovtseva. 1991. Role of the immune system in the athlete's adaptive processes. *Teoriya i Praktika Phyzicheskoyi Kulturyi* 8:27-37.

83. Surkina, I.D., E. P. Gotovtseva, O.N. Vatagina, A.I. Golovachev, M.R. Shurin, N.V. Kost, P.N. Uchakin, V.P. Chemizov, A.A. Zozulya. 1993. The effect of immune-interferongenesis stimulation on adaptation to exercise: participation of the opioid system. *Vestnik Sportivnoi Medicinyi Rossii*, 1: 11-16.

84. Surkina, I.D., E. P. Gotovtseva, O.N. Vatagina, A.A. Vorobyev, N.S. Dudov, L.V. Kostina, N.N. Ozolin, P.N. Uchakin. 1994. The effect of immune-interferongenesis stimulation on adaptation to exercise: participation of the endocrine system. *Vestnik Sportivnoi Medicinyi Rossii* 1: 10-16.

85. Tabor, J.M. 1986. Production-purification of interferons: recombinant technology. In *Clinical application of interferons and their inducers*, ed. D.A. Stringfellow, 61-82. New York: Marcel Dekker.

86. Tonew, M., E. Tonew, R. Mentel. 1977. The antiviral activity of dipyridamole. *Acta Virologica* 21: 146-150.

87. Tonew, E., M.K. Indulen, D.R. Dzeguze. 1982. Antiviral action of dipyridamole and its derivates against influenza virus A. *Acta Virologica* 26: 125-129.

88. Tvede, N., M. Kappel, K. Klarlund, S. Duhn, J. Halkjær-Kristensen, M. Kjær, H. Galbo, B.K. Pedersen. 1994. Evidence that the effect of bicycle exercise on blood mononuclear cell proliferative responses and subsets is mediated by epinephrine. *International Journal of Sports Medicine* 15: 100-104.

89. Tvede, N., J. Steensberg, B. Baslund, J. Halkjær-Kristensen, B.K. Pedersen. 1991. Cellular immunity in highly trained elite racing cyclists during periods of training with high and low intensity. *Scandinavian Journal of Medicine and Science in Sports* 1: 163-166.

90. Uchakin, P.N. 1993. Production of lymphokines (interleukin-2 and gamma-interferon) in stress modelings in humans. PhD diss., The N.F. Gamaleya Research Institute of Epidemiology and Microbiology, Russian Academy of Medical Sciences.

91. Van Erp-Baart, A.M.J., W.M.H. Saris, R.A. Binkhorst, J.A. Vos, J.W.H. Elvers. 1989. Nationwide survey on nutritional habits in elite athletes. Part II. Mineral and vitamin intake. *International Journal of Sports Medicine* 10: S11-S16.

92. Viru, A. 1985. *Hormones in muscular activity*. Vol. II: Boca Raton, FL: CRC Press.

93. Weiss, M. 1994. Anamnestic, clinical and laboratory data of 1300 athletes in a basic medical check with respect to the incidence and prophylaxis of infectious diseases. *International Journal of Sports Medicine* 6: 360.

94. Weicker, H., E. Werle. 1991. Interaction between hormones and the immune system. *International Journal of Sports Medicine* 12: S30-S37.

95. Weidner, T.G. 1994. Literature review: upper respiratory illness and sport and exercise. *International Journal of Sports Medicine* 15: 1-9.

96. Werner, G.H. 1987. Immunostimulants: the western scene. In *Immunostimulants now and tomorrow*, eds. I. Azuma, G. Jolles, 3-39. Berlin: Springer-Verlag.

97. Wolfarth, B., M. Richter, E. Jakob, J. Keul. 1994. Clinical-immunological course observations during an 8-week training cycle in endurance athletes. *International Journal of Sports Medicine* 15: 360-361.

98. Wybran, J., A. Governs. 1977. Levamisole and human lymphocyte surface markers. *Clinical and Experimental Immunology* 27: 319-321.

Nutritional Considerations of Overreaching and Overtraining

CHAPTER 13

Energy Intake, Diet, and Muscle Wasting

Jacqueline R. Berning, PhD, RD

Introduction

The body maintains or at least strives to maintain a constant environment (homeostasis) even in periods of heavy training or overreaching. A key aspect of maintaining homeostasis during periods of heavy work or training is satisfying the body's required energy intake as well as other various nutrients. Muscle damage, dietary intake, and overtraining interrelate. Specifically, some athletes who are overtrained appear to have decreased hunger and appetite, which can lead to reduced energy intake and less than optimal intake of carbohydrates and other nutrients; these dietary deficiencies could lead to reduced muscle glycogen levels and therefore affect performance, mood, and normal bodily functions. This chapter will focus on nutrition factors that could be affected during periods of prolonged training, that when left unattended, may contribute to the syndrome known as overtraining or overreaching.

Energy Intake

Energy intake is dependent on kilocalorie input. Many factors influence when and what a person eats. For example, some athletes who train vigorously may not be hungry for hours after a hard workout, while other athletes who experience mood disturbances (16) may eat too much. Social issues also influence how much a person eats. For example, many athletes train away from home or in unfamiliar surroundings and may be forced to eat by themselves. Athletes who do not like to eat alone may not eat enough food or may make lousy food choices. On the other hand, athletes who are more social and comfortable with their surroundings and teammates may have a tough time passing up the dessert after a filling meal with their teammates.

Numerous physiological and social forces govern how much an individual eats or what foods the individual picks to eat. Hunger and appetite both influence food intake. Many of the hunger-related and appetite-related forces overlap each other. It is thought that these forces are redundant because nutrients are vital to the body, and it is critical that many factors encourage eating for maintenance of health and well being. Overall, the system is not perfect. However, there is agreement that powerful internal and social forces greatly influence energy intake.

Hunger-Related Forces

Hunger

Hunger has been defined as a primarily physiological drive to find and eat food. It is mostly driven by negative internal forces such as low blood glucose and fatty acids, and low levels of neurotransmitters and circulating hormones (see table 13.1) For example, several hours after eating the concentrations of glucose, amino acids and fatty acids in the blood have dropped, leaving the body to rely upon the liver for energy reserves and causing hunger to return to the person. Hunger can also be triggered by gastric contractions, an absence of nutrients in the small intestines, and gastrointestinal hormones. Two compounds that may be increased during overtraining may lead to increased feedings—increased endorphins and cortisol levels have been shown to increase food intake (37). When the body is responding to stress, either physical or mental, which result in increased endorphins or cortisol, the body responds by increasing hunger signals to meet the energy demands of the stress. In general then, as physiological markers of overtraining increase in the athlete one would expect hunger responses to increase to meet the extra energy demands of prolonged training. However, there are other factors that decrease hunger cues. For instance, environmental factors such as temperature and humidity, as well as the availability of food, influence hunger negatively (37). Increased body temperatures have been known to decrease hunger in patients with fevers or diseases in whom core temperatures have been elevated (37).

Table 13.1 Factors That Influence Hunger

• Stomach distention	• Environmental factors
• Glucose level	— temperature
• Insulin	— humidity
• Glucagon	• Emotional status
• Fatty acid levels	— stress
• Neurotransmitters	— mood
— sertonin	

It is possible that athletes who are involved in heavy training programs in hot and humid environmental conditions could have elevated body temperatures throughout the day. This could result in abnormal hunger cues and failure to eat enough food to sustain the energy expended during training despite other physiological signs such as increased endorphins and cortisol. Food availability is certainly a factor influencing energy intake. If an athlete can not afford to grocery shop or consume enough food necessary to meet energy demands despite physiological signs to increase energy intake, weight loss will occur, that could impede training or compound the situation during overtraining.

Emotional factors such as stress and mood also affect hunger. While these responses differ in each individual, these factors certainly should not be overlooked in the athlete who is training hard for prolonged periods of time.

Appetite

Appetite is defined as the psychological and environmental forces that influence eating behaviors. Emotional factors such as stress, mood, and personal beliefs can affect the desire to eat. For example, depressed individuals may exhibit either decreased or increased food intake. The particular response could be as a result of experiences in the past, when eating or not eating made them feel better or worse during a stressful period or crisis. There may be a link to overtraining and the desire to eat (appetite) through mood. Indeed, studies in which mood was monitored in athletes who have trained for prolonged periods of time have shown that moods can change in a negative manner very rapidly with overtraining (27). These findings suggest that there could be a link between overtraining and the desire to eat. It has been shown that as training advances, athletes develop dose-related mood disturbances, meaning those who are at levels of overtraining exhibit a low score for vigor and increasing scores for negative moods (27). Many researchers have anecdotally commented on the fact that loss of appetite is a factor in overtraining. Therefore, monitoring profiles of mood scores (POMS) can help to predict athletic success and prevent overtraining in several types of athletes (3, 20, 33). In athletes who are prone to overtraining, mood should be monitored to not only be aware of overtraining but to monitor food intake as well.

Environmental factors, such as availability of food, time of day, social obligations, insufficient sleep, and temperature and humidity can influence eating behavior (37). Athletes who train rigorously and who have other commitments such as school or part-time jobs often do not have an appetite nor the time to sit and consume the amount of kilocalories necessary to maintain bodily reserves of nutrients and energy. Other factors such as bodily appearance, social customs, and peers can influence the desire to eat. Concern about appearance in a swim suit or gymnastics outfit can influence food choices. A female athlete who is concerned about her weight may choose smaller portions of food or none at all despite the fact that she is hungry and has physiological signs of hunger.

Hunger is primarily the physiological or internal desire to find and eat food, while appetite is influenced by external forces like mood, social customs, time of day, palatability, and cultural habits. While both hunger and appetite play major roles in determining what and how much a person eats, thus determining energy intake, most Americans probably respond more to external, appetite-related forces than to hunger-related ones in choosing when and what to eat. Table 13.2 lists the influence of hormones and certain chemical compounds produced in the body and their relationship to both hunger and appetite.

Table 13.2 Hormones and Other Factors That Affect Feeding Behavior

Increase food intake	Decrease food intake
• Insulin	• Insulin
• Endorphins	• CPK
• Norephinephrine	• Dopamine
• Cortisol	• Somatostatin
• Growth hormone	• Histamine
• Progesterone	• Vasopressin
• Some tranquilizers	• Amphetamines
	• Lepin
	• Nicotine

Nutritional, Balanced Diets for Athletes

Nutritional guidelines for athletes have been gathered and agreed upon from a number of sophisticated studies throughout the years. However, the nutritional requirements for athletes who are in the process of being overtrained or who have reached overtraining or overreaching are not well known. While the overtraining literature is replete with signs of the nutritional implications of overtraining, no specific studies have been designed to investigate specific nutrients and their role in overtraining. For example, Sherman and Maglischo (35) alluded to the nutritional implications of "loss of appetite" and "weight loss" as examples of symptoms of chronic fatigue in overtrained swimmers. Costill and his colleagues (12) found that swimmers who were overtrained failed to consume enough carbohydrate on a daily basis, compounding the fatigue factor found with heavy training; and Randy Eichner (16) concluded that proper nutrition is crucial in the prevention of overtraining. Most individuals in the fields of endurance and resistance training would agree that proper nutrition is a must for the prevention of overtraining. However, what specific role carbohydrates, proteins, fats, or vitamins and minerals might play has not been extensively studied using an overtraining model.

Carbohydrate

In one of the few studies dealing with carbohydrate intake and overtraining, Costill and colleagues (12) overtrained 12 collegiate-level swimmers by doubling the volume of work done by each swimmer for 10 days. As a group, performance tests were not altered by the 10 days of increased work, while cortisol and creatine kinase levels elevated significantly during the increased work. However, it was noted that four of the swimmers (group A) had the greatest difficulty completing workouts and they were chosen by the researchers for closer evaluations and comparison to the eight swimmers (group B), who had less difficulty with the training. Dietary records kept the last 2 days of training revealed that group A consumed fewer calories than the swimmers in group B (a difference of almost 1,000 kcal). In terms of carbohydrate (CHO), group B was close to the recommended 7-10 grams of CHO per kilogram of body weight (g/kg/d) consuming about 8.2 grams of CHO/kg/d, while group A consumed only 5.3 grams of CHO/kg/d. The authors also measured glycogen content and noted that the swimmers in group A had significantly lower glycogen values both before and after the 10 day regimen compared to the swimmers in group B. The authors concluded that the swimmers in group A were insensitive to changes in their daily energy expenditure and may have had difficulty maintaining a calorie and carbohydrate balance. Their conclusions were based on food records performed at the beginning of the experiment showing that group A consistently ate fewer kilocalories per kilogram of body weight than group B. While the swimmers were not classified as typical overtrained athletes, some of them did have difficulty finishing the workouts. These swimmers may have become chronically fatigued as a result of improper nutrition. The symptoms described by Costill et al. (12), such as fatigue and difficulty completing workouts associated with heavy volumes of work, were reported in about 30% of the swimmers in the study. These same symptoms have been alluded to over and over again in the overtraining literature; however, this is one of the few studies where the nutritional status of the athlete has been investigated and linked to the possibility of the athlete becoming chronically fatigued. If this phenomenon is happening in about 30% of swimmers who have increased their volume of work, might it not happen in other athletes? See chapter 14 for more information regarding the relationship of carbohydrate availability and overtraining.

Protein

Because protein and especially the branched-chain amino acids can be used as a metabolic fuel during exercise, much controversy surrounds the need for protein in exercising individuals (5, 22). Concerns about the quality of the protein consumed, as well as the energy intake of the subjects, may have led to incorrect interpretations of the early protein studies determined by classic nitrogen balance techniques. Protein requirements in exercise are not only dependent upon the quality and quantity of the protein but also on energy intake. Gail Butterfield and colleagues (7) have shown that feeding as much as 2 g protein/kg/d to men running five or 10 miles per day at 65-75% of their VO_2max is insufficient to maintain nitrogen balance when energy intake is inadequate by as little as 100 kilocalories per day. Furthermore, when Butterfield et al. (6), investigated protein requirements in women running 3-5 mi/d at about 65% of VO_2max, protein requirements were estimated to be 1.1 g/kg/d in women who consumed 35 kcal/kg/d to maintain body weight.

Indeed, when athletes are in a negative energy balance protein requirement will need to increase, while athletes with a positive energy balance appear to need less protein to maintain health and lean body mass.

Protein Needs for Endurance Athletes.   One interesting note for endurance athletes is that there appears to be a drop in nitrogen balance in response to initiation of a moderate endurance exercise program (18, 39). However, this decline is corrected within two weeks of the start of exercise without any dietary manipulation (17). Butterfield and Calloway (8) verified this transient drop in nitrogen balance and found that nitrogen balance was more positive after the adaptation than before. This suggests that the protein requirement for nitrogen equilibrium in individuals participating in moderate endurance exercise may actually be lower than that of the sedentary population, provided that energy intake is adequate.

Protein Catabolism.   Despite the negative energy balance that has been shown by Calloway and Spector (9) and Butterfield (7), athletes need to maintain blood glucose output. If necessary, in order to supply the needed glucose, muscle tissue may be used as an energy source. Amino acids derived mainly from the catabolism of muscles are transaminated or deaminated and transported to the liver and used as a source of glucose production via gluconeogenesis (muscle wasting). These amino acids are transported to the liver primarily in the form of alanine and glutamine. In the case of glutamine, ammonium ions are coupled with glutamate to yield glutamine. The glutamine is released from the muscle to the blood and taken up by the kidneys and the intestines. Normal levels of glutamine range from 0.5-0.6 mmol/L (19).

Recently, attention has been directed toward glutamine levels and the overtrained athlete. Parry-Billings et al. (31) found that overtrained runners exhibit slightly, but significant, lower plasma glutamine concentrations compared to nonovertrained runners. Glutamine is required as a nitrogen source for nucleotide synthesis by lymphocytes (19). This implies that low glutamine levels may be linked to immunologic responses and thus put overtrained athletes at higher risk for infections. However, Mackinnon and Hooper (24) studied swimmers whose workouts were intensified over the course of four weeks to determine whether changes in glutamine concentrations were related to the appearance of upper respiratory tract infections. Eight of the twenty-four swimmers were classified as overtrained based on decrements in timed trial performances and self-reported persistent fatigue ratings. The other sixteen subjects were considered the well-trained group. Plasma glutamine levels were determined before, midway, and postovertraining period. For all of the subjects, plasma glutamine levels increased during the overtraining period. However, the overtrained group had lower plasma glutamine levels than the well-trained swimmers, although both groups were well above the normal levels of glutamine of 0.5-0.6 mmol/L (19).

While the authors found no significant differences in glutamine levels and the incidence of upper respiratory infections, this study supports the idea that athletes who are subjected to large volumes of work or overtraining may be at risk for protein-energy malnutrition and muscle wasting, particularly if the athlete is not consuming enough food (especially carbohydrates) to spare protein and allow it to be used for protein synthesis. While further research needs to be done to explore this issue, many of the complaints that the subjects listed in the Mackinnon and Hooper study (24) could result from not consuming enough calories (carbohydrates). If carbohydrate intake was insufficient, protein would have been used as an energy source, thereby wasting lean body mass, that could certainly affect performance and explain the high levels of glutamine seen in all the swimmers during the intensive training period.

Implications for the Overtrained Athlete.   The implications that protein will be used as a fuel source for athletes who do not maintain energy balance during periods of overtraining or overreaching are enormous. Although no studies have investigated the protein status or nitrogen balance of overtrained athletes, one can assume that due to an increased volume of work with its possible effects on energy intake via hunger and appetite, some athletes may experience the negative nitrogen balance that may lead to a loss of lean body mass. Moreover, these reactions, in which protein is used as a substrate for glucose production via gluconeogenesis, are accelerated by high glucagon-to-insulin ratios (as seen in low carbohydrate intakes or low energy intake) and by gluococorticoids such as cortisol (19), which is found in athletes who are overtrained (12, 16)

Nitrogen Balance.   Athletes who are involved with heavy training or overtraining need to maintain a positive nitrogen balance to eliminate the wasting of lean body mass that occurs when protein is used as an energy source. The data on protein requirements for the maintenance of nitrogen balance is controversial at best. Celejowa and Homa (10) found no increase in lean tissue when consuming 1.85 g protein/kg/d while energy intake balanced energy output within 100 kcal/day. Other studies (2, 25) suggest that an energy surplus may be required for positive nitrogen balance. Butterfield and Tremblay (6) found that when energy intake exceeded need by 400 kcal, protein intake had no significant effect on nitrogen retention. The improvement in nitrogen balance seen with increased energy and protein intake was explained by the energy contribution of the protein. These data suggest that it is far more important to have athletes who are involved with heavy training to consume adequate amounts of kilocalories than to overconsume on protein or to take protein supplements at the expense of other energy-providing nutrients.

Protein Recommendations.   The recommended dietary allowance for protein for individuals in the United States is based on the weight of the individual. The suggested range of protein intake for athletes is 0.8-2.0 g/kg/d (1). Thus, a 17-year-old male swimmer who weighs 165 pounds (70 kg) needs 60-150 grams of protein per day. The typical American diet supplies about 100 grams of protein per day, for an average of 1.4 grams of protein per kilogram for the reference man. It has been suggested that athletes who consume 1.5-2.0 g protein/kg/d would meet the demands of protein synthesis and maintain nitrogen balance as long as they remain in energy balance. Athletes who are consuming enough kilocalories and have reasonably balanced diets do not need to consume extra protein. The exceptions would be athletes who are on restricted kilocalorie diets or those who make lousy food choices. Their higher protein requirement is not unexpected due to their low energy intake.

Fat Consumption

Some researchers have proposed that consuming more fat (fat loading) may help with endurance training (28, 32). It is generally accepted that an increased supply and subsequent oxidation of free fatty acids inhibits carbohydrate utilization (15). This observation may be of value to the athlete, since during sustained physical exercise depletion of body carbohydrate stores is a major factor leading to exhaustion. In fact, it has been found that an increased availability of free fatty acids reduces the rate of glycogen utilization during exercise and delays the onset of exhaustion (11).

Several investigators (28, 32) have found that consuming a high fat diet can cause an increase in the utilization of fatty acids during exercise. While these studies found no difference in endurance time between high fat and moderate fat diets in spite of diminished carbohydrate stores, other studies have found different responses. Johansson et al. (21) fed seven moderately fit subjects a high fat diet in either a liquid or solid form (76% fat, 14% protein, 10% carbohydrate), or an isocaloric high carbohydrate diet in a solid form (10% fat, 14% protein, 76% carbohydrate) over four days of no exercise. After the feedings the subjects were exercised by running on a treadmill until exhaustion. Subjects ran at 70% of VO_2max using a 30 min:10 min work:rest ratio. Subjects on the high carbohydrate diet ran significantly longer than those on either the solid or liquid high fat diet (106 + 5 min vs. 64 min + 6 min and 59 min + 6 min, respectively). Blood glucose levels dropped following the high fat diets and all subjects exhibited symptoms of low blood glucose at exhaustion. The authors concluded that four days of a high fat diet produces premature fatigue compared to a high carbohydrate diet.

In another study, Simonsen et al. (36) studied 22 subjects (12 men and 10 women) during 28 days of twice-daily intensive workouts, that consisted of training for 40 min at 70% peak O_2 consumption in the morning and either three 500 m time trials to assess power output or interval training at 70-90% VO_2max in the afternoon. Mean daily training was 65 min at 70% VO_2max and 38 min at >90% VO_2max. On the seventh day each week a VO_2max test was performed and subjects rowed for 35 minutes at 70-80% of VO_2max. Subjects consumed either 5 g carbohydrate/kg/d or 10 g carbohydrate/kg/d. Protein intake was 2 g/kg/d and fat intake was adjusted to maintain body mass. Muscle glycogen content increased significantly for the high carbohydrate group compared to the low carbohydrate group. Mean power output in the timed trials increased in both groups, but more significantly in the high carbohydrate group (+11%) compared to the low carbohydrate group (+2%). The authors concluded that diets with 10 g/kg/d of carbohydrate promote greater muscle glycogen concentration and a greater power output during training compared to a diet containing 5 g/kg/d of carbohydrate. However, the authors did state that a carbohydrate intake of 5 g/kg/d did not lead to glycogen depletion or impairment of power output during training.

While the thought of increasing performance by consuming high fat diets to spare muscle glycogen is interesting there appears to be insufficient evidence at this time to support this concept. A definite concern is the association of a high fat diet with a higher risk for heart disease and certain types of cancer. As a result, most high fat diets are not recommended.

Female Athlete Triad

In the past twenty years, the number of women participating in strenuous sports has increased dramatically. Many questions and concerns have arisen with the changing role of the female athlete (13). One observation that has emerged is that young female athletes who are driven to excel in their sport, may be at risk of developing a potentially lethal triad of medical disorders: disordered eating, amenorrhea, and osteoporosis (30) (see figure 13.1). Alone, each disorder is worrisome and can yield considerable disability, but in combination, the triad disorders are potentially fatal (14).

Female athlete triad

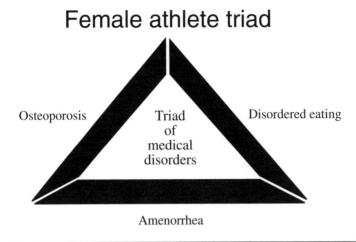

Figure 13.1   Schematic diagram of the triad of medical disorders seen in some young female athletes.

Etiology of Amenorrhea in Female Athletes

The current thinking on the irregular menstrual cycles seen in active and athletic women is that they may be due in part to periods of energy deficiency (energy drain). Female athletes may have increased energy expenditures due to both the physical and psychological stresses of training (14, 15). This increase in energy expenditure as well as restricted eating practices may lead to a greater kilocalorie deficit than is found with just dieting alone. The combination of increased kilocalorie expenditure and restricted eating can influence the secretion of reproductive hormones in female athletes. For example, Schweiger et al. (34) found a positive correlation (p < .01) between energy intake and progesterone levels during the luteal phase of the menstrual cycle. They also found that psychological stress (subjective ratings of stress) from partners, family, and friends correlated negatively with luteal phase progesterone levels as well as estrogen production during the luteal phase (p < .01). Their data supports the concept that nutritional status and both emotional and physical stress play a crucial role in the etiology of menstrual dysfunction in athletes.

Further support of the energy drain theory comes from studies in which thyroid measurements and metabolic rates were determined on amenorrheic athletes. In general, when the body is in a negative calorie balance, thyroid hormones and resting metabolic rate are decreased.

If an amenorrheic athlete is restricting kilocalorie intake, the possibility exists that thyroid hormones and resting metabolic rate will be decreased. Two recent studies investigated the level of thyroid hormones in amenorrheic women (23, 38). The researchers found reduced free T4, free T3, reverse T3, and lower T3 uptake. In addition to measuring thyroid hormones both studies investigated the energy intakes of their amenorrheic subjects and found different results. Ann Loucks and her associates (23) found reduced energy intake in their amenorrheic subjects, while Wilmore and colleagues (38) did not report lower energy intakes in their subjects. Other researchers (29) have not only found lower energy intakes and reduced thyroid hormones but have also documented lower resting metabolic rates, which may demonstrate that these female athletes are in a state of energy drain.

Amenorrheic Athletes as Overtraining Models

Female athletes who present with the symptoms of the female athlete triad could be viewed as a model for overtraining. The rationale for this statement is in the fact that most of the women in this triad of disorders have training regimens with excessive workloads and an emphasis on restricting calories in order to maintain thinness or leanness. Other support for this model comes from a recent case study performed by Dueck et al. (14). The researchers investigated the effect of a 15-week diet and exercise intervention program on energy balance, hormonal profiles, body composition, and menstrual function on an amenorrheic runner. The subject had lost 9 kg of body weight and had not had a normal menstrual cycle for three months after switching from sprinting events to distance running. Her exercise regimen before the intervention consisted of double workouts seven days a week. She complained of chronic fatigue, poor performance, and a high frequency of illness and injury. Baseline data on body composition, fasting hormonal profiles, and bone density were collected before the intervention and again at 15 weeks. The intervention consisted of both a dietary component and an exercise component. The subject was supplemented with one 11-oz liquid meal replacement consisting of 360 kcals/serving (59 g CHO, 17 g protein, 7 g fat). The training component consisted of adding one complete day of rest to her weekly training regimen. Weighed dietary records were determined before the intervention and showed that the athlete was in a negative energy balance by 155 kcals. After the intervention the subject experienced a positive energy balance (+600), an increase in body fat from 8.2 to 14.4%, increased fasting luteinizing hormone from 3.9 to 7.3 mIU/ml, and decreased cortisol from 41.2 to 33.2 ug/dl.

It is important to note that the athlete had no eating disorder or any adverse feelings toward food; she did, however, in her previous track season have difficulty maintaining weight, felt tired, and experienced chronic fatigue during the track season. After using the dietary supplement and with the subsequent reduction in training, the athlete set more personal records, broke two school records and qualified for the National Junior Collegiate Athletic Association track and field meet in several events. Obviously her increase in body fat did not hinder her performance. The authors concluded that using a nonpharmacological approach (increased kilocalories and a reduction in training) for the treatment of athletic amenorrhea can contribute to the resumption of a more normal menstrual hormone profile and improve performance. While this study was not designed as an overtraining or overreaching study, it does have some similar factors: weight loss, chronic fatigue, decrease in performance, higher incidence of illness and injury, and high levels of circulating cortisol, all classic signs and symptoms of overtraining. This study does seem to support the hypothesis that help with the muscle wasting and poor performances seen in overtrained athletes is provided by a nutrition intervention program that increases calorie consumption in combination with a reduction in training.

Conclusion

While no definitive studies have been performed on the specific role of nutrition in overtraining or overreaching, it appears that caloric intake and the nutrient density of the calories consumed during periods of overtraining may play a role in athletic performance and the ensuing health of the athlete (see figure 13.2). Athletes involved in heavy periods of training need to make sure that they are consuming enough carbohydrates and calories so as not to catabolize amino acids from muscle as a fuel source for the exercise. Trainers, coaches, and athletes must also pay special attention to the hunger and appetite sensations in overtrained individuals who are involved with extensive training programs. Many times athletes may not be hungry immediately after a prolonged workout, and therefore wait to eat. However, if food is not available or if the athlete has no appetite or has some mood disturbances due to the results of a performance, the athlete may go without eating, which may result in muscle wasting and poor performances. Monitoring food intake, body weight, POMS, and hormonal indicators of overtraining may help to keep athletes from entering into the overtraining state.

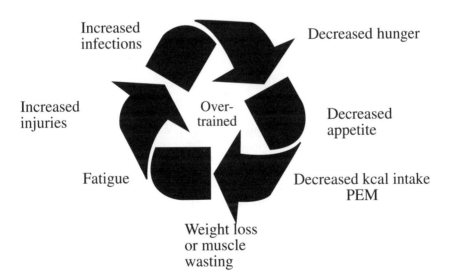

Role of nutrition in overtraining

Figure 13.2   Schematic diagram of the interaction of nutrition and exercise in the overtrained athlete.

References

1. American Dietetic Association. 1987. Nutrition and physical fitness and athletic performance. *Journal of the American Dietetics Association* 87: 933-939.
2. Bartels, R.L., D.R. Lamb, V.M. Vivian, J.T. Snook, K.F. Rinehart, J.P. Delaney, K.B. Wheeler. 1989. Effects of chronically increasing consumption of energy and carbohydrate on anabolic adaptations to strenuous weight training. Report of the Ross Symposium. In *The theory and practice of athletic nutrition: bridging the gaps*, eds. A.C. Grandjean, J. Storlie, 70-80. Columbus, OH: Ross Laboratories.
3. Berglund, B., H. Safstrom. 1994. Psychological monitoring and modulation of training load of world-class canoeists. *Medicine and Science in Sports and Exercise* 26: 1036-1040.
4. Brownell, K.D., J. Rodin, J.H. Wilmore. 1992. Eating, body weight and performance in athletes. Philadelphia: Lea & Febiger.
5. Butterfield, G.E. Amino acids and high protein diets. 1991. In Perspectives in exercise science and sports medicine, vol. 4, *Ergogenics: enhancement of performance in exercise and sport*, eds. D.R. Lamb, M.H. Williams, 87-122: Ann Arbor, MI: Brown and Benchmark.
6. Butterfield, G.E. and A. Tremblay. 1990. Physical activity and nutrition in the context of fitness and health. In *Exercise, fitness, and health; a consensus of current knowlege,* ed. C. Bouchard. Ann Arbor, MI: Books on Demand Publisher.
7. Butterfield, G.E. 1987. Whole body protein utilization in humans. *Medicine and Science in Sports and Exercise* 19: S157-S165.
8. Butterfield, G.E., D.H. Calloway. 1984. Physical activity improves protein utilization in young men. *British Journal of Nutrition* 51: 171-184.
9. Calloway, D.H., H. Spector. 1954. Nitrogen balance as related to caloric and protein intake in active young men. *American Journal of Clinical Nutrition* 2: 405-411.
10. Celejowa, I., M. Homa. 1970. Food intake, nitrogen and energy balance in Polish weight lifters during a training camp. *Nutrition and Metabolism* 12: 259-274.
11. Costill, D.L., E. Coyle, G. Dalsky, W. Evans, W. Fink, Hoopes. 1977. Effects of elevated FFA and insulin on muscle glycogen usage during exercise. *Journal of Applied Physiology* 43: 695-699.
12. Costill, D.L., M.G. Flynn, J.P. Kirwan, J.A. Houmard, J.B. Mitchell, R. Thomas, S.H. Park. 1988. Effects of repeated days of intensified training on muscle glycogen and swimming performance. *Medicine and Science in Sports and Exercise* 20: 249-254.
13. Dueck, C.A., K.S. Matt, M.M. Manore, J.S. Skinner. 1996. A diet and training intervention program for the treatment of athletic amenorrhea. *International Journal of Sports Nutrition* 6: 24-40.
14. Dueck, C.A., M.M. Manore, K.S. Matt. 1996. Role of energy balance in athletic menstrual dysfunction. *International Journal of Sports Nutrition.* 6: 165-190.
15. Dyck, D.J., C.T. Putman, G.J.F. Heighenhauser, E. Hultman, L.L. Spriet. 1993. Regulation of fat-carbohydrate interaction in skeletal muscle during intense aerobic cycling. *American Journal of Physiology* 265: E852-E859.
16. Eichner, E.R. 1995. Overtraining: consequences and prevention. *Journal of Sport Science* 13: S41-S48.

17. Gontzea, I., P. Sutzesco, S. Dumitrache. 1975. The influence of adaptation to physical effort on nitrogen balance in man. *Nutrition Reports International* 11: 231-236.

18. Gontzea, I., P. Sutzesco, S. Dumitrache. 1974. The influence of muscular activity on nitrogen balance and on the need of man for protein. *Nutrition Reports International* 10: 35-43.

19. Groff, J.L., S.S. Gropper, S.M. Hunt. 1995. Advanced nutrition and human metabolism. 2d ed. St. Paul: West.

20. Gutmann, M.C., M.L. Pollack, C. Foster, D. Schmidt. 1984. Training stress in Olympic speed skaters: a psychological perspective. *Physician and Sportsmedicine* 12: 45-57.

21. Johannsson, A., C. Hagen, H. Galbo. 1981. Prolactin, growth hormone, thyrotropin, 3, 5, 3'-triiodothyronine, and throxyine responses to exercise after fat- and carbohydrate-enriched diet. *Journal of Clinical Endocrinology and Metablolism* 52: 56-61.

22. Lemon, P.W.R. 1991. Protein and amino acid needs of the strength athlete. *International Journal of Sports Nutrition* 1: 127-145.

23. Loucks, A.B., G.A. Laughlin, J.F. Mortola, L. Girton, J.C. Nelson, S.S.C. Yen. 1992. Hypothalamic-pituitary-thyroidal function in eumenorrheic and amenorrheic athletes. *Journal of Clinical Endocrinology and Metabolism* 75: 514-518.

24. Mackinnon, L.T., S.L. Hooper. 1996. Plasma glutamine and upper respiratory tract infection during intensified training in swimmers. *Medicine and Science in Sports and Exercise* 28: 285-290.

25. Marable, N.L., N.L. Kehrberk, J.T. Judd, E.S. Prather, C.E. Bodwell. 1988. Caloric and selected nutrient intakes and estimated energy expenditures for adult women: identification of non-sedentary women with low energy intakes. *Journal of the American Dietetics Association* 88: 687-693.

26. Montgomery, R., T.W. Conway, A.A. Spector, D. Chappell. 1996. *Biochemistry: a case oriented approach*. 6th ed., 256. St. Louis: Mosby.

27. Morgan, W.P., D.R. Brown, J.S. Raglin. 1987. Psychological monitoring of overtraining and staleness. *British Journal of Sports Medicine* 21: 107-114.

28. Muoio, D.M., J.J. Leddy, P.J. Horvath, A.B. Awad, D.R. Pendergast. 1994. Effect of dietary fat on metabolic adjustments to maximal VO$_2$ and endurance in runners. *Medicine and Science in Sports and Exercise* 26: 81-88.

29. Myerson, M., B. Gutin, M.P. Warren, M.T. May, I. Contento, M. Lee, F.X. Pi-Sunyer, R. N., Pierson, J. Brooks-Gunn. 1991. Resting metabolic rate and energy balance in amenorrheic and eumenorrheic runners. *Medicine and Science in Sports and Exercise* 23: 15-22.

30. Nattiv, A., R. Agostini, B. Drinkwater, K. Yeager. 1994. The female athlete triad: the interrelatedness of disordered eating, amenorrhea, and osteoporosis. *Clinical Sports Medicine* 13: 405-418.

31. Parry-Billings, M., R. Budgett, Y. Koutedakis. 1992. Plasma amino acid concentrations in the overtraining syndrome: possible effects on the immune system. *Medicine and Science in Sports and Exercise* 24: 1353-1358.

32. Phinney, S.D., B.R. Bistrian, W.J. Evans, E. Gervino, G.L. Blackburn. 1983. The human metabolic response to chronic ketosis without caloric restriction and preservation of submaximal exercise capabilities with reduced carbohydrate oxidation. *Metabolism* 32: 769-776.

33. Raglin, J.S., W.P Morgan, A. E. Luchsinger. 1990. Mood and self-motivation in successful and unsuccessful female rowers. *Medicine and Science in Sports and Exercise* 22: 849-853.

34. Schweiger, U., F. Herrmaan, R. Laessie, W. Riedel, M. Schweiger, K.M. Pirke. 1988. Caloric intake, stress, and menstrual function in athletes. *Fertility and Sterility* 49: 447-450.
35. Sherman, W.M., E.W. Maglischo. 1991. Minimizing chronic athletic fatigue among swimmers: special emphasis on nutrition. In *Sports science exchange* 35: 4. Chicago: Gatorade Sports Science Exchange.
36. Simonsen, J.C., W.M. Sherman, D.R. Lamb, A.R. Dernbach, J.A. Doyle, R. Strauss. 1991. Dietary carbohydrate, muscle glycogen, and power output during rowing training. *Journal of Applied Physiology* 70: 1500-1505.
37. Wardlaw, G.M., P.M Insel. 1996. *Perspectives in nutrition*. 3d ed. St. Louis: Mosby.
38. Wilmore, J.H., K.C. Wambsgans, M. Brenner, C.E. Broeder, I. Paijmans, J.A. Volpe, K.M. Wilmore. 1992. Is there energy conservation in amenorrheic compared with eumenorrheic distance runners? *Journal of Applied Physiology* 72: 15-22.
39. Yoshimura, H. 1961. Adult protein requirements. *Federal Proceedings* 20: 103-110.

CHAPTER 14

Carbohydrate Metabolism During Endurance Exercise

W. Michael Sherman, PhD, Kevin A. Jacobs, MS, and Nicole Leenders, MS

Metabolism of Fuels

The adaptation to intense training is facilitated in part through the maintenance of bodily energy stores. Training intensity, duration, amount of recovery, and consumption of macronutrients and micronutrients all play roles in this process. An imbalance among these factors, however, may contribute to the failure of an athlete to adapt to the stresses of training, a characteristic of overreaching and subsequent overtraining (18, 30, 32). It has previously been suggested that acute depletion or a chronic reduction in carbohydrate stores may play a role in the onset of overreaching and subsequent overtraining (44). In examining the role of carbohydrate metabolism and availability in overreaching and overtraining, it is important to also consider the influences and interactions of protein and lipid metabolism.

Also of importance to the discussion of the metabolism of fuels in the context of overreaching and overtraining is whether bodily fuel stores undergo net catabolism or anabolism. The study of the balance of anabolism and catabolism as it relates to overreaching and overtraining has primarily stemmed from research in the area of neuroendocrinology. Recently, researchers have examined the ratio of plasma free testosterone to cortisol (2, 5, 28). Although there is evidence to the contrary (53), it has been suggested that a 30% decrease in the plasma free testosterone to cortisol ratio be used as an indicator of a state of overreaching and overtraining. An imbalance between the anabolic and catabolic states in response to training would possibly have its greatest effects during recovery, where the anabolic state normally predominates (30). A net catabolic state may primarily affect protein metabolism; however, due to the interrelationship of the fuel systems, lipid and carbohydrate metabolism may also be affected by a net catabolic state. On the other hand, it is possible that training itself protects against the prevalence of the catabolic response to acute stress (52), and even very intense daily work such as that which occurs with orienteering does not appear to produce the initial indications of overreaching when adequate carbohydrate is consumed to normalize muscle glycogen (28).

Protein

Protein represents a substantial source of potential energy, accounting for approximately 17% of bodily energy stores (21). Skeletal muscle represents the largest source of protein. Protein also has a role in the transport and regulation of the metabolism of various fuels. Protein's structural and functionally important roles are thought to limit protein use as a primary and limiting substrate for energy production during exercise.

It is generally accepted that during exercise, the rate of protein synthesis declines while the rate of degradation may increase leading to a greater pool of amino acids, which are available for catabolism. Much of the research on amino acid oxidation by skeletal muscle during exercise has focused on the branched-chain amino acids (BCAA)—leucine, valine, and isoleucine. It appears that BCAA are oxidized by skeletal muscle at a greater rate than other amino acids (1, 3, 19). The rate of oxidation of BCAA is influenced by exercise intensity and duration; however, the total contribution of the oxidation of BCAA is thought not to exceed 3-4% of the total energy expenditure of exercise (24).

Another important factor that influences the rate of protein catabolism during exercise is the availability of carbohydrate. Lemon and Mullin (33) examined urea excretion in urine and sweat in subjects cycling for one hour at 60% VO_2max in either a carbohydrate-loaded or carbohydrate-depleted condition. The breakdown of protein at a rate of 14 g/h contributed 10% of the total energy expenditure during exercise in the carbohydrate-depleted state, whereas the contribution of protein oxidation to the total energy expenditure was only 4% (6 g/hr) in the carbohydrate-loaded state. Lemon and Mullin (33) concluded that carbohydrate-loading has a protein-sparing effect during prolonged exercise.

Wagenmakers et al. (55) examined the activation of the branched-chain 2-oxoacid dehydrogenase complex (BC-complex) during exercise after glycogen depletion or carbohydrate loading. The BC-complex is the rate-limiting enzyme in the degradation of BCAA and measurements of the activity of the BC-complex have been used as an indicator of the breakdown of these amino acids. Subjects exercised on a bicycle ergometer for up to two hours between 70-75% of the maximal workload. This intensity was gradually lowered to 50% of the maximal workload for most of those in the carbohydrate-depleted group. Exercise led to a 4-fold increase in the activity of the BC-complex in the carbohydrate-depleted group with no significant change in the BC-complex in the carbohydrate-loaded group. Thus it appears that carbohydrate loading prevents the oxidation of BCAA.

While the BCAA are primarily oxidized by skeletal muscle for energy production, other amino acids such as alanine, glutamine, and aspartate are utilized by the liver to make glucose through gluconeogenesis. Gluconeogenesis is thought to contribute 25-35% of hepatic glucose production in the fasted state (6). Alanine itself may contribute 20-50% of hepatic glucose production from gluconeogenesis in the fasted state (3, 6). During long duration, low-intensity exercise, however, gluconeogenesis contributes only 8% of the total energy expenditure per hour (3).

Also, as bodily carbohydrate stores are lowered during prolonged intense exercise, the rate of gluconeogenesis cannot maintain the blood glucose concentration, the blood glucose concentration falls, and fatigue occurs.

Because prolonged exercise can lead to the degradation of bodily protein stores, it has been suggested that endurance athletes consume 1.2-1.4 g protein/kg/d (34). Consumption of a diet containing 12-15% of total energy intake as protein will be adequate to meet the protein requirement as long as total energy intake is adequate to maintain body weight (16). As discussed previously, the extent of protein degradation during exercise is heavily influenced by the availability of bodily carbohydrate stores. The combination of intense training, inadequate recovery, and inadequate carbohydrate intake may lead to a persistent state of suboptimal bodily carbohydrate levels. This would likely result in an increased utilization of protein for energy production and possibly produce a net catabolic state. In addition to the negative effects this would have on skeletal muscle, there is the possibility a net catabolic state may also compromise the other fuel systems that rely on structures comprised primarily of protein. Free fatty acids (FFA) liberated from adipocytes are bound to the protein albumin for transport in the blood. The movement of FFAs from the blood across the sarcolemma is thought to be mediated by a fatty acid-binding protein (50). Glucose entry into adipose, skeletal muscle, and cardiac cells is mediated by Glut 4, a glucose transporter protein. The enzymes that catalyze the reactions involved in the metabolism of fuels are also proteins. Therefore, a net catabolic state related to proteins alone could have far-reaching consequences, all of which may play roles in the onset of overreaching and overtraining.

Triglyceride

Triglyceride (TG) is the major storage form of lipids in the body. Triglyceride represents the largest source of potential energy, accounting for approximately 80% of total bodily energy stores (20). Beyond the role of TG as a vast source of energy, TG also provides thermal insulation, serves as a cushioning protection for organs, acts as a carrier for fat-soluble vitamins, and is a structural component of many cell membranes.

With the onset of exercise there is normally a transient fall in plasma free fatty acid (FFA) concentration as skeletal muscle uptake of FFA exceeds FFA production from lipolysis in adipocytes. Within 20-30 minutes of light- and moderate-intensity exercise, however, the plasma FFA concentration begins to rise as epinephrine stimulates hormone-sensitive TG lipase, stimulating the degradation of TG into FFA and glycerol that are released into the blood. The FFA is bound to albumin in the blood and is carried by the circulation to peripheral tissues. Upon reaching muscle tissue, FFAs are thought to be transported across the sarcolemma by a carrier-mediated diffusion process (50). Once inside the cell FFA can either be converted to TG or converted to a fatty acyl-CoA and transported into the mitochondria by carnitine acyl-transferase where FFA undergoes beta-oxidation and energy transduction.

The use of lipids as a significant energy source during exercise is thought to be limited to low- to moderate-intensity activities ($< 60\%$ VO_2max). It was once thought that lipid metabolism during exercise was limited by the low concentration of TG in muscle and that FFA must be transported in the blood to the muscle. However, the supply of FFA in the blood normally far exceeds its uptake in the muscle even during moderate-intensity exercise (40). The finding that FFA oxidation is increased with endurance training indicates that the limiting factors for lipid metabolism during exercise likely reside at the level of the muscle cell. Turcotte et al. (51) examined the FFA metabolism of trained and untrained subjects during three hours of knee extensions at 60% of maximal dynamic knee extension capacity. Although arterial plasma FFA concentration increased similarly in both groups, the fractional uptake of FFA across the thigh of the trained subject remained at 15% while that of the untrained subject decreased from 15% to 7%, especially during the last hour of exercise. The FFA uptake increased linearly with FFA delivery in the trained, while FFA uptake became saturated in the untrained. It is quite possible that the carrier-mediated diffusion of FFA across the sarcolemma represents a rate-limiting step in lipid metabolism during exercise (50) that can be increased via endurance training.

The relationship between lipid metabolism and overreaching and overtraining has not been thoroughly examined. Lehmann et al. (31) investigated the effects of more than doubling the training volume of experienced middle- and long-distance runners within a 3 wk period on performance, catecholamines, and energy metabolism. The subjects trained six days each week and their volume of training was increased from 85.9 km/wk in week one to 174.6 km/wk in week four. Resting serum TG, low density lipoprotein, and very low density lipoprotein all decreased significantly. Resting and maximal exercise concentrations of serum FFA and glycerol also decreased significantly. While these results are preliminary and the mechanisms responsible for these changes are not known, it appears that lipid metabolism may be affected by overreaching. Thus, overreaching and overtraining may coincide with a catabolic state in which protein degradation may decrease serum albumin and thus reduce FFA transport in the blood.

Carbohydrate

Carbohydrates are a very limited source of energy compared to the energy stored as protein and TG. The average carbohydrate stores represent approximately 2,000 kcal of energy, comprising only 1-2% of the total bodily energy stores (21). Carbohydrate is stored in skeletal muscle (79% of total carbohydrate) and liver (14% of total carbohydrate) in the form of glycogen, and in the blood (7% of the total carbohydrate) in the form of glucose (41).

As with protein and lipid metabolism, the utilization of carbohydrate stores is sensitive to exercise intensity. At intensities less than 60% of VO_2max, although carbohydrates are oxidized, a substantial amount of energy is provided by oxidation of lipids. However, carbohydrates appear to be the preferred fuel for oxidation at intensities between 65-85% of VO_2max (23, 25, 39). Muscle glycogen represents an immediate source of energy and it is during the first 20-30 minutes of moderate-intensity exercise that muscle glycogen degradation is most pronounced. The rate of muscle glycogenolysis then slows as blood glucose supplies a larger proportion of the substrate for oxidation during exercise. The blood glucose concentration reflects the balance between muscle glucose uptake and hepatic glucose production (44). Blood glucose levels normally remain constant for the first 60-90 minutes of exercise, after which blood glucose begins to decline as liver glycogen levels become depleted (4). Exhaustion during moderate intensity exercise (65-85% of VO_2max) is strongly associated with the depletion of muscle glycogen and in some cases, a lowering of the blood glucose concentration.

With carbohydrate as the preferred fuel for oxidation at the intensities at which athletes train and compete, and because bodily carbohydrate content is limited and can be depleted, adequate intake of carbohydrates coupled with adequate recovery periods is important (41) for training and performance. Because many athletes do not consume the recommended amount of energy as carbohydrate, it is thought that the possible resulting acute reduction or chronic depletion of carbohydrate reserves may play a role in the onset of overreaching and subsequent overtraining (44). As discussed earlier, there is also the possibility that a net catabolic state may effect glucose utilization by compromising the number and activity of glucose transport proteins.

Bodily Carbohydrate and Training

In sporting activities such as running, swimming, and cycling, daily training regimens at 65-85% VO_2max for 90-120 min are not uncommon. It is well known that such moderate-intensity prolonged exercise causes a substantial decrease in bodily carbohydrate stores. Furthermore, it has been suggested that lowered bodily carbohydrate stores may lead to a decline in the athlete's ability to train and perform at high intensities. Thus, to maximize exercise capacity on a daily basis, the intramuscular and hepatic glycogen stores must be replenished between training sessions. Considering that daily energy expenditure for an endurance athlete is approximately 1.5 to 3 times higher than for the average individual, it is important that endurance athletes consume an adequate amount of energy in their meals during all phases of the training and competitive season to maintain the balance between energy expenditure and energy intake. Also, because of the strong relationship between the amount of carbohydrate consumed (188, 325, or 525 g) in the 24 h period after strenuous exercise and the amount of muscle glycogen synthesized (r = .84) (12), athletes are advised to consume up to 8-10 g carbohydrate/kg body weight/d (41, 44) to restore the body's carbohydrate stores between training sessions.

Obviously, endurance athletes must ensure that adequate amounts of carbohydrate are ingested on a daily basis during all phases of training and the competitive season to avoid a reduction in training and maximal exercise performance capacity due to insufficient bodily carbohydrate. Thus, adequate energy and carbohydrate consumption may be an important factor in the prevention of the onset of the overreaching syndrome during a short-term period (< 2 weeks) when the training load is increased (44, 48). The overreaching syndrome as a result of training or nontraining stress results in a short-term decrement in performance that can be accompanied by physiological and psychological symptoms, with restoration of performance capacity in several days or weeks (see figure 14.1). Furthermore, in the long term (> 2 weeks), sufficient energy and dietary carbohydrate intake may also play a preventive role in the onset of the overtraining syndrome during which the restoration of performance capacity may take several weeks or months (see figure 14.2).

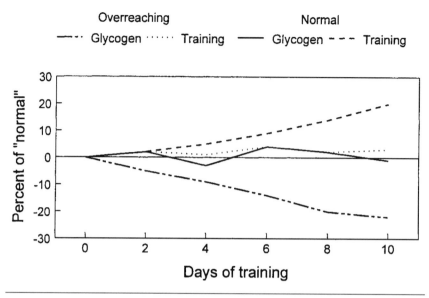

Figure 14.1   Schematic representation of (1) the normal relationship between muscle glycogen and the response to training, and (2) the results published in the literature of the relationship between muscle glycogen and the response to training over the short term in the context of overreaching.
Based on data from references 13, 29, 42, and 48.

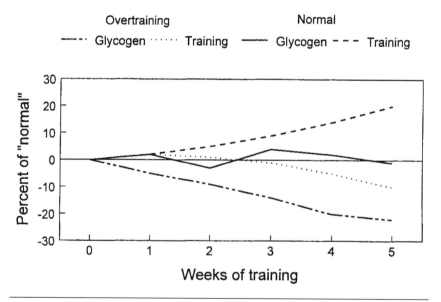

Figure 14.2   Schematic representation of (1) the normal relationship between muscle glycogen and the response to training, and (2) the results published in the literature of the relationship between muscle glycogen and the response to training over the long term in the context of overtraining.
Based on data from references 22 and 47.

Carbohydrate Consumption and Training Over the Short Term (< 2 wk)

Several studies have examined the effects of short-term nutritional manipulations on muscle glycogen concentrations, training, and maximal exercise capacity in endurance athletes when training volume is maintained or suddenly increased (see table 14.1).

After a standardized 7 d period for training and diet to ensure equal bodily carbohydrate stores, Sherman et al. (42) fed cyclists and runners either five or 10 g carbohydrate/kg body weight/d for a 7 d training period. All of the food was provided to match total daily energy expenditure. During the 7 d period, subjects exercised under supervision for 1 h at 75% peak VO_2 followed by five, 1 min sprints at 100% peak VO_2 interrupted by 1 min rest intervals. Muscle biopsies for the analysis of muscle glycogen concentration were obtained before training on days 1, 3, 5, and 7 of the training period. Maximal exercise performance capacity, consisting of two maximal performance trials at 80% peak VO_2 until exhaustion were measured after the normal training session on day 7. For the runners and the cyclists consuming the moderate carbohydrate diet (5 g/kg body weight/d), muscle glycogen was significantly reduced by 30-36% during days 5-7 of the training period. Consuming a diet of 10 g carbohydrate/kg body weight/d, however, maintained muscle glycogen concentration during days 5-7 of the training period. Despite the decline in muscle glycogen stores for the cyclists and runners consuming the moderate carbohydrate diet, training and performance capacity were not impaired.

Costill et al. (13) examined the effect of increased training volume on muscle glycogen concentration, performance, and several psychological parameters in swimmers. During a 10 d period swimmers doubled their daily training volume from 4,266 to 8,970 m/d in two 1.5 hr training sessions, five days a week, while swimming intensity was maintained at 94% VO_2max. The increased training volume was associated with an estimated caloric expenditure of 4,667 kcal/d. Swimming performance tests were performed one day before, and after five and 11 days of increased training volume. Subjects self-selected a diet that was analyzed from the subject's dietary recall records over the final 2 d of the 10 d training period. Six swimmers consumed a self-selected diet containing 4,682 kcal/d and 8.2 g carbohydrate/kg body weight/d while four swimmers consumed only 3,631 kcal/d and 5.3 g carbohydrate/kg body weight/d. Overall, all the swimmers reported local muscular fatigue and difficulty completing the training sessions. However, endurance swimming performance, sprinting, and swimming power were not affected by the increased training volume. Biopsies obtained at the beginning and end of the 10 d period from the deltoid muscle revealed that muscle glycogen had declined 20% by the end of the 10 d training period.

The swimmers that consumed only 3,631 kcal/d had the greatest difficulty completing the training sessions and muscle glycogen was significantly lower before and after the 10 d period compared to the other swimmers with the higher daily energy intake. The swimmers who failed to increase their energy intake to maintain the balance between energy expenditure and energy intake had a 20% reduction in muscle glycogen concentration, experienced local muscle fatigue, and were forced to significantly reduce daily swimming velocity. Similarly, Kirwan et al. (29) had runners double their daily training distance over a 5 d period while consuming either 3.9 or 8.0 g carbohydrate/kg body weight/d. The muscle glycogen concentration was also lower for the moderate carbohydrate diet when compared to the high carbohydrate diet (81 mmol/kg versus 121 mmol/kg, respectively) on day five of the study. Nevertheless, the reduced muscle glycogen did not adversely affect training capability in the runners.

Table 14.1 Studies That Have Manipulated Diet and Increased Training Volume or Intensity for Up to 15 d to Assess Effects on Muscle Glycogen and Subsequent Training and Performance Capabilities

Reference	Carbohydrate consumption (g/kg/d)	Study duration	Percent change in muscle glycogen	Percent change in performance	Evidence of overreaching
Costill et al. (13)	8.2	10	−15	±0	no
Kirwan et al. (29)	3.9	5	−27	±0	no
	8.0	5	−32	±0	no
Snyder et al. (48)	7.4	15	±0	±0	no

Snyder et al. (48) had cyclists increase their training volume from 12.5 h/wk during a normal 7 d training period (norm period) to 18 h/wk during a training period that lasted 15 days (over period). During the 15 d period, cyclists trained outdoors and training intensity was controlled by monitoring heart rate. Maximal performance capacity was measured by means of an incremental test to fatigue twice during the norm and once during the over period. Immediately after each training session, subjects consumed a beverage containing 160 g of carbohydrates (100% maltodextrins) to optimize glycogen synthesis. Biopsies from the vastus lateralis for the determination of muscle glycogen concentration were obtained at the end of the norm and over periods. Dietary intake was standardized and was estimated based upon the subjects' dietary recall food records. Despite the fact that training volume substantially increased as did daily energy expenditure during the over period compared to norm, total energy intake and carbohydrates consumed during over and norm were similar. Muscle glycogen concentrations for both over and norm periods were similar. No significant changes in body weight were observed and thus either energy expenditure during the over period was not substantially different when compared to the norm period, or food intake was not accurately reported. While the authors indicated that the athletes could be identified as overtrained based upon both physiological and psychological markers, it is obvious that overtraining was not associated with reduced muscle glycogen and a decrement in maximal work performed during the performance test. Importantly, the lack of supervision of diet and training and the lack of a control group makes interpretation of the results difficult (48). Thus, additional research is necessary to explore the relationship between dietary carbohydrate intake, muscle glycogen concentration, and the onset of the overreaching syndrome when training volume is suddenly increased for a short period of time.

Based on these studies (13, 29, 42) it can be suggested that a diet containing at least 8-10 g carbohydrate/kg body weight/d (see table 14.2) will likely maintain the muscle glycogen concentration, while the muscle glycogen concentration will be reduced over days of training when carbohydrate intake is 5 g carbohydrate/kg body weight/d. Although these studies did not demonstrate impaired training and maximal exercise performance capacity, the lowered bodily carbohydrate reserves might eventually make the athlete more susceptible to the overtraining syndrome. Thus, to prevent the onset of the overreaching state and eventually overtraining during heavy training or intensified training, athletes should attempt to consume 8-10 g carbohydrate/kg body weight/d. However, most athletes involved in strenuous training have a busy lifestyle and it is difficult to consume 10 g carbohydrates kg/body weight/d in three daily meals. Indeed, athletes rarely achieve the goal of consuming 8-10g/kg carbohydrate/kg body weight/d. Therefore, athletes gradually increase the risk that over the short term and the long term, bodily carbohydrate stores may decline to a level that eventually may lead to a decline in training and performance capacity. Furthermore, lowered bodily carbohydrate reserves may play a factor in the onset of the overreaching or overtraining syndromes. Therefore, to avoid inadequate dietary carbohydrate consumption, athletes may benefit from consuming a solid or liquid carbohydrate supplement before, during, and immediately after training or competition with the objective to maintain optimal levels of liver and muscle glycogen concentration and energy consumption to match total daily energy expenditure. Consequently, this may result in the ability to maintain maximal training capability, theoretically optimize performance (41, 44), and reduce the potential for the onset of the overreaching syndrome.

Table 14.2 Daily Carbohydrate Consumption for Variously Sized Athletes Undertaking Heavy Training; Based Upon 65% of Daily Energy Consumption as Carbohydrate and Maintaining Carbohydrate Consumption at 10 g/kg/d

Body mass (kg)	Total daily energy consumption (kcal)*	Daily carbohydrate consumption (g/d)
45	1800	450
68	3650	675
91	4200	912

*To convert to scientific units (MJ), divide the kcal unit by 238.92.
Based on data from Sherman and Wimer 1991.

Muscle Glycogen Supercompensation

Because of the positive relationship between the preexercise muscle glycogen concentration and the length of time that moderate-intensity exercise can be maintained, methods to supercompensate the muscle glycogen concentration (i.e., carbohydrate load) have been examined by manipulating training, nutritional status, and exercise patterns. Supraelevating the muscle glycogen concentration either improves performance (37) by a quantity effect, or if the high glycogen concentrations increase the rate of glycogenolysis, supercompensation may occur in the liver and this may enhance endurance performance by delaying the onset of the decline of the blood glucose concentration.

Although it has been suggested that endurance athletes who undergo strenuous twice-daily training and consume a high carbohydrate diet may be supercompensated on a regular basis, athletes may benefit from the taper and glycogen supercompensation protocol before an important endurance event (41). In the week before an important competition, athletes might gradually decrease training volume from 90, 40, 40, 20, 20 min at a hard to moderate intensity (75% VO_2max) and rest the day before competition. This taper regimen, accompanied by a diet consisting of three days of a moderately high carbohydrate diet (350 g carbohydrate/d) and three days of a high carbohydrate diet (500-600 g carbohydrate/d), either as liquid or solid carbohydrate sources, will increase muscle glycogen by 20-40% above normal (i.e., supercompensated) levels (46).

Recently, Tarnopolsky et al. (49) advocated the perspective that women do not increase muscle glycogen during a carbohydrate loading and taper protocol to the same extent as men. Eight female and seven male athletes underwent a taper and supercompensation protocol prior to cycling at 75% peak VO_2 for 1 h followed by a time trial to fatigue at 85% peak VO_2. Subjects were fed in a counterbalanced fashion either a high carbohydrate diet (75% of energy consumption as carbohydrate) or a low carbohydrate diet (55-50% of energy consumption as carbohydrate). Muscle biopsies were obtained from the vastus lateralis before and after each performance trial. For the women, preexercise muscle glycogen did not differ significantly between the two diets, while muscle glycogen concentration increased by 41% when the men consumed the high carbohydrate diet. Because muscle glycogen concentration did not differ between the two diets for the women, exercise performance was not changed. The men, however, experienced a significantly longer time to fatigue (+25%) on the high carbohydrate diet compared with the low carbohydrate diet.

It has been recommended that the amount of carbohydrate consumed by athletes should be calculated relative to body weight (g/kg bw/d). In addition, the amount of carbohydrate necessary to optimally replenish muscle glycogen likely exceeds 500 carbohydrate g/d. The female subjects in the Tarnopolsky et al. study (49) consumed 6.4 g carbohydrate/kg body weight/d on the high carbohydrate diet which is obviously less than the recommended amount. Further, on this basis the men consumed more carbohydrate than did the women. Thus, this study probably does not unequivocally support the contention that under "identical" circumstances, women do not supercompensate compared to men. Therefore, the study that controls all aspects of dietary carbohydrate intake and muscle glycogen supercompensation in women awaits completion.

Although, this study requires replication these preliminary results may suggest that female athletes undergoing strenuous training may be more susceptible to the overreaching and overtraining syndrome than men as a result of lowered bodily carbohydrate stores.

Preexercise Carbohydrate Consumption

Preexercise carbohydrate feeding may increase liver (36) and muscle glycogen (14) and thus increase the bodily carbohydrate stores during the hours before the exercise. This nutritional manipulation may enhance the ability to maintain training at higher intensity and presumably optimize performance. Several investigators (43, 45, 56) have examined the effects on performance of consuming different amounts and types of carbohydrates and of consuming carbohydrates at different time intervals before a strenuous exercise session in either moderately- or highly-trained cyclists. When moderately-trained cyclists consumed 312 g of carbohydrate (4.5 g of carbohydrate/kg body weight) 4 h before intermittent endurance exercise lasting 95 min followed by a performance trial, performance was increased by 15% when compared to either a placebo, 45 g of carbohydrate, or 156 g of carbohydrate (0, 0.6, or 2 g carbohydrate/kg body weight) consumed 4 h before exercise (45). In a similar experiment, Sherman et al. (43) fed trained cyclists either 1.1 g liquid carbohydrate/kg body weight or 2.2 g carbohydrate/kg body weight 1 h before exercise at 70% VO_2max lasting 90 min that was followed by a time trial. All subjects completed the time trial faster (by 12-13%) after consuming the preexercise carbohydrate feeding when compared with the placebo.

In another study, using a similar carbohydrate feeding protocol, Wright et al. (56) fed well-trained cyclists either 5 g carbohydrate/kg body weight or a placebo 3 h before exercise at 70% VO_2max until fatigue. After 20 min of exercise subjects consumed either 0.2 g carbohydrate/kg body weight or a placebo every 20 min thereafter. The carbohydrate feedings consumed before and during the exercise increased time to exhaustion by 44% and total work output by 46% compared with the placebo. Thus, consuming carbohydrates before and during the exercise significantly improved performance compared to consuming carbohydrate either only preexercise or during exercise. Although preexercise feedings may produce a decline in blood glucose at the onset of the exercise in some subjects, few subjects experienced fatigue during the initial lowering of blood glucose caused by the preexercise carbohydrate feedings.

It appears that preexercise carbohydrate feedings have the potential to improve training and performance capacity due to enhanced glucose availability and oxidation when glycogen stores are depleted. Furthermore, this dietary manipulation may help athletes consume adequate amounts of carbohydrate in order to consume 8-10 g of carbohydrate/kg body weight/d during strenuous training periods.

Carbohydrate Ingestion During Exercise

Several investigators have demonstrated that carbohydrate feedings during exercise improve performance in exercise sessions lasting longer than 90 min performed at intensities > 70% VO_2max by preventing a decline in blood glucose concentration and facilitating glucose oxidation late in exercise. Bosch et al. (8) demonstrated that carbohydrate ingestion during prolonged exercise at 70% VO_2max for 3 h presumably spares liver glycogen. When carbohydrate is ingested during exercise and blood glucose remains high, athletes maintain reliance upon glucose oxidation for energy and fatigue is delayed by up to 1 h (15). Glucose, sucrose, and maltodextrins are equally effective when 30-65 g carbohydrate/h are consumed beginning early in the exercise (15). Performance can also be improved by ingesting a large amount (i.e., 200 g) of carbohydrate late in the exercise, but only when the carbohydrate is consumed at least 30 min before the blood glucose concentration starts to decline (11). Carbohydrate feedings given throughout prolonged intermittent exercise lasting longer than 90-120 min when the intensity varies enhances performance presumably due to a slower rate of muscle glycogenolysis (54).

Thus, when 4-5 g carbohydrate/kg body weight are consumed 3-4 h before exercise or when 1-2 g carbohydrate/kg body weight are consumed 1 h before exercise, each in combination with carbohydrate ingestion during exercise (40-65 g) in either solid or liquid form (35), blood glucose will be maintained and carbohydrate oxidation will be maintained late in the exercise. Using these manipulations might improve training capabilities and thus facilitate maximal endurance exercise performance. These dietary manipulations may also protect against a possible reduction in bodily carbohydrate stores and thus presumably against the onset of the overreaching and overtraining syndromes.

Ingestion of Carbohydrate After Exercise

Because many endurance athletes undertake strenuous endurance training that substantially reduces the bodily carbohydrate stores, it is important to consume sufficient amounts of carbohydrates to enhance recovery of bodily carbohydrate. The rate of muscle glycogen synthesis in the first few hours after glycogen-depleting exercise can be influenced by the timing of ingestion of carbohydrate after exercise, the amount of carbohydrate consumed, and the type of carbohydrate that is ingested after exercise (15).

Ivy et al. (26) demonstrated that the rate of glycogen synthesis after glycogen-depleting exercise is 6 mmol/kg/h when athletes consume 1.5 g carbohydrate solution/kg body weight (70% maltodextrin, 15% glucose, 15% sucrose) immediately, and every 2 h after completing the exercise. When carbohydrate consumption is delayed for 2 h after exercise, muscle glycogen synthesis is 47% slower. Blom et al. (7) and Ivy et al. (27) determined that consuming 0.75-3 g glucose/kg every 2 h for the 4-6 h after glycogen-depleting exercise results in higher rates of muscle glycogen synthesis when compared to consuming 0.35 g glucose/kg every 2 h (the rates of glycogen synthesis were 5.7, 5.8, and 2.1 mmol/kg/hr, for 0.35 g, 0.75 g, and 3 g glucose/kg consumed, respectively). However, consuming carbohydrates at more frequent intervals results in a higher rate of muscle glycogen synthesis (8-10 mmol/kg/h) (17, 57). Apparently, consuming 1.6 g maltodextrin/kg body weight/h at 15-min intervals during the first 4 h after exercise may provide an even greater stimulus for muscle glycogen synthesis (17) after glycogen-depleting exercise. This higher rate of glycogen synthesis appears to be related to the higher overall blood insulin concentration that is maintained by the more frequent ingestion of carbohydrate (17).

Several investigators have examined the influence of the type of carbohydrate consumed (sucrose vs. glucose vs. fructose), the form of carbohydrate ingested (solid vs. liquid), and the glycemic index of foods on muscle glycogen synthesis. During the first 4 h after strenuous exercise the consumption of either solid or liquid carbohydrate (1.5 g carbohydrate/kg body weight) immediately after and at 2 h intervals after strenuous exercise will result in similar rates (5.6 mmol/kg/h) of muscle glycogen synthesis (38). Blom et al. (7) demonstrated that during the first 6 h after strenuous exercise the rate of muscle glycogen synthesis is similar when the same amounts of glucose or sucrose are ingested (1.5 g/kg body weight) immediately after and every 2 h thereafter. However, when fructose is ingested in the same manner, the rate of muscle glycogen synthesis is 50% lower. This leads to the suggestion that to obtain a high rate of glycogen synthesis in the initial hours after the exercise, an athlete weighing 70 kg might consume a large amount of carbohydrate in smaller portions per feeding but at more frequent intervals during the initial 4 h after exercise (28 g carbohydrate every 15 min, or 448 g carbohydrate over 4 h).

When adequate dietary carbohydrate is consumed in the initial hours after strenuous exercise, approximately 10-20 hours are required to recover muscle glycogen stores to normal. Burke et al. (9) determined the rate of muscle glycogen synthesis in the 24 h period after prolonged strenuous exercise when subjects consumed a combination of carbohydrate, fat, and protein (489 g of carbohydrate or 7.8 g/kg/d) or a matched energy diet (829 g of carbohydrate or 12 g/kg/d). After 24 h, the rate of muscle glycogen synthesis was similar for the two diets, suggesting that when carbohydrate intake is adequate, muscle glycogen synthesis is optimized.

Costill et al. (12) compared the rate of muscle glycogen synthesis during a 48 h period after glycogen-depleting exercise when subjects consumed a 70% carbohydrate diet consisting of simple sugars or complex carbohydrates. After completion of a 10 mile run at 80% VO_2max and five 1 min sprints at 130% VO_2max with three minute rest intervals, subjects consumed 648 g carbohydrate during the first 24 h and 415 g carbohydrate during the second 24 h after exercise. There was no difference in the rate of muscle glycogen synthesis during the first 24 h; however, during the second 24 h the complex carbohydrate diet produced a higher rate of muscle glycogen synthesis. Today, carbohydrates are classified according to the extent to which they increase the blood glucose concentration (i.e., glycemic index).

Burke et al. (9) examined the effect of the glycemic index of meals on the muscle glycogen concentration after strenuous exercise. After 24 h, the increase in muscle glycogen was greater for the high glycemic diet compared to the low glycemic diet (both diets were isocaloric). More research, however, should be performed in this area. Athletes who consume predominantly simple carbohydrates or low glycemic index foods who also undertake increased training volume may predispose themselves to inadequate replenishment of bodily carbohydrate stores over the short-and long-term period that may promote overreaching or overtraining. On the other hand, over the long term athletes may ensure optimal replenishment of the bodily carbohydrate stores by consuming predominantly foods with at least a moderate glycemic index or complex carbohydrates. Presumably this nutritional approach will result in greater muscle glycogen concentrations, replenish the bodily carbohydrate stores, and maintain optimal training and performance capacity.

Thus, it appears that the most critical dietary manipulation for optimal restoration of bodily carbohydrate reserves after strenuous exercise is adjustment of the total amount and the type of carbohydrate consumed. Therefore, athletes can play a preventive role in the onset of the overreaching syndrome by paying close attention to the amount and type of carbohydrate they consume after exercise.

Carbohydrate Consumption and Training Over the Long Term

Based upon the preceding studies over the short-term period, it is apparent that moderate dietary carbohydrate intake (5 g carbohydrate/kg body weight/d) produces a lower muscle glycogen concentration compared to higher carbohydrate diet (10 g carbohydrate/kg body weight/d). Although these acute changes in muscle glycogen have not directly affected training capabilities and performance, it has been suggested that chronic inadequate dietary carbohydrate intake may influence muscle glycogen even more and consequently impair performance and this may play a role in the onset of the overtraining syndrome. In one study (47), 28 rowers underwent high-intensity exercise twice daily for a 4 wk period. During the 4 wk period rowers consumed either 5 g or 10 g carbohydrate/kg body weight/d with protein content constant at 2 g/kg body weight/d. The rowers that consumed the diet consisting of 10 g/kg body weight/d consumed a carbohydrate beverage as a supplement to the total energy and carbohydrate intake, while the rowers in the other group consumed a placebo beverage. The rowers trained in the laboratory daily, six days a week for 100 min/d at > 70% VO_2max. Also during each week, on three afternoons time trials consisting of 3×2500 m were included in the training sessions to assess the effects of the diets on performance capacity. Muscle biopsies were obtained once each week. Over the four weeks, muscle glycogen increased by 65% for rowers consuming 10 g carbohydrate/kg body weight/d while the muscle glycogen remained constant (116 mmol/kg) for the rowers consuming 5 g carbohydrate/kg body weight/d. Performance capacity increased significantly more in the high carbohydrate group (11%) compared to the group on the moderate carbohydrate diet (2%) (see figure 14.3). Thus, over the 4 wk period the rowers consuming the high carbohydrate diet improved their training and performance capacity, while moderate carbohydrate intake did not increase muscle glycogen nor was there an impaired training or performance capacity. Furthermore, neither dietary regimen had an effect on either psychological or self-reported physiological states that could have been associated with the onset of the overtraining syndrome (10).

The fact that a diet containing a large amount of energy from carbohydrate (546 g/d vs. 177 g/d) compared to the amount of energy from fat (75 g/d vs. 217 g/d) promotes a greater adaptive response to training was also demonstrated by Helge et al. (22) in untrained subjects (N = 20) who trained under either diet for seven weeks. Over this period of time the diets were well supervised. Daily training was well supervised and monitored and consisted of exercise between 50 and 85% of maximal aerobic power 3-4 times per week for 60-75 min per session. Whereas

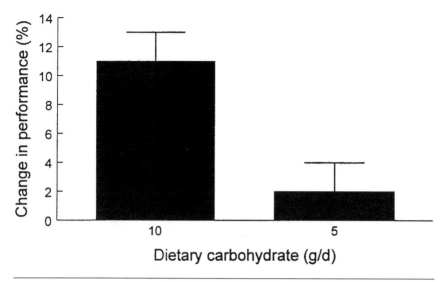

Figure 14.3 Percent change in training capability over 28 days of twice-daily rowing training when fed high (10 g/kg bw/d) and moderate (5 g/kg bw/d) amounts of carbohydrate. Significant difference between groups.
Based on data from Simonsen et al. 1991.

both groups increased their maximal aerobic power by 11%, the group that consumed the high carbohydrate diet improved endurance time to exhaustion at 81% of the pretraining maximal aerobic power by 191% (from 34.2 to 102.4 min), whereas the low carbohydrate group only improved by 83% (from 35.7 to 65.2 min) (see figure 14.4). Importantly, the degree of increase in endurance capability between groups was significantly different ($p < .05$). While both groups derived the benefit of muscle glycogen sparing from the training program (rate of glycogenolysis decreased from 6.3 to 3.0 mmol/kg/min), the high carbohydrate group apparently was able to derive a greater percentage of the total exercise energy expenditure from the oxidation of fat. Interestingly, plasma norepinephrine concentration and heart rate were higher at exhaustion in the low carbohydrate group suggesting a greater cardiovascular stress during exercise when bodily carbohydrate availability was low; this suggests that chronic training while consuming less than optimal amounts of carbohydrate over time may contribute to overtraining (30).

Summary

Based upon this review of the literature, it should be apparent that no study has directly linked a reduction in muscle glycogen to the phenomenon of overreaching. One might also argue that the literature has failed to unequivocally support the presumption that a phenomenon such as overreaching exists. Further, it should also be apparent that no study has directly linked a reduction in muscle glycogen to the phenomenon of overtraining. One might also argue that the literature has failed to unequivocally support the presumption that a phenomenon such as overtraining exists. Nevertheless, if one is to assess the relationship between muscle

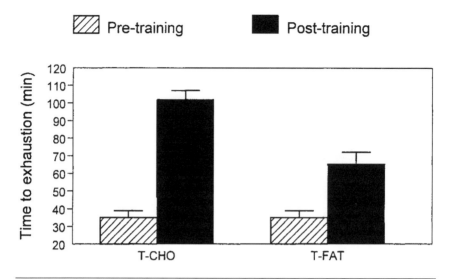

Figure 14.4 Time to exhaustion at 81% of pretraining VO_2max before and after training 3-4 times/wk for 7 wk when consuming moderate carbohydrate (T-CHO: 6.8 g/kg bw/d) and low carbohydrate (T-FAT: 2.4 g/kg bw/d) diets. Significant increase in time to exhaustion for both groups. The increase in time to exhaustion was significantly greater for T-CHO compared to T-FAT.
Based on data from Helge, Richter, and Kiens 1996.

glycogen and training or performance capabilities over the short or long term, tests that can detect relatively small changes in performance should be utilized. Additionally, to more likely ensure the potential to assess the relationship between muscle glycogen and training over the long term to study overtraining, studies lasting longer than 4-6 wk are required.

References

1. Abidi, S.A. 1976. Metabolism of branched-chain amino acids in altered nutrition. *Metabolism* 25: 1287-1302.
2. Adlercreutz, H., M. Härkönen, K. Kuoppasalmi, I. Huhtaniemi, H. Tikkanen, K. Remes, A. Dessypris, J. Karvonen. 1986. Effect of training on plasma anabolic and catabolic steroid hormones and their response during exercise. *International Journal of Sports Medicine* 7:S27-S28.
3. Ahlborg, G., P. Felig, L. Hagenfeldt, R. Hendler, J. Wahren. 1974. Substrate turnover during prolonged exercise in man. *Journal of Clinical Investigation* 53: 1080-1090.
4. Ahlborg, G., P. Felig. 1982. Lactate and glucose exchange across the forearm, legs, and splanchnic bed during and after prolonged leg exercise. *Journal of Clinical Investigation* 69: 45-54.
5. Alén, A., A. Pakarinen, K. Häkkinen, P. Komi. 1988. Responses of serum androgenic-anabolic and catabolic hormones to prolonged strength training. *International Journal of Sports Medicine* 9: 229-233.

6. Bjorkman, O., J. Wahren. Glucose homeostasis during and after exercise. 1988. In *Exercise, nutrition, and energy metabolism*, eds. E.S. Horton, R.L. Terjung, 100-115. New York: Macmillan.

7. Blom, P.C.S., A.T. Hostmark, O. Vaage, K.R. Kardel, S. Maehlum. 1987. Effect of different post-exercise sugar diets on the rate of muscle glycogen synthesis. *Medicine Science Sports and Exercise* 19: 491-496.

8. Bosch, A.N., A.C. Dennis, T.D. Noakes. 1994. Influence of carbohydrate ingestion on fuel substrate turnover and oxidation during prolonged exercise. *Journal of Applied Physiology* 76: 2364-2372.

9. Burke, L.M., G.R. Collier, M. Hargreaves. 1993. Muscle glycogen storage after prolonged exercise: effect of the glycemic index of carbohydrate feedings. *Journal of Applied Physiology* 75: 1019-1023.

10. Cogan, K., P.S. Highlen, T.A. Petrie, W.M. Sherman, J. Simonsen. 1991. Psychological and physiological effects of controlled intensive training and diet on collegiate rowers. *International Journal of Sport Psychology* 22: 165-180.

11. Coggan, A.R., E.F. Coyle. 1987. Reversal of fatigue during prolonged exercise by carbohydrate infusion and ingestion. *Journal of Applied Physiology* 63: 2388-2395.

12. Costill, D.L., W.M. Sherman, W.J. Fink, C. Maresh, M. Witten, J.M. Miller. 1981. The role of dietary carbohydrates in muscle glycogen resynthesis after strenuous running. *American Journal of Clinical Nutrition* 34: 1831-1836.

13. Costill, D.L., M.G. Flynn, J.P. Kirwan, J.A. Houmard, J.B. Mitchell, R. Thomas, S.H. Park. 1988. Effects of repeated days of intensified training on muscle glycogen and swimming performance. *Medicine Science Sports and Exercise* 20: 249-254.

14. Coyle, E.F., A.R. Coggan, M.K. Hemmert, R.C. Lowe, T.J. Walters. 1985. Substrate usage during prolonged exercise following a preexercise meal. *Journal of Applied Physiology* 55: 230-235.

15. Coyle, E.F. 1991. Timing and method of increased carbohydrate intake to cope with heavy training, competition and recovery. *Journal of Sports Sciences* 9: 29-52.

16. Dohm, G.L. 1984. Protein nutrition for the athlete. *Clinics in Sports Medicine* 3: 595-604.

17. Doyle, J.A., W.M. Sherman, R.L. Strauss. 1993. Effects of eccentric and concentric exercise on muscle glycogen replenishment. *Journal of Applied Physiology* 74: 1848-1855.

18. Fry, R.W., A.R. Morton, D. Keast. 1991. Overtraining in athletes: an update. *Sports Medicine* 12: 32-65.

19. Goldberg, A.L., R. Odessey. 1972. Oxidation of amino acids by diaphragms from fed and fasted rats. *American Journal of Physiology* 223: 1384-1391.

20. Gollnick, P.D., B. Saltin. 1988. Fuel for muscular exercise: role of fat. In *Exercise, nutrition, and energy metabolism*, ed. E.S. Horton, R.L. Terjung, 72-88. New York: Macmillan.

21. Goodman, M.N. 1988. Amino acid and protein metabolism. In *Exercise, nutrition, and energy metabolism*, eds. E.S. Horton, R.L. Terjung, 89-99. New York: Macmillan.

22. Helge, J.W., E.A. Richter, B. Kiens. 1996. Interaction of training and diet on metabolism and endurance during exercise in man. *Journal of Physiology* 492: 293-306.

23. Hermansen, L., E. Hultman, B. Saltin. 1967. Muscle glycogen during prolonged severe exercise. *Acta Physiologica Scandinavica* 71: 129-139.
24. Hood, D.A., R.L. Terjung. 1990. Amino acid metabolism during exercise and following endurance training. *Sports Medicine* 9: 23-35.
25. Hultman, E., P. Greenhaff. 1991. Skeletal muscle energy metabolism and fatigue during intense exercise in man. *Science Progress* 75: 361-370.
26. Ivy, J.L., A.L. Katz, C.L. Cutler, W.M. Sherman, E.F. Coyle. 1988. Muscle glycogen synthesis after exercise: effect of time of carbohydrate ingestion. *Journal of Applied Physiology* 64: 1480-1485.
27. Ivy, J.L., M.C. Lee, J.T. Brozinick Jr., M.J. Reed. 1988. Muscle glycogen storage after different amounts of carbohydrate ingestion. *Journal of Applied Physiology* 65: 2018-2023.
28. Johansson, C., L. Tsai, E. Hultman, T. Tegelman, A. Pusette. 1990. Restoration of anabolic deficit and muscle glycogen consumption in competitive orienteering. *International Journal of Sports Medicine* 11: 204-207.
29. Kirwan, J.P., D.L. Costill, J.B. Mitchell, J.B. Houmard, M.G. Flynn, W.J. Fink, J.D. Beltz. 1988. Carbohydrate balance in competitive runners during successive days of intense training. *Journal of Applied Physiology* 65: 2601-2606.
30. Kuipers, H., H.A. Keizer. 1988. Overtraining in elite athletes: review and directions for the future. *Sports Medicine* 6: 79-92.
31. Lehmann, M., H.H. Dickhuth, G. Gendrisch, W. Lazar, M. Thum, R. Kaminski, J.F. Aramendi, E. Peterke, W. Wieland, J. Keul. 1991. Training-overtraining: a prospective, experimental study with experienced middle- and long-distance runners. *International Journal of Sports Medicine* 12: 444-452.
32. Lehmann, M., C. Foster, J. Keul. 1993. Overtraining in endurance athletes: a brief review. *Medicine and Science in Sports and Exercise* 25: 854-862.
33. Lemon, P.W., J.P. Mullin. 1980. Effect of initial muscle glycogen levels on protein catabolism during exercise. *Journal of Applied Physiology* 48: 624-629.
34. Lemon, P.W. 1995. Do athletes need more dietary protein and amino acids? *International Journal of Sports Nutrition* 5: S39-S61.
35. Lugo, M., W.M. Sherman, G.S. Wimer, K. Garleb. 1993. Metabolic responses when different forms of carbohydrate energy are consumed during cycling. *International Journal of Sport Nutrition* 3: 398-407.
36. Nilsson, L.H., E. Hultman. 1973. Liver glycogen in man: the effect of total starvation or a carbohydrate-poor diet followed by carbohydrate feeding. *Scandinavian Journal of Clinical Laboratory Investigation* 32: 325-330.
37. Rauch, L.H.G., I. Rodger, G.R. Wilson, J.D. Belonje, S.D. Dennis, T.D. Noakes, J.A. Hawley. 1995. The effects of carbohydrate loading on muscle glycogen content and cycling performance. *International Journal of Sport Nutrition* 5: 25-36.
38. Reed, M.J., J.T. Bronzinick Jr., M.C. Lee, J.L. Ivy. 1989. Muscle glycogen storage postexercise: effect of mode of carbohydrate administration. *Journal of Applied Physiology* 66: 720-726.
39. Saltin, B., J. Karlsson. 1971. Muscle glycogen utilization during work of different intensities. In *Ergogenic aids in sports*, eds. B. Pernow, B. Saltin, 289-300. New York: Plenum Press.
40. Saltin, B., P.O. Åstrand. 1993. Fatty acids and exercise. *American Journal of Clinical Nutrition* 57: S752-S757.

41. Sherman, W.M. 1995. Metabolism of sugars and physical performance. *American Journal of Clinical Nutrition* 62: S228-S241.

42. Sherman, W.M., J.A. Doyle, D.R. Lamb, R.H. Strauss. 1993. Dietary carbohydrate, muscle glycogen, and exercise performance during 7 d of training. *American Journal of Clinical Nutrition* 57: 27-31.

43. Sherman, W.M., M.C. Peden, D.A. Wright. 1991. Carbohydrate feedings 1 h before exercise improves cycling performance. *American Journal of Clinical Nutrition* 54: 866-870.

44. Sherman, W.M., G.S. Wimer. 1991. Insufficient dietary carbohydrate during training: does it impair athletic performance? *International Journal of Sport Nutrition* 1: 28-44.

45. Sherman, W.M., G. Brodowicz, D.A. Wright, W.K. Allen, J. Simonsen, A. Dernbach. 1989. Effects of 4 h preexercise carbohydrate feedings on cycling performance. *Medicine Science Sports and Exercise* 21: 589-604.

46. Sherman, W.M., D.L. Costill, W.J. Fink, J.M. Miller. 1981. The effect of exercise and diet manipulation on muscle glycogen and its subsequent use during performance. *International Journal of Sports Medicine* 2: 114-118.

47. Simonsen, J.C., W. M. Sherman, D.R. Lamb, A.R. Dernbach, J.A. Doyle, R.H. Strauss. 1991. Dietary carbohydrate, muscle glycogen, and power output during rowing training. *Journal of Applied Physiology* 70: 1500-1505.

48. Snyder, A.C., H. Kuipers, B. Cheng, R. Servais, E. Fransen. 1995. Overtraining following intensified training with normal muscle glycogen. *Medicine Science Sports and Exercise* 27: 1063-1070.

49. Tarnopolsky, M.A., S.A. Atkinson, S.M. Phillips, J.D. MacDougall. 1995. Carbohydrate loading and metabolism during exercise in men and women. *Journal of Applied Physiology* 78: 1360-1368.

50. Turcotte, L.P., B. Kiens, E.A. Richter. 1991. Saturation kinetics of palmitate uptake in perfused skeletal muscle. *FEBS Letters* 279: 327-329.

51. Turcotte, L.P., E.A. Richter, B. Kiens. 1992. Increased plasma FFA uptake and oxidation during prolonged exercise in trained vs. untrained humans. *American Journal of Physiology* 262: E791-E799.

52. Vasankari, T.J., U.M. Kujala, O.J. Heinonen, I.T. Huhtaniemi. 1993. Effects of endurance training on hormonal responses to prolonged physical exercise in males. *Acta Endocrinologica* 129: 109-113.

53. Vervoorn, C., A.M. Boelens-Quist, L.M. Vermulst, W.M. Erich, W.R. de Vries, J.H. Thijssen. 1991. The behaviour of the plasma free testosterone/cortisol ratio during a season of elite rowing training. *International Journal of Sports Medicine* 3: 257-263.

54. Yaspelkis III, B.B., J.G. Paggerson, P.A. Anderla, Z. Ding, J.L. Ivy. 1993. Carbohydrate supplementation spares muscle glycogen during variable intensity exercise. *Journal of Applied Physiology* 75: 1477-1485.

55. Wagenmakers, A.M., E.J. Beckers, F. Brouns, H. Kuipers, P.B. Soeters, G.J. van der Vusse, W.M. Saris. 1991. Carbohydrate supplementation, glycogen depletion, and amino acid metabolism during exercise. *American Journal of Physiology* 260: E883-E890.

56. Wright, D.A., W.M. Sherman, A.R. Dernbach. 1991. Carbohydrate feedings before, during, or in combination improve cycling endurance performance. *Journal of Applied Physiology* 71: 1082-1088.

57. Zachwieja, J.J., D.L. Costill, D.D. Pascoe, R.A. Robergs, W.J. Fink. 1991. Influence of muscle glycogen depletion on the rate of resynthesis. *Medicine and Science in Sport and Exercise* 23: 44-48.

CHAPTER 15

Central Fatigue Hypothesis and Overtraining

Richard B. Kreider, PhD

Introduction

Amino acids are the foundation of protein in the body and are essential for the synthesis of tissue, specific proteins, hormones, enzymes, and neurotransmitters (8, 26, 30, 31, 38). Amino acids are also involved in the synthesis of energy through gluconeogenesis and regulation of numerous metabolic pathways (19, 22, 24, 31, 56). Over the last number of years, there has been significant interest in determining the role of specific amino acids on metabolic, physiological, and psychological responses to exercise that may affect acute fatigue and the physiological and psychological responses to intense training. In addition, exercise-induced alterations in specific amino acids have been suggested to be associated with chronic fatigue, overtraining, and immunosuppression (31, 36, 42-44, 57).

The purpose of this chapter is to discuss the potential relationship that exercise-induced alterations in amino acid concentrations may have on central fatigue, overreaching, and overtraining, and moreover, to discuss possible nutritional strategies that may lessen the impact of alterations in amino acid concentrations during exercise on central fatigue, overreaching, and overtraining.

Central Fatigue Hypothesis

Effects of Exercise on Plasma Amino Acid Concentrations

Several amino acids appear to play an important role in the metabolic, physiological, and psychological responses to exercise. These amino acids include leucine, isoleucine, valine, tryptophan, and glutamine. During prolonged exercise, the branched-chained amino acids leucine, isoleucine, and valine (BCAA) and glutamine are taken up by the muscle rather than the liver in order to contribute to oxidative metabolism (31, 42). The source of BCAA and glutamine for muscular oxidative metabolism during exercise is the plasma amino acid pool that is replenished through the catabolism of whole body proteins during prolonged exercise. However, since the oxidation of BCAA and glutamine in the muscle may exceed the catabolic capacity to increase BCAA and glutamine availability, plasma BCAA and glutamine levels may decline during prolonged exercise (31).

For example, Blomstrand and colleagues (4) reported that an army training regimen (16 km run and 30 min of circuit weight training) decreased plasma BCAA levels by 24%, and that runners performing a 42.2 km marathon significantly decreased plasma BCAA concentrations by 19% while increasing free tryptophan (fTryp) by 146%. Further, Blomstrand et al. (6) reported that plasma leucine (–33%), isoleucine (–36%), and valine (–24%) levels significantly decreased while plasma fTryp (+45%) levels significantly increased in females playing a standard soccer match despite ingestion of a carbohydrate drink during the match. In another study, Blomstrand and associates (7) reported that valine (–26%), leucine (–29%), and isoleucine (–33%) levels were significantly decreased during a 30 km run, and that valine (–17%), leucine (–19%), and isoleucine (–26%) levels significantly decreased during a marathon run.

Lehmann and coworkers (37) reported that the sum concentration of 25 amino acids decreased by 18% in response to an ultraendurance triathlon lasting approximately 23 hours. More specifically, leucine concentrations remained essentially unchanged (–2%), isoleucine levels decreased by 28%, valine levels declined by 24%, and fTryp concentrations increased by 74% despite an approximately 7.6% increase in plasma volume. Davis and colleagues (16) reported that plasma BCAA levels remained essentially unchanged (1-3% decrease) while fTryp levels markedly increased (almost 3-fold) in trained subjects cycling at approximately 68% of VO_2max for up to 255 min. Finally, Schena et al. (48) reported that plasma BCAA levels decreased by 16% following 90 min of running at 75% of VO_2max.

Exercise-induced alterations in plasma glutamine associated with exercise-induced immunosuppression have also been reported. Parry-Billings and associates (45) reported that overtrained athletes had significantly lower (–8.5%) plasma glutamine levels than nonovertrained athlete controls. Schena et al. (48) reported that plasma glutamine levels decreased by 10% following 90 min of running at 75% of VO_2max. In addition, Kargotich and coworkers (28) reported that plasma glutamine levels decreased by 16% in male swimmers who performed 15 × 100 m freestyle swims at 95% of maximal capacity compared to nonexercising, nontrained controls. Moreover, the decline in plasma glutamine levels peaked 6 h postexercise. Interestingly, no change in plasma glutamine levels was observed when performing 15 × 100 m sprints at 70% of maximal capacity. The researchers suggested that the decline in plasma glutamine following exercise may be related to postexercise immunosuppression (30, 31, 42-44). Collectively, these reports indicate that exercise bouts ranging from high-intensity repeated sprint exercise to prolonged endurance exercise may significantly reduce plasma BCAA and glutamine levels while increasing free tryptophan levels.

Relationship of Alterations in Amino Acid Concentrations on 5-HT Synthesis

The significance of the exercise-induced alterations in plasma BCAA and fTryp described previously is the basis for the central fatigue hypothesis (4-7,16, 18, 31, 42). In this regard, Newsholme and colleagues (42) have reported that fTryp and BCAA compete for entry into the brain via the same amino acid carrier. A decrease in the level of BCAA in the blood during exercise or an increase in the level of fTryp increases the ratio of fTryp to BCAA (fTryp/BCAA), facilitating entry of tryptophan into the brain. An increased concentration of tryptophan in the brain promotes the formation of the neurotransmitter 5-hydroxytryptamine (5-HT) or serotonin. Increased synthesis of 5-HT has been reported to induce sleep, depress motor neuron excitability, alter autonomic and endocrine function, and suppress appetite (4, 42, 47, 57). Consequently, exercise-induced increases in 5-HT have been suggested to affect tiredness, psychological perception of fatigue, muscle power output, and hormonal regulation during exercise (i.e., central fatigue hypothesis). It has also been hypothesized (31, 42-44) that chronic elevations in 5-HT concentrations, that may occur in athletes maintaining high volume training, may explain many of the reported signs and symptoms of the overtraining syndrome including postural hypotension, hypercardia, anemia, amenorrhea, immunosuppression, appetite suppression, weight loss, depression, and decreased performance capacity.

Influence of Substrate Availability on 5-HT Synthesis

The etiology of the changes in plasma fTryp and BCAA during exercise appear to be related to changes in substrate availability (18, 31, 42). As carbohydrate availability declines during prolonged exercise or if exercise was initiated with low glycogen availability, there is an increased utilization of free fatty acids (FFA) and BCAA as metabolic fuels. The levels of FFA in blood have been reported to be correlated with the concentration of fTryp (16, 42). In this regard, most tryptophan in the blood is bound to albumin. The proportion of tryptophan bound to albumin is influenced by the availability of long-chained fatty acids. As FFA levels increase during exercise (particularly above 1 mmol/L), the amount of tryptophan bound to albumin is reduced, thus increasing the concentration of fTryp in the blood. At the same time, the rate of BCAA and glutamine oxidation is also related to carbohydrate availability. Consequently, as glycogen availability declines during prolonged exercise or if exercise is initiated with low glycogen stores, the rate of BCAA oxidation increases resulting in a gradual decline in plasma BCAA levels. Collectively, the increase in plasma fTryp and the decline in BCAA alters the ratio of fTryp/BCAA resulting in an increased synthesis of 5-HT in the brain (see figure 15.1).

Relationship of 5-HT and Fatigue

In the last 10 years, a number of studies have been conducted on animals and humans to investigate the effects of exercise on tissue concentrations of 5-HT, the

%Change

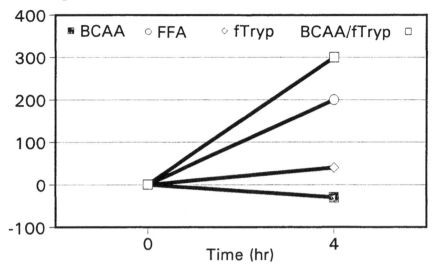

Figure 15.1   Exercise-induced changes in plasma BCAA, FFA, fTryp, and the ratio of fTryp/BCAA during prolonged exercise.

relationship of 5-HT to fatigue, and whether pharmacological or nutritional manipulations affect the ratio of fTryp/BCAA, 5-HT synthesis, and fatigue during exercise. Initial studies by Chaouloff and colleagues (12, 13) reported that 1-2 hours of treadmill running in conditioned rats did not significantly alter plasma total tryptophan. However exercise produced a significant increase in plasma fTryp and concentrations of tryptophan and the primary metabolite of 5-HT, 5-hydroxyindoleacetic acid (5-HIAA) in the brain. Chaouloff and colleagues (14) also reported that exercise significantly increased the concentrations of tryptophan and 5-HIAA in the cerebrospinal fluid of rats and that these levels returned to normal within one hour postexercise. In a similar study, Blomstrand et al. (5) reported that plasma fTryp and regional brain tryptophan, 5-HT, and 5-HIAA were increased at exhaustion in trained and untrained rats. Moreover, Bailey and colleagues (2) reported that 5-HT, 5-HIAA, dopamine, and DOPAC (the primary metabolite of dopamine) significantly increase in various regions of the rat brain after 1 h of exercise (60-65% VO_2max) and that levels peaked at exhaustion (approximately 3 h). Collectively, these animal data suggest that exercise affects plasma fTryp levels as well as tissue concentrations of 5-HT and 5-HIAA and that increases in fTryp, 5-HT, and 5-HIAA appear to be associated with fatigue.

In addition to these studies, several pharmacological studies designed to stimulate or inhibit 5-HT synthesis in rats and humans have been conducted in order to evaluate the relationship of exercise-induced changes in 5-HT and fatigue. Bailey and colleagues (3) reported that rats administered titrated doses of the 5-HT agonist m-chlorophenyl piprazine (m-CPP) caused decreased run time to exhaustion in a dose-ordered relationship. In addition, Bailey and coworkers (1) reported that rats administered another 5-HT agonist (quipazine dimaleate, QD) reduced run time to exhaustion whereas administration of the 5-HT antagonist (LY-53,857) delayed fatigue. Finally, Wilson and Maughan (55) and Davis et al. (17) reported that humans administered the 5-HT agonists paroxetine and fluoxetine prior to prolonged running and cycling at 70% VO_2max decreased time to fatigue (55) and increased rating of perceived exertion (17) during exercise in comparison to placebo control trials. These data provide strong evidence that alterations in 5-HT concentrations affect physiological and psychological responses to exercise and fatigue.

Nutritional Strategies Designed to Delay Central Fatigue

Theoretical Rationale

Since substrate availability affects the ratio of fTryp/BCAA, alterations in the fTryp/BCAA ratio appear to affect 5-HT synthesis, and increases in 5-HT have been associated with fatigue, significant research attention has recently been focused on potential nutritional strategies to delay central fatigue. Theoretically, preventing or minimizing the normal exercise-induced alterations in the fTryp/BCAA ratio would serve to minimize 5-HT synthesis and delay fatigue. Currently, three nutritional strategies have been proposed to influence central fatigue. First, carbohydrate supplementation during exercise has been suggested to be a nutritional means of blunting FFA mobilization and BCAA oxidation, minimizing increases in fTryp and reductions in BCAA, minimizing increases in 5-HT, and thereby delaying fatigue. Second, BCAA supplementation has been proposed as a means of maintaining or increasing plasma BCAA levels during exercise, minimizing or eliminating increases in the fTryp/BCAA ratio during exercise, minimizing synthesis of 5-HT, and thereby delaying fatigue. Finally, a balanced approach of carbohydrate and BCAA supplementation has been proposed as an effective means of minimizing increases in the fTryp/BCAA ratio, 5-HT synthesis, and delaying fatigue. While there is general agreement that exercise-induced alterations in the fTryp/BCAA ratio may affect physiological and psychological responses to exercise and that nutrition may play an important role in minimizing these changes, there is some disagreement regarding which nutritional strategies are effective. There are two primary viewpoints regarding the effects of carbohydrate and BCAA supplementation on central fatigue. First, some suggest that carbohydrate supplementation is effective to delay central fatigue and that BCAA supplementation provides no additional benefit (18, 54). Further, supporters of this viewpoint note that BCAA supplementation increases plasma NH_3 levels (50, 54), which theoretically may impair oxidative metabolism and performance (54) and additionally may be potentially dangerous because NH_3 is toxic to the brain (18, 54). Finally, some researchers state that BCAA supplementation is impractical because large quantities of BCAA must be consumed in order to promote the physiological change in the fTryp/BCAA ratio, which may promote gastrointestinal distress and slowing of fluid absorption from the gut during exercise (18).

On the other hand, a number of researchers suggest that BCAA supplementation with or without carbohydrate is a safe and effective means of minimizing changes in the fTryp/BCAA ratio during exercise, thereby affecting hormonal, physiological and psychological responses to exercise and possibly improving performance (4, 6, 7, 29, 31). The following discussion will examine the relevant literature supporting each of these views as well as comment on their validity.

Carbohydrate Supplementation

There are several studies that serve as the basis for the viewpoint that carbohydrate supplementation is effective in delaying central fatigue and that BCAA supplementation provides no additional benefit. First, studies indicate that dietary availability of carbohydrate and muscle glycogen content affect BCAA oxidation and ammonia production during exercise. In this regard, Greenhaff and colleagues (23) investigated the effects of dietary availability of carbohydrate on plasma NH_3 accumulation and acid-base status in response to incremental cycling for 5 min at 30%, 50%, 70%, and 95% of VO_2max. Subjects performed the exercise test following 3 d of ingesting either their normal diet (49% carbohydrate, 38% fat, 13% protein), a low carbohydrate diet (3% carbohydrate, 69% fat, 28% protein), or a high carbohydrate diet (84% carbohydrate, 5% fat, 11% protein). Results showed that following the low carbohydrate diet, mild acidosis was observed in all subjects, and plasma NH_3 levels tended to be higher than observed in the normal and high carbohydrate diet trials at each workload. No differences were observed between the normal and high carbohydrate trials. The investigators concluded that dietary availability of carbohydrate affects plasma NH_3 accumulation during exercise and that use of NH_3 as a marker of fatigue should be viewed with caution.

In a similar study, MacLean and coworkers (39) investigated the effects of diet on plasma amino acid and NH_3 responses to altered dietary intakes prior to prolonged exercise. Six recreational cyclists rode to exhaustion at 75% of VO_2max following 3 d on either a low carbohydrate diet (4% carbohydrate, 64% fat, 32% protein), a normal mixed diet (49% carbohydrate, 36% fat, 18% protein), or a high carbohydrate diet (66% carbohydrate, 22% fat, 12% protein) assigned in a random order. Blood samples were collected prior to exercise and at 15 min intervals until exhaustion. Results revealed that time to exhaustion was significantly different among trials (low carbohydrate 58.8 ± 4 min, mixed 112.1 ± 7 min, high carbohydrate 152.8 ± 10 min) and that increases in plasma NH_3 levels were significantly greater in the low carbohydrate trial. In addition, resting BCAA levels were significantly greater and decreased to a greater degree during exercise in the low carbohydrate trial. Results indicated that alterations in dietary patterns affect plasma NH_3 and BCAA concentrations during exercise.

To further evaluate the relationship of carbohydrate availability and BCAA oxidation, Wagenmakers and colleagues (54) investigated the effects of glycogen status on BCAA oxidation during prolonged exercise. Eight trained cyclists completed 2 h of cycling at 70-75% of VO_2max in glycogen-depleted and glycogen-loaded states. Results revealed that plasma NH_3 levels increased more rapidly and plasma alanine, glutamine, and glutamate were lower in the glycogen-depleted trial. Exercise caused a 3.6-fold increase in the proportion of B-C, 2-oxoacid dehydrogenase (BC) activity in the glycogen-depleted trial, while no changes were seen in the glycogen-loaded trial. Further, NH_3 levels were inversely related to muscle glycogen content. It was concluded that carbohydrate loading abolishes increases in BCAA oxidation and BC activity, and that NH_3 levels are related in part to deamination of amino acids.

van Hall and colleagues (52) investigated the influence of preexercise muscle glycogen content on NH_3 production, adenine nucleotide breakdown, and amino acid metabolism during prolonged exercise. Six subjects performed one-leg knee extensions at a work rate corresponding to 60-65% of maximal power for 90 min with a glycogen-depleted leg and with a nonglycogen-depleted leg. The NH_3 concentrations increased gradually during the first 60 min of exercise and then plateaued with a total NH_3 production of 9.1 ± 0.4 and 9.5 ± 1.4 mmol/kg dry muscle in the normal leg and glycogen-depleted leg, respectively. No differences were observed in levels of phosphocreatine, total adenine nucleotides, or inosine monophosphate, and only minor differences were observed in muscle concentrations of glutamine, alanine, and BCAA. Further, muscle glutamate concentrations decreased primarily during the first 10 min of exercise, whereas NH_3 production increased gradually during the exercise bout suggesting that deamination of glutamate was not the primary contributor to NH_3 production. The researchers concluded that NH_3 production does not appear to be affected by preexercise muscle glycogen content, that NH_3 production exceeds the breakdown of adenine nucleotides and inosine monophosphate suggesting alternate sources, and that deamination of amino acids during prolonged one-leg exercise is probably the source of NH_3 production.

Collectively, most of these studies indicate that dietary availability of carbohydrate and muscle glycogen availability affect BCAA oxidation and NH_3 accumulation during exercise. However, they do not directly address the second rationale in this viewpoint, i.e., that BCAA supplementation provides no additional benefit to carbohydrate supplementation in affecting the fTryp/BCAA ratio and central fatigue; that BCAA supplementation may increase NH_3 accumulation to toxic levels and possibly impair performance; and BCAA supplementation may cause gastrointestinal distress. The following studies from the Wagenmakers (50-54) and Davis (16, 20) research groups, are often cited as evidence for these contentions.

Wagenmakers et al. (53) and Vandewalle et al. (50) evaluated the effects of leucine and BCAA supplementation on plasma amino acid concentrations, NH_3 accumulation, and performance in glycogen-depleted subjects. Seven male subjects performed a glycogen depleting cycling protocol. Thirty minutes later, subjects ingested 2 g/kg of skimmed yogurt containing either 0.12 g/kg each of leucine, isoleucine, and valine (25-30 g of BCAA), or 0.35 g/kg of leucine (25-30 g of leucine), or a skimmed yogurt control containing no BCAA or leucine (placebo). Subjects rested for 90 min and then performed a cycling protocol (5 min at 50% VO_2max followed by 30 min at 65% VO_2max and then to exhaustion at 75% of VO_2max). Results revealed that the sum concentration of BCAA was significantly higher in the BCAA and leucine trials prior to and following exercise (control 0.42 ± 0.1 to 0.37 ± 0.13; BCAA 3.0 ± 1 to 2.5 ± 0.8; leucine 2.0 ± 0.8 to 1.4 ± 0.6 mmol/L) and that NH_3 accumulation was significantly greater in the BCAA and leucine trials (control 15 ± 10 to 68 ± 33; BCAA 23 ± 14 to 139 ± 55; leucine 33 ± 8 to 201 ± 87 mmol/L). In addition, no significant differences were observed in time to exhaustion between the BCAA group (34.9 ± 6 min) and placebo group (31.2 ± 7 min). These researchers concluded that BCAA supplementation results in significant elevations in plasma NH_3 and that BCAA supplementation does not provide ergogenic benefit (despite an 11.5% increase in time to exhaustion in the BCAA trial).

van Hall and coworkers (51), investigated the effects of BCAA and tryptophan supplementation on plasma amino acid concentrations and performance. Ten endurance-trained males cycled at 70-75% of VO_2max while ingesting in a random and double-blind manner either a 6% glucose electrolyte solution (GES), or the 6% GES with 3 g/L tryptophan, or the 6% GES with 6 g/L BCAA, or the 6% GES with 18 g/L BCAA. Results revealed that tryptophan and BCAA supplementation significantly increased plasma tryptophan and BCAA. Based on kinetic parameters of transport of human brain capillaries, researchers estimated that BCAA supplementation reduced brain tryptophan uptake by 8-12%, while tryptophan supplementation resulted in a 7- to 20-fold increase in brain tryptophan uptake.

However, no significant differences were observed in performance times. The investigators concluded that nutritional manipulation of tryptophan supply to the brain either has no additional effect upon serotogenic activity or that serotogenic activity does not contribute to fatigue during prolonged exercise.

Galiano and colleagues (20) reported the effects of ingesting 0.5 g/L of BCAA mixed in a 6% GES on the fTryp/BCAA ratio and hormonal responses during endurance cycling performed at 70% of VO_2max for up to 255 min. Results revealed that plasma BCAA levels were maintained throughout the BCAA supplement trial while plasma BCAA levels decreased significantly towards the end of exercise in the placebo trial. However, no significant differences were observed in heart rate, RPE, oxygen uptake, fTryp, cortisol, adrenal corticotropic hormone (ACTH), norepinephrine, and epinephrine. Likewise, no significant differences were reported in time to exhaustion between the placebo (235 ± 10 min) and BCAA trials (220 ± 11 min). It was concluded that the addition of small quantities of BCAA to a typical sport drink may serve to maintain plasma BCAA levels throughout the duration of prolonged exercise, but the addition does not appear to have any affect on physiological, endocrine, or performance responses during prolonged cycling.

Based on this study and previous observations from Wagenmakers and colleagues (50, 53, 54), Davis and colleagues (16) hypothesized that BCAA supplements in the quantities necessary to significantly alter the fTryp/BCAA ratio did not appear to be a viable nutritional strategy. Rather, they sought to influence the fTryp/BCAA ratio through carbohydrate feedings in an attempt to blunt FFA and fTryp levels. In this study, eight male cyclists cycled for up to 255 min at an intensity corresponding to VO_2 at lactate threshold (LT) on four occasions. Following familiarization, subjects were randomly fed 5 ml/kg of either a flavored placebo, a 6% GES, or a 12% GES every 30 min during exercise. Results revealed that power output was significantly greater in the GES groups (6% GES 238 ± 9, 12% GES 237 ± 13, placebo 191 ± 10 W) and that fatigue occurred earlier in the placebo group. In addition, FFA levels and the fTryp/BCAA ratio increased sooner and to the greatest degree in the placebo group. The GES supplement increased glucose and insulin levels while attenuating increases in FFA and the fTryp/BCAA ratio. However, no differences were observed between GES treatments. Finally, the fTryp levels observed correlated well (r = .86) to FFA levels. The researchers concluded that carbohydrate feedings during prolonged exercise attenuates exercise-induced increases in FFA, fTryp, and the fTryp/BCAA ratio, which may explain in part the enhanced performance capacity associated with carbohydrate feedings (see figure 15.2).

In summary, these studies suggest that dietary (23, 39) and muscle glycogen (54) availability affect plasma amino acid levels (23, 39, 54) and thereby may affect NH_3 accumulation during exercise. Further, BCAA and leucine supplementation in glycogen-depleted subjects (50, 53) significantly increased resting and postexercise BCAA concentrations promoting significant increases in NH_3 concentrations with no significant effect on performance. Deamination of BCAA appears to be a primary source of the NH_3 production (52). Supplementation of 3 g/L of tryptophan causes marked increases in blood levels of tryptophan and increased estimated brain uptake of tryptophan by 7- to 20-fold, while BCAA supplementation (6 g/L and 18 g/L) increased plasma BCAA levels and decreased estimated tryptophan uptake by 8-12% (51). Low-dose BCAA supplementation (0.5 g/L) attenuated exercise-induced reductions in BCAA levels but did not, however, affect physiological or psychological responses to exercise or time to exhaustion

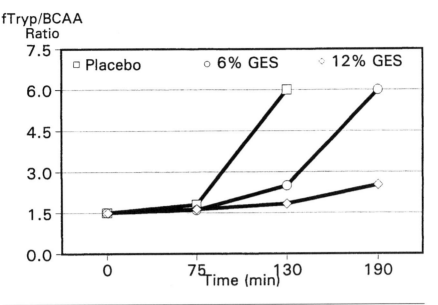

fTryp/BCAA Ratio

Figure 15.2 Effects of endurance cycling on the ratio of fTryp/BCAA following administration of a flavored placebo, 6% GES, or 12% GES.
Based on data from Davis et al.1992.

(20). Finally, ingestion of a 6% or 12% glucose electrolyte solution attenuated exercise-induced increases in FFA and fTryp, serving to minimize increases in the fTryp/BCAA ratio and delay fatigue during prolonged exercise (16).

While it is clear that carbohydrate supplementation should be recommended for athletes engaged in prolonged exercise and that carbohydrate availability may delay central and peripheral fatigue, the following points should be noted regarding concerns that have been raised with respect to BCAA supplementation during exercise. First, we are not aware of any studies (including those to be cited below) that have reported gastrointestinal distress or slowed fluid absorption when ingesting 2-30 g of BCAA prior to, during, or following exercise. Second, not all studies have reported significant increases in plasma NH_3 following BCAA supplementation. In fact, some studies report no effect or that plasma NH_3 or other markers of protein degradation are decreased with BCAA supplementation. In addition, of the studies reporting elevations in plasma NH_3 concentrations, the NH_3 levels reported were well within normal values reported for athletes following intense exercise or training. Finally, there is no evidence that BCAA supplementation prior to or during exercise impairs performance. Rather, as will be reviewed, several studies indicate that exercise performance may be improved in comparison to carbohydrate or noncarbohydrate placebos.

BCAA Supplementation

As previously stated, a number of researchers have suggested that BCAA supplementation with or without carbohydrate prior to and during exercise may affect plasma amino acid concentrations, the fTryp/BCAA ratio, central fatigue, physiological responses to exercise, psychological responses to exercise, and performance capacity. The following studies have evaluated these hypotheses.

Blomstrand and associates (7) investigated the effects of BCAA supplementation on endurance performance and psychological responses to a 30 km and a 42 km run. In the 30 km trial, runners ingesting a flavored-water placebo had significant reductions in plasma valine (–26%), leucine (–29%), and isoleucine (–33%) levels. Conversely, subjects administered a drink containing a total of 7.5 g BCAA during the race (2.5-3 g/h) had significant increases in valine (+82%) and nonsignificant increases in leucine (+22%) and isoleucine (+4). In addition, mental performance was significantly improved following the 30 km run in the BCAA trial, yet not improved in the placebo trial.

In the 42 km marathon run, valine (–17%), leucine (–19%), and isoleucine (–26%) levels were significantly decreased in the flavored-water placebo trial, whereas BCAA supplementation (total of 16 g or 4-5 g/h) resulted in a significant increase in valine (+177%), leucine (+80%), and isoleucine (+95%) levels. Interestingly, tyrosine levels increased by 22% in the placebo trial and only 3% in the BCAA trial. Since tyrosine has been suggested to be an indicator of net protein degradation (8, 19), these data suggest that protein degradation was less in the BCAA trial. No significant differences between the placebo and BCAA trials were observed in final performance times of all study participants. However, when the subjects were divided into subgroups of runners who completed the run in less than 3.05 hr or between 3.05-3.30 hr, performance time ratios were significantly lower in the slower runners who received BCAA indicating that run performance was improved in this group.

In a similar study, Blomstrand et al. (6) evaluated the effects of BCAA supplementation on mental performance in six female soccer players following a standard Swedish national soccer match lasting about 2.5 h. During the soccer match, players were given either a 6% GES placebo drink or a 6% GES drink with 7.5 g (3-3.5 g/h) of BCAA (40% valine, 35% leucine, and 25% isoleucine) in a double-blind, randomized, and crossover manner. Results revealed that in the placebo trial, plasma leucine (–33%), isoleucine (–36%), and valine (–24%) levels significantly decreased while plasma fTryp (+45%) levels significantly increased, thereby increasing the ratio of fTryp/BCAA. However, when the subjects ingested the carbohydrate with BCAA drink, plasma leucine (+168%), isoleucine (+198%), and valine (+148%) levels were significantly increased, while plasma fTryp levels were not significantly increased by 20%. Therefore, the ratio of fTryp/BCAA was significantly lower in the BCAA supplement trial. Following the soccer match, mental performance was significantly improved in the players who ingested the BCAA, whereas mental performance was unchanged in all players in the GES placebo drink group. The authors concluded that an intake of BCAA in addition to a standard carbohydrate drink during exercise appears to affect mental alertness during and following exercise.

To test this hypothesis, Kreider and associates (29) investigated the effects of BCAA supplementation on ultraendurance triathlon performance. In a double-blind, randomized, and crossover manner, five nationally competitive triathletes were supplemented with 6-8 g/d of a commercially available supplement containing either BCAA and glutamine, or a lactose placebo for 17 days. On day 14 of supplementation, subjects performed a competitive laboratory simulation of an Ironman half triathlon (2 k swim, 90 km bike, and a 21 km run). Subjects ingested 2.25 g of the BCAA and glutamine supplement 30 min prior to the triathlon and 0.75 g/h during the triathlon in addition to normal refueling practices, which were recorded and replicated in the subsequent trial. Physiological and psychological data were collected prior to, during, and for 3 d after the triathlon. Subjects then observed a 14 d washout period and repeated the experiment while ingesting the remaining lactose placebo supplement.

Results revealed that the BCAA and glutamine supplement increased serum insulin levels by 16% and decreased serum NH_3 (−23%) and sweat urea nitrogen (−13%). No differences were observed in swim- or bike-segment times. However, subjects' average run splits were significantly decreased in the BCAA and glutamine trial (from 9.71 ± 2 to 8.75 ± 1 min) resulting in a decrease in total run time (from 126.2 ± 22 to 113.8 ± 10 min) and total triathlon time (from 328.6 ± 26 to 315.7 ± 17 min). No significant differences were noted in RPE or pre-and postexercise profile of mood-states (POMS). Recovery analysis indicated that the athletes experienced less protein degradation and maintained a more optimal protein balance following the triathlon. These findings support the hypothesis that BCAA and glutamine supplementation may decrease net protein degradation during ultraendurance exercise and possibly improve performance in events lasting more than 3 h.

Petruzzello and associates (46) investigated the effects of a BCAA supplement on mood changes related to prolonged exercise. Nine trained males performed two prolonged cycling bouts (2 h) at 65% of VO_2max following 14 d of BCAA or lactose placebo supplementation administered in a double-blind manner. Subjects ingested according to the manufacturer's guidelines 2-4 capsules containing 0.75 g of BCAA per capsule prior to and following exercise training bouts. The profile of mood-states (POMS) psychological inventory and a critical flicker fusion test to assess CNS status were performed prior to and following the exercise bout. While fatigue did not significantly increase during the exercise bout, no significant differences were observed between groups in POMS, the CNS test, or performance times.

Carli and associates (10) investigated the effects of BCAA supplementation on hormonal responses in 14 male distance runners on two separate occasions before, during, and after 60 min of running at a constant running speed corresponding to 2 mmol/L on a 400 m track. In the first trial, subjects were administered a commercial carbohydrate and protein drink containing 5.1 g of leucine, 2.6 g of isoleucine, and 2.6 g of valine. In the second trial, subjects ingested the carbohydrate and protein drink with no BCAA added (placebo). Results revealed that in comparison to the placebo trial, human growth hormone, and prolactin levels, and the ratio of testosterone to sex hormone binding globulin were significantly lower following exercise in the BCAA trial while levels of insulin, testosterone, sex hormone binding globulin, and the ratio of testosterone to cortisol were significantly increased before and after exercise in the BCAA trial. No significant differences were observed between trials in levels of androstenedione, glucose, or serum proteins. These findings support contentions that protein or amino acid ingestion with carbohydrate increases serum insulin concentrations (11, 21, 27, 58) and may suppress endogenous amino acid oxidation (7, 29) through a hormonally-induced anticatabolic effect.

Schena and coworkers (48) investigated the effects of BCAA supplementation on amino acid metabolism during endurance exercise. In this study, 11 male runners performed two 90 min runs on a treadmill at 75% of VO_2max. Subjects received in a double-blind and randomized manner either a placebo or five, 5 g doses of BCAA beginning 30 min prior to exercise and concluding 30 min following exercise. Blood samples were measured for plasma amino acid concentrations.

Results showed that plasma BCAA ($-16 \pm 3\%$) and glutamine ($-10 \pm 2\%$) levels significantly decreased in the placebo trial, while BCAA supplementation resulted in a $125 \pm 12\%$ increase in plasma BCAA and a $7 \pm 2\%$ increase in glutamine levels. Total typtophan and fTryp levels were increased in a similar pattern in both trials. Consequently, results revealed that BCAA supplementation was effective in minimizing the normal exercise-induced alterations in the fTryp/BCAA ratio.

In another study, Coombes and McNaughton (15) investigated the effects of BCAA supplementation on indicators of muscle damage after prolonged exercise. Sixteen male subjects were divided into two equal groups and randomly and blindly assigned to ingest either 12 g/d of BCAA or a placebo, in addition to their normal diet for 14 days. Baseline creatine kinase (CK) and lactate dehydrogenase (LDH) levels were determined every other day for seven days preceding supplementation. No differences were observed between groups in levels of CK and LDH prior to supplementation. Subjects cycled for 120 min at 70% VO_2max and were required to keep activity to a minimum during the recovery period. Postexercise blood samples were taken at 0, 1, 2, 3, and 4 h postexercise as well as at 1, 2, and 4 days postexercise. Results showed that LDH and CK levels significantly increased in response to exercise in both groups. However, LDH levels in the BCAA supplement group were significantly lower than in the control group at 3 and 4 h postexercise and 1 d following the exercise bout. In addition, CK levels in the BCAA group were significantly lower than in the control group at 4 h postexercise and 1 and 2 days following exercise. The researchers concluded that BCAA supplementation significantly reduced serum CK and LDH levels following prolonged exercise, suggesting that BCAA supplementation may reduce exercise-induced muscle damage.

Nemoto and colleagues (41) investigated the effects of BCAA supplementation on endurance capacity and RPE levels in 30 active but nontrained college females. Lactate threshold (LT) and onset of blood lactate accumulation (OBLA) were determined during an incremental bicycle ergometer test. Subjects then performed submaximal exercise tests at LT, 75% of OBLA, or 90% of OBLA. Thirty minutes prior to performing submaximal exercise tests, subjects were administered either 11 g of BCAA or a placebo in a double-blind and randomized manner. Lactate, NH_3, and RPE were determined prior to, during, and following exercise. Results showed no significant differences between groups performing at LT. However, subjects performed 13% more work to OBLA with BCAA supplementation (153 ± 14 vs. 135 ± 18 W). Further, although NH_3 levels were consistently higher with the BCAA supplement, lactate and RPE levels were significantly lower. The researchers concluded that BCAA supplementation may improve endurance capacities and RPE during prolonged exercise of moderate intensity (from LT to 90% OBLA) despite accumulation of NH_3.

Mittleman and associates (40) investigated the effects of BCAA supplementation on gender-specific responses to prolonged exercise in the heat. Four male and four female subjects performed two performance trials consisting of 30 min of rest, 2 h resting heat exposure, and walking to exhaustion at 40% VO_2max in the heat. In a double-blind, randomized, counterbalanced manner, subjects received either a drink containing 5 ml/kg of GES or a drink containing BCAA (amount not specified). Results revealed that the BCAA supplement significantly increased time to exhaustion by 23.5% for females (168 ± 20 vs. 136 ± 18 min) but not for males (151 ± 34 vs. 152 ± 32 min), and that no significant differences were observed in serum FFA levels. These findings suggest a possible gender influence in the response to BCAA supplementation during prolonged exercise.

Finally, Hefler and associates (25) investigated the effects of BCAA supplementation on performance capacity in competitive cyclists. In a double-blind, randomized, and crossover design with a two-week washout period, 10 competitive cyclists were randomly assigned to ingest either 16 g/d of a glucose placebo or 16 g/d of BCAA (50% leucine, 30% valine, 20% isoleucine) for 14 days. On day two and day 14 of each treatment, subjects completed a 40 km time trial using their racing bicycles attached to a wind trainer with a cyclocomputer mounted to measure distance. Results revealed no significant differences in performance times between day two (58.1 ± 3 min) and day 14 (56.7 ± 2 min) when the subjects ingested the placebo. However, the subjects performed the 40 km time trial significantly faster when ingesting BCAA (58.2 ± 5 vs. 51.4 ± 3 min for days two and 14, respectively) which represented an 11.7% improvement in performance time. In addition, while no significant differences existed between groups at day two of supplementation, subjects rode significantly faster on day 14 of supplementation in the BCAA group. These data provide the best evidence to date that BCAA supplementation may provide an ergogenic benefit during intense endurance exercise.

In summary, studies indicate that BCAA supplementation (0.5-10 g/h) with or without carbohydrate serves to significantly increase plasma BCAA levels and thereby minimizes increases in the fTryp/BCAA ratio (6, 7, 16, 20, 48, 50-54) in comparison to water or carbohydrate placebos. In addition, BCAA supplementation has been reported to decrease exercise-induced protein degradation and serum muscle enzyme release (7, 10, 15, 29) possibly by promoting an anticatabolic hormonal profile (10, 29), to improve mental performance following endurance exercise (6, 7), to increase power output during exercise (29, 41), and to improve performance capacity in endurance and ultraendurance events (25, 29, 40, 41). Finally, it should be noted that no studies have reported gastrointestinal distress, toxic ammonia levels, or impaired performance associated with these supplementation regimens as suggested by some researchers (18, 54). Although additional research is necessary, these data provide preliminary support that BCAA supplements may provide additional benefit to normal carbohydrate feedings during exercise.

Implications on Overreaching and Overtraining

Most of the available literature in this area has investigated the effects of carbohydrate or BCAA supplementation on acute exercise responses following relatively short periods of supplementation (0-14 d). Theoretically, if exercise-induced alterations in the fTryp/BCAA ratio serotonergically contribute to fatigue during single bouts of prolonged exercise, maintaining a training program that involves frequent, intense, or prolonged training bouts may result in chronic elevations in the fTryp/BCAA ratio and chronic fatigue. This possible relationship was first noted by Newsholme and colleagues (42) and Parry-Billings and coworkers (43, 44). Unfortunately, few studies have examined the effects of carbohydrate and BCAA supplementation during intense training. However, the following two studies provide rationale as to adding protein or BCAA to carbohydrate supplements during training.

Zawadzki and associates (58) investigated the effects of ingesting a carbohydrate and protein complex supplement on serum substrates, hormones, and muscle glycogen storage following endurance cycling. Nine fasted male subjects cycled at 70% of VO_2max for 2 h on three occasions in order to deplete muscle glycogen levels. The participants were then fed a supplement containing 112 g carbohydrate, or 40.7 g protein, or the combination of 112 g carbohydrate and 40.7 g protein immediately after exercise and following 2 h of recovery. Plasma substrates and hormones were assessed for 6 h following exercise, and postexercise muscle glycogen levels were determined immediately after exercise and 4 h postexercise. Results revealed that serum insulin levels were significantly higher following ingestion of the combined carbohydrate and protein supplement as compared to either the carbohydrate or the protein trial. In addition, the rate of muscle glycogen resynthesis was 38% greater in the combined carbohydrate and protein trial compared to the remaining treatments. While the carbohydrate trial increased serum insulin levels and muscle glycogen storage in comparison to the protein trial, the rate of muscle glycogen resynthesis in the carbohydrate trial was significantly less than when subjects consumed the combined carbohydrate and protein supplement (see figure 15.3). The investigators suggested that the enhanced rate of muscle glycogen storage observed following the combined carbohydrate and protein supplementation may have been mediated by a carbohydrate- and protein-stimulated increase in plasma insulin levels promoting glycogen restoration. Since maintenance of muscle glycogen levels is important during intense training, this nutritional approach would seemingly enhance muscle glycogen restoration.

Cade and associates (9) investigated the effects of carbohydrate and protein supplementation following swim training in 20 male and 20 female intercollegiate swimmers. Swimmers were randomized into one of five supplementation groups while maintaining their normal training diets. One group of swimmers consumed water during exercise and was not administered a supplement following training sessions. A second group consumed water during exercise bouts and

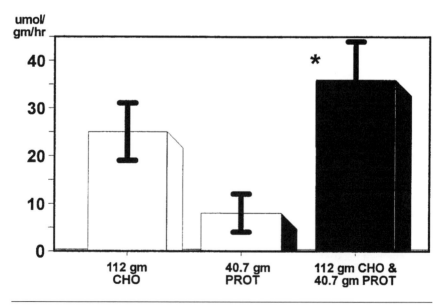

Figure 15.3   Effects of carbohydrate, protein, or combined carbohydrate and protein ingestion following endurance exercise on rate of muscle glycogen resynthesis.

Based on data from Zawadzki et al. 1992.

was supplemented with 80 g of sucrose immediately following each training session. The third group ingested water during exercise, followed by consuming 80 g of sucrose with 15 g of milk protein following exercise. The remaining two groups ingested a 6% GES during swim training followed by ingesting either 80 g of sucrose, or 80 g of sucrose with 15 g of milk protein after training. Serum enzymes were monitored during recovery in order to examine the effects of these nutritional strategies on muscle enzyme release and recovery during swim training. Results revealed that regardless of ingesting water or GES during exercise, creatine kinase (CK) levels were lower in the groups receiving the combined sucrose and protein supplement following exercise. In addition, CK levels returned to baseline sooner in the group receiving sucrose and protein following exercise (3 h) in comparison to carbohydrate supplementation alone (8 h). These findings suggest that ingestion of protein with carbohydrate following exercise may hasten the rate of muscle recovery during intense training.

To our knowledge, our study investigating the effects of BCAA supplementation during 25 weeks of intercollegiate swim training is the only data available in published form that has evaluated the effects of BCAA supplementation on markers of overtraining (30, 32-34). In this study, 10 intercollegiate swimmers (5 male, 5 female) were matched based on training volume, intensity, and event specificity to 10 swim team counterparts (5 male, 5 female), and 10 apparently healthy control subjects (5 male, 5 female) who represented a cross section of sedentary to moderately active people. During weeks 1-10 of training, participants were randomly and blindly assigned to ingest either 2.175 g of a supplement containing leucine (850 mg), isoleucine (625 mg), valine (625 mg), glutamine (50 mg), and carnitine (25 mg) or 2.175 g of a lactose placebo with 250 ml of water containing 22.5 g of maltodextrin prior to and following each swim workout (9/wk). During weeks 11-25, subjects ingested 2.9 g of the BCAA/glutamine supplement or the lactose placebo with the maltodextrin drink prior to and following workouts. In addition, the athletes supplemented their diet with 3.2 g of a supplement containing 20 amino acids or 3.2 g of the lactose placebo four times daily (12.8 g/d). Fasting blood samples were obtained at weeks 0, 4, 10, 14, 15, 22, and 25 of training. The swimmers performed a standardized 90 min swim workout following each training phase in order to assess the effects of BCAA/glutamine supplementation on physiological and psychological responses to swim training.

Results revealed that serum ammonia, cortisol, and the ratio of cortisol to testosterone were significantly lower throughout 25 weeks of swim training in the BCAA/glutamine supplemented group (BG) than in the carbohydrate supplemented group (CG) (32). In addition, analysis of personal strain questionnaires revealed that physical strain (CG, 23.3 ± 6; BG, 19.0 ± 5) indicating fewer physical symptoms (i.e., colds, aches and pains, erratic eating patterns, weight changes, sleep disturbance, lethargy and apathy) and psychological strain (CG, 26.1 ± 8; BG, 22.7 ± 6) indicating less depression and anxiety were significantly lower in the BCAA/glutamine group during training compared to the carbohydrate supplemented trial (see figures 15.4 and 15.5). Moreover, POMS anger scores (CG, 15.7 ± 9; BG, 11.2, p = .07) and total mood disturbance scores (CG, 47.1 ± 32; BG, 34.3 ± 23, p = .06) tended to be lower in the BCAA/glutamine group (34). Finally, analysis of fatigue inventories completed daily and four weeks posttraining revealed that subjects in the BCAA/glutamine supplemented group experienced significantly less mental fatigue, arm and shoulder fatigue, leg fatigue, general tiredness, and tiredness of school (34).

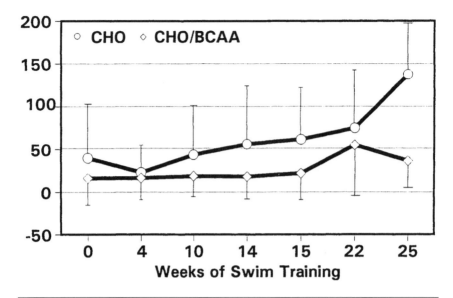

Figure 15.4   Fasting cortisol to testosterone ratio during 25 weeks of intercollegiate swimming in the carbohydrate (CHO) and carbohydrate/BCAA/glutamine (CHO/BCAA) supplemented groups.

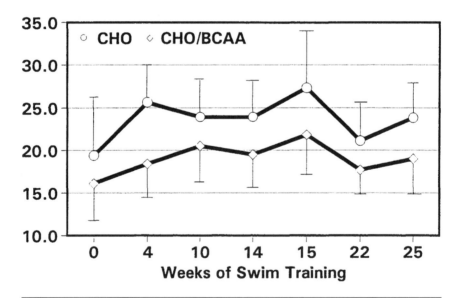

Figure 15.5   Personal strain questionnaire psychological stress responses observed during 25 weeks of intercollegiate swimming in the carbohydrate (CHO) and carbohydrate/BCAA/glutamine (CHO/BCAA) supplemented groups.

Analysis of immune data throughout the 25 wk training period (30, 31, 35) revealed that the fasting CD4/CD8 ratio (control 2.57 ± 1.4 mg/dl; carbohydrate 2.46 ± 1.2 mg/dl; BCAA/glutamine 3.49 ± 1.8 mg/dl) and serum IgA levels (control 123 ± 54 mg/dl, carbohydrate 135 ± 60 mg/dl, BCAA/glutamine 192 ± 69 mg/dl) were significantly greater in the BCAA/glutamine group (see figures 15.6 and 15.7). No significant differences were observed among control, carbohydrate placebo, or BCAA/glutamine groups lymphocyte proliferation response induced by pokeweed (PWM) or phytohemagglutinin (PHA) mitogens. Analysis of retrospectively reported symptoms of infection revealed that the BCAA/glutamine group experienced significantly lower frequency of chest congestion and tightness, coughing, stomach upset, and degree of muscle and joint pain. However, no significant differences were reported in the total number or severity of sicknesses among groups.

Data from this study, which were presented at the International Conference on Overreaching and Overtraining in Sport indicated that the percentage of fasting monocytes (control 8.0 ± 1.3%, carbohydrate 7.7 ± 1.1%, BCAA/ glutamine 9.5 ± 2.9%) and plasma IgG levels (control 828 ± 343 mg/dl, carbohydrate 948 ± 382 mg/dl, BCAA/glutamine 1,045 ± 364 mg/dl) tended to be higher (p < .10) in the BCAA/glutamine group (49). Further, analysis of leukocytic responses to the standardized 90 min swim revealed that subjects in the BCAA/glutamine group experienced a 21%, 18%, and 29% less change in percentage of lymphocytes, neutrophils and the Ne/Ly ratio, respectively, following the exercise stimulus throughout the season (35).

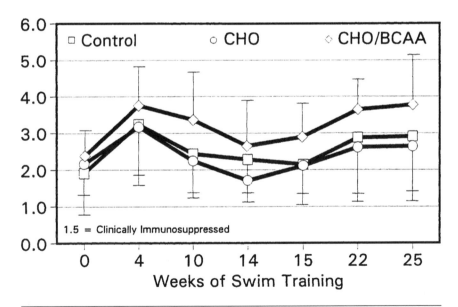

Figure 15.6   Fasting CD4/CD8 ratio observed during 25 weeks of intercollegiate swimming in the nontrained controls, the carbohydrate supplemented group (CHO), and the carbohydrate/BCAA/glutamine (CHO/BCAA) supplemented group.

mg/dl

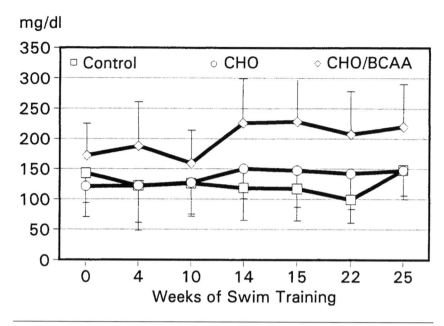

Figure 15.7 Plasma IgA levels observed during 25 weeks of intercollegiate swimming in the nontrained controls, the carbohydrate supplemented group (CHO), and carbohydrate/BCAA/glutamine (CHO/BCAA) supplemented group.

Collectively, these findings support the hypothesis that BCAA/glutamine supplementation during intense training may affect hormonal, psychological, and immunological responses to training. Further, the addition of small quantities of BCAA/glutamine to carbohydrate drinks may lessen the physiological, psychological, and immunological responses to heavy training to a greater degree than carbohydrate alone. However, additional research is necessary to determine the effects of carbohydrate and BCAA supplementation during intense training on markers of overtraining before definitive conclusions can be drawn.

Summary

Review of the literature and preliminary reports, that have investigated the effects of BCAA supplementation on endurance and ultraendurance exercise and training has revealed some interesting physiological and psychological phenomena, that will provide areas of research interest for some time. While there are data to support the hypothesis that BCAA supplementation with or without carbohydrate may affect the physiological and psychological responses to endurance exercise, many questions remain to be answered regarding the hypothesized mechanisms of action, and the physiological and psychological effects on endurance exercise performance and training.

However, based on the available literature, the following conclusions can be drawn. First, it is clear that dietary manipulations of carbohydrate affect substrate utilization during exercise. Athletes initiating exercise with low muscle and liver glycogen levels oxidize FFA and BCAA to a greater degree than athletes initiating exercise with normal glycogen levels. As glycogen availability declines during prolonged exercise, FFA and fTryp levels in the blood increase while levels of essential amino acids decline (particularly BCAA and glutamine). These alterations result in significant increases in the ratio of fTryp/BCAA. Increases in the fTryp/BCAA ratio in the blood increases 5-HT synthesis in the brain, which has been reported to affect psychological and physiological factors affecting fatigue in animal and human studies.

There appear to be two nutritional strategies that may be effective in delaying the exercise-induced changes in the fTryp/BCAA ratio, and thereby theoretically delaying fatigue. First, carbohydrate supplementation (6-12% GES) has been shown to attenuate increases in FFA, minimize increases in fTryp, and enhance power output during prolonged cycling. Second, BCAA supplementation (2-10 g/h) with or without carbohydrate has been shown to increase plasma BCAA levels, minimize increases in the fTryp/BCAA ratio, and affect psychological and physiological responses to exercise. The effects of BCAA supplementation on endurance performance remain equivocal. However, no study has reported a negative impact on performance.

Less is known regarding the effects of BCAA supplementation during training. However, evidence is accumulating that protein and BCAA supplementation with carbohydrate prior to, during, and following exercise may lessen physiological, psychological, and immunological responses to training. Theoretically, these adaptations may lessen the incidence of overtraining. However, additional well-controlled research is necessary to investigate these hypotheses before definitive conclusions can be drawn.

Based on a review of literature in this area as well as sport nutrition in general, the following nutritional recommendations can be made for athletes engaged in heavy training:

1. Maintain a high carbohydrate (55-65% CHO, 15% protein, <30% fat), energy balanced, nutrient dense diet.
2. Ingest a preexercise meal 4-6 hours prior to exercise in order to saturate muscle and liver glycogen levels prior to exercise.
3. Ingest a light carbohydrate and protein meal (e.g., 30-50 g carbohydrate as fructose, 5-10 g of protein) 30-60 min prior to intense or prolonged exercise.
4. If exercise is an intense (>70% VO_2max), intermittent, or sustained exercise bout (>60 min), ingest a 4-8% GES during exercise.
5. BCAA and glutamine supplementation (2-10 g/h) prior to and during intense or prolonged exercise bouts (>60 min) in addition to normal carbohydrate feedings may be beneficial. However, additional research is necessary to investigate this hypothesis.
6. Ingestion of a carbohydrate and protein meal (e.g., 120 g CHO, 40 g protein) within 2 h following intense exercise will enhance glycogen resynthesis.

References

1. Bailey, S.P., J.M. Davis, E.N. Ahlborn. 1992. Effect of increased brain serotogenic (5-HT$_{IC}$) activity on endurance performance in the rat. *Acta Physiologica Scandinavica* 145: 75-76.

2. Bailey, S.P., J.M. Davis, E.N. Ahlborn. 1993. Neuroendocrine and substrate responses to altered brain 5-HT activity during prolonged exercise to fatigue. *Journal of Applied Physiology* 74: 3006-3012.

3. Bailey, S.P., J.M. Davis, E.N. Ahlborn. 1993. Brain serotogenic activity affects endurance performance in the rat. *International Journal of Sports Medicine* 6: 330-333.

4. Blomstrand E., F. Celsing, E.A. Newsholme. 1988. Changes in plasma concentrations of aromatic and branch-chain amino acids during sustained exercise in man and their possible role in fatigue. *Acta Physiologica Scandinavica* 133: 115-121.

5. Blomstrand E., D. Perrett, M. Parry-Billings, E.A. Newsholme. 1989. Effect of sustained exercise on plasma amino acid concentrations and on 5-hydroxy-tryptamine metabolism in six different brain regions in the rat. *Acta Physiologica Scandinavica* 136: 473-481.

6. Blomstrand E., P. Hassmen, E. Newsholme. 1991. Effect of branch-chain amino acid supplementation on mental performance. *Acta Physiologica Scandinavica* 143: 225-226.

7. Blomstrand E., P. Hassmen, B. Ekblom, E.A. Newsholme. 1991. Administration of branch-chain amino acids during sustained exercise—effects on performance and on plasma concentration of some amino acids. *European Journal of Applied Physiology* 63: 83-88.

8. Butterfield, G. 1991. Amino acids and high protein diets. In Perspectives in exercise science and sports medicine, vol. 4; *Ergogenics: enhancement of performance in exercise and sport*, eds. D. Lamb, M. Williams, 87-122. Indianapolis: Brown & Benchmark.

9. Cade, J.R., R.H. Reese, R.M. Privette, N.M. Hommen, J.L. Rogers, M.J. Fregly. 1991. Dietary intervention and training in swimmers. *European Journal of Applied Physiology and Occupational Physiology* 63: 210-215.

10. Carli, G., M. Bonifazi, L. Lodi, C. Lupo, G. Martelli, A.Viti. 1992. Changes in exercise-induced hormone response to branched chain amino acid administration *European Journal of Applied Physiology and Occupational Physiology* 64: 272-277.

11. Castellino, P., L. Luzi, D.C. Simonson, M. Haymond, R.A. DeFronzo. 1987. Effect of insulin and plasma amino acid concentrations of leucine metabolism in man. *Journal of Clinical Investigations* 80: 1784-1793.

12. Chaouloff, F., J.L. Elgohozi, Y. Guezennec, D. Laude. 1985. Effects of conditioned running on plasma, liver and brain tryptophan and on brain 5-hydroxytryptamine metabolism of the rat. *British Journal of Pharmacology* 86: 33-41.

13. Chaouloff, F., G.A. Kennett, B. Serrurier, D. Merina, G. Curson. 1986. Amino acid analysis demonstrates that increased plasma free tryptophan causes the increase of brain tryptophan during exercise in the rat. *Journal of Neurochemistry* 46: 1647-1650.

14. Chaouloff, F., D. Laude, Y. Guezennec, J.L. Elgohozi. 1986. Motor activity increases tryptophan, 5-hydroxyinoleacetic acid, and homovanillic acid in ventricular cerebrospinal fluid of the conscious rat. *Journal of Neurochemistry* 46: 1313-1316.

15. Coombes, J., L. McNaughton. 1985. The effects of branched chain amino acid supplementation on indicators of muscle damage after prolonged strenuous exercise. *Medicine and Science in Sports and Exercise* 27: S149.

16. Davis, J.M., S.P. Bailey, J.A. Woods, F.J. Galiano, M.T. Hamilton, W.P. Bartoli. 1992. Effects of carbohydrate feedings on plasma free tryptophan and branched-chain amino acids during prolonged cycling. *European Journal of Applied Physiology and Occupational Physiology* 65: 513-519.

17. Davis, J.M., S.P. Bailey, D.A. Jackson, A.B. Stansner, S.L. Morehouse. 1993. Effects of a serotonin (5-HT) agonist during prolonged exercise to fatigue in humans. *Medicine and Science in Sports and Exercise* 25: S78.

18. Davis, J.M.. 1995. Carbohydrates, branched-chain amino acids, and endurance: the central fatigue hypothesis. *International Journal of Sport Nutrition* 5: S29-S38.

19. Dohm, G.L. 1986. Protein as a fuel for endurance exercise. *Exercise Sport Science Reviews* 14: 143-173.

20. Galiano, F..J., J.M. Davis, S.P. Bailey, J.A. Woods, M. Hamilton, W.P. Bartoli. 1992. Physiological, endocrine and performance effects of adding branch chain amino acids to a 6% carbohydrate electrolyte beverage during prolonged cycling. *Medicine and Science in Sports and Exercise* 23: S14.

21. Garlick, P.J., I. Grant. 1986. Amino acid infusion increases the sensitivity of muscle protein synthesis in vivo to insulin. *Biochemistry Journal* 254: 579-584.

22. Goldberg, A.L., T.W. Chang. 1978. Regulation and significance of amino acid metabolism in skeletal muscle. *Federal Proceedings* 37: 2301-2307.

23. Greenhaff, P.L., J.B. Leiper, D. Ball, R.J. Maughan. 1991. The influence of dietary manipulation on plasma ammonia accumulation during incremental exercise in man. *European Journal of Applied Physiology* 63: 338-344.

24. Haralambie, G., A. Berg. 1976. Serum urea and amino nitrogen changes with exercise duration. *European Journal of Applied Physiology* 36: 39-48.

25. Hefler, S.K., L. Wildman, G.A. Gaesser, A. Weltman. 1993. Branched-chain amino acid (BCAA) supplementation improves endurance performance in competitive cyclists. *Medicine and Science in Sports and Exercise* 25: S24.

26. Hood, D.A., R.L Terjung. 1990. Amino acid metabolism during exercise and following endurance training. *Sports Medicine* 9: 23-35.

27. Hutton, J.C., A. Sener, W.J. Malaisse. 1980. Interaction of branched-chain amino acids and keto acids upon pancreatic islet metabolism and insulin secretion. *Journal of Biological Chemistry* 255: 7340-7346.

28. Kargotich, S., D.G. Rowbottom, D. Keast, C. Goodman, A.R. Morton 1996. Plasma glutamine changes after high intensity exercise in elite male swimmers. *Medicine and Science in Sports and Exercise* 28: S133.

29. Kreider, R.B., G.W. Miller, M. Mitchell, C.W. Cortes, V. Miriel, C.T. Somma, S.R. Sechrist, D. Hill. 1992. Effects of amino acid supplementation on ultraendurance triathlon performance. In *Proceedings of the first world congress on sport nutrition*, 488-536. Barcelona, Spain: Enero.

30. Kreider, R.B., B. Leutholtz. 1993. Nutrition for the immune system: the role of amino acids. *Journal of Optimal Nutrition* 2: 278-291.

31. Kreider, R.B., V. Miriel, E. Bertun. 1993. Amino acid supplementation and exercise performance: proposed ergogenic value. *Sports Medicine* 16: 190-209.

32. Kreider, R.B., V. Miriel, E. Bertun, T. Somma, S. Sechrist. 1993. Effects of amino acid and carnitine supplementation on markers of protein catabolism and body composition during 25 weeks of swim training. *Southeast American College of Sports Medicine Abstracts* 20: 45.

33. Kreider, R.B., R. Ratzlaff, E. Bertun, J. Edwards, V. Miriel, G. Lloyd, J. Gentry. 1993. Effects of amino acid and carnitine supplementation on immune status during an intercollegiate swim season. *Medicine and Science in Sports and Exercise* 25: S123.

34. Kreider, R.B., C.W. Jackson. 1994. Effects of amino acid supplementation on psychological status during an intercollegiate swim season. *Medicine and Science in Sports and Exercise* 26: S115.

35. Kreider, R., V. Miriel, D. Tulis, E. Bertun. 1996. Effects of amino acid supplementation during a 25-week intercollegiate swim season on leukocytic response to swimming. *International Conference on Overreaching and Overtraining in Sport, Conference Abstracts* 1: 76.

36. Lancranjan, I.A., A. Wirz-Justice, W. Puhringer, E. Del Pozo. 1977. Effect of L-5-hydroxytryptophan infusion of growth hormone and prolactin secretion in man. *Journal of Endocrinology and Metabolism* 45: 588-593.

37. Lehmann, M., M. Huonker, F. Dimeo, N. Heinz, U. Gastmann, N. Treis, J.M. Steinacker, J. Keul, R. Kajewski, D. Häussinger. 1995. Serum amino acid concentrations in nine athletes before and after the 1993 Colmar ultratriathlon. *International Journal of Sports Medicine* 16: 155-159.

38. Lemon, P.W.R. 1991. Protein and amino acid needs of the strength athlete. *International Journal of Sport Nutrition* 1: 127-145.

39. MacLean, D.A., L.L. Spriet, T.E. Graham. 1992. Plasma amino acid and ammonia responses to altered dietary intakes prior to prolonged exercise in humans. *Canadian Journal of Physiology and Pharmacology* 70: 420-427.

40. Mittleman, K., C. Miller, M. Ricci, L. Fakhrzadeh, S.P. Bailey. 1995. Branched-chain amino acid (BCAA) supplementation during prolonged exercise in heat: influence of sex. *Medicine and Science in Sports and Exercise* 27: S148.

41. Nemoto, I., A. Tanaka, Y. Kuroda. 1996. Branched-chain amino acid (BCAA) supplementation improves endurance capacities and RPE. *Medicine and Science in Sports and Exercise* 28: S37.

42. Newsholme, E.A., M. Parry-Billings, M. McAndrew, R. Budgett. 1991. Biochemical mechanism to explain some characteristics of overtraining. In Medical Sports Science, vol. 32, *Advances in nutrition and top sport*, ed. F. Brouns, 79-93. Basel: Karger.

43. Parry-Billings, M., E. Blomstrand, B. Leighton, G.D. Dimitradis, E.A. Newsholme. 1990. Does endurance exercise impair glutamine metabolism? *Canadian Journal of Sport Science* 13: 13P.

44. Parry-Billings, M., E. Blomstrand, N. McAndrew, E.A. Newsholme. 1990. A communicational link between skeletal muscle, brain and cells of the immune system. *International Journal of Sports Medicine* 11: S122-S128.

45. Parry-Billings, M., R. Budgett, Y. Koutedakis, E. Blomstrand, S. Brooks, C. Williams, P.C. Calder, S. Pilling, R. Baigrie, and E. Newsholme. 1992. Plasma amino acid concentrations in the overtraining syndrome: Possible effects on the immune system. *Medicine and Science in Sports and Exercise* 24: 1353-1358.

46. Petruzzello, S.J., D.M. Landers, J. Pie, J. Billie. 1992. Effect of branched-chain amino acid supplements on exercise-related mood changes. *Medicine and Science in Sports and Exercise* 24: S2.

47. Rang, H.P., M.M. Dale. 1987. *Pharmacology*, 80-92. Edinburgh: Churchill Livingstone.

48. Schena, F., F. Guerrine, P. Tregnaghi. 1993. Effects of branched-chain amino acid supplementation on amino acid metabolism during endurance exercise. *Medicine and Science in Sports and Exercise* 25: S24.
49. Taylor, T., R. Kreider, L. Ramsey, H. Yamashito, V. Miriel, D. Tulis, E. Bertun. 1996. Effects of amino acid supplementation during a 25-week intercollegiate swim season on fasting immunoglobulins and leukocytes. *International Conference on Overreaching and Overtraining in Sport, Conference Abstracts* 1: 76.
50. Vandewalle, L., A.J.M. Wagenmakers, K. Smets, F. Brouns, W.H.M. Saris. 1991. Effect of branched-chain amino acid supplements on exercise performance in glycogen depleted subjects. *Medicine and Science in Sports and Exercise* 23: S116.
51. van Hall, G., J.S. Raymakers, W.H. Saris. 1995. Ingestion of branched-chain amino acids and tryptophan during sustained exercise in man: failure to affect performance. *Journal of Physiology* 486: 789-794.
52. van Hall, G., G.J. van der Vusse, K. Soderlund, A.J. Wagenmakers. 1995. Deamination of amino acids as a source for ammonia production in human skeletal muscle during prolonged exercise. *Journal of Physiology* 489: 251-261.
53. Wagenmakers, A.J.M., K. Smets, L. Vandewalle, F. Brouns, W.H.M. Saris. 1991. Deamination of branched-chain amino acids: a potential source of ammonia production during exercise. *Medicine and Science in Sports and Exercise* 23: S116.
54. Wagenmakers, A.J.M., E.J. Bechers, F. Brouns, H. Kuipers, P.B. Soeters, G.J. van der Vusse, W.H.M. Saris. 1991. Carbohydrate supplementation, glycogen depletion, and amino acid metabolism during exercise. *American Journal of Physiology* 260: E883-E890.
55. Wilson, W.M., R.J. Maughan. 1992. Evidence for a possible role of 5-hydroxytryptamine in the genesis of fatigue in man: administration of paroxetine, a 5-HT re-uptake inhibitor, reduces the capacity to perform prolonged exercise. *Journal of Physiology* 77: 921-924.
56. Wolfe, R.R., M.H. Wolfe, E.R. Nadel, J.H.F. Shaw. 1984. Isotopic determination of amino acid-urea interactions in exercise in humans. *Journal of Applied Physiology* 56: 221-229.
57. Young, S.N. 1986. The clinical psychopharmacology of tryptophan. In *Nutrition and the brain*, eds. Wurtman, Wurtman, 49-88. New York: Raven Press.
58. Zawadzki, K.M., B.B. Yaspelkis, J.L. Ivy. 1992. Carbohydrate-protein complex increases the rate of muscle glycogen storage after exercise. *Journal of Applied Physiology* 72: 1854-1859.

.

Psychological Considerations of Overreaching and Overtraining

CHAPTER 16

A Systemic Model for Understanding Psychosocial Influences in Overtraining

Andrew W. Meyers, PhD, and James P. Whelan, PhD

Introduction

In the pre-Atlanta Olympic Games frenzy taking place in the summer of 1996, personal stories about elite athletes appeared in our newspapers and other media on an almost daily basis. As we prepared this chapter, one particular story captured our attention. Recently, articles in the sports pages announced that Kent Bostick, a 42-year-old American cyclist known fondly to his competitors as "Bostisaurus" in reference to his relatively advanced age, qualified for the U.S. Olympic team by winning the men's 4,000-meter individual pursuit at the Olympic trials.

Mr. Bostick, who turned 43 years of age just three weeks after the trials, had been an alternate on three previous Olympic teams, and he was quoted as saying "Each time that I've been an alternate, I learned that I needed to do a little more to get there." The article continued, "Realizing that this was his final chance to reach the Olympics, Bostick stuck to an exhausting training regimen. In addition, Bostick . . . holds down a full-time job as a groundwater hydrologist" (71). So how does a 42-year-old fellow with a full-time job and an exhaustive training schedule deal with these demands and prevent staleness and burnout? We are told that much of his training took place on a 40-mile round-trip bicycle commute to work. These commutes were followed by evening rides with his wife, also a competitive cyclist, which he described as their "social time." Finally, his 84-year-old mother, who attended the Olympic trials, claimed responsibility for Bostick's rigorous, organically-based diet that helped him to maintain "the body of a 20 year old" (5). We have, in Mr. Bostick, a thoughtful, resilient athlete with an accommodating career and a supportive wife and family, and the result is a positive response to intense training. Mr. Bostick's story presages the issues that we examine in this chapter.

We begin with a brief overview of several important issues in the overtraining literature in sport psychology. While we do not intend to cover exhaustively this growing body of literature, we do hope to represent its strengths and weaknesses and to focus on its points of conflict. We will then present what we believe is the first comprehensive, systemic model of overtraining. We will attempt to illustrate the model both by description and example. Finally, we will discuss the implications of this model for the prevention and treatment of the overtraining phenomenon.

Review of the Literature

Perhaps the single best review of the psychological issues in overtraining is Raglin's (74) chapter in Singer, Murphey, and Tennant's book, *Handbook of Research on Sport Psychology*, and a more detailed review of this literature can be found there (78). The first issue that demands our attention is the apparent contradictory elements of the hypothesized relationship between physical exercise and emotional well-being. One contradiction is illustrated by our call here to discuss psychological deterioration in response to intense exercise levels, yet there is a common assumption that some forms of exercise and physical activity confer mental health benefits (63, 93). There is an accumulating body of research that supports the benefits of exercise for a variety of psychological states and problems including depression (38, 64), anxiety (83), and low self-esteem (73). Furthermore, Morgan and Goldston (63) have asserted that athletes who engage in regular and rigorous physical training possess a positive mental health profile. These athletes self-report high vigor and low tension, depression, anger, fatigue, and confusion on the Profile of Mood States (55), a pattern Morgan and Goldston have labeled the Iceberg Profile (see figure 16.1).

We will not directly address the issue of exercise and mental health here and we assume that the physical activity levels examined in studies of exercise and emotional well-being are typically much less demanding than those relevant to overtraining. Indeed, in today's competitive athletic world, athletes, in their search for excellence, are often called on to train at high levels of intensity for extended periods of time. This intense training is typically part of a periodic training cycle where progressive, systematic increases in training intensity and volume are designed to maximally stress the athlete (35). The assumption here is that the athlete is forced to adapt to the intense training stress and that such an overload extends his or her physical and psychological boundaries. These gains are then solidified during a period of active rest prior to competition (23, 77). Failure to adapt adequately to the training stress produces decrements in performance and psychological functioning arbitrarily labeled overtraining, staleness, or burnout (61, 77, 91).

Unfortunately, not all athletes profit from the intense training experience. A second apparent contradiction in the relationship between exercise and psychological well-being is demonstrated by the finding that competitors with similar physical skills and capabilities, exposed to almost identical training regimens, may demonstrate widely variable performance outcomes (27, 48). The body of this chapter is devoted to building a framework for understanding why this may occur.

An additional underlying problem in the psychological overtraining literature is one of definition. One aspect of this problem is the failure of investigators in this

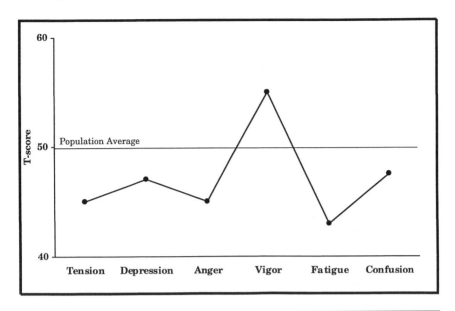

Figure 16.1   The Iceberg Profile from the profile of mood states (POMS).
Based on data from McNair et al. 1971.

field to adopt shared definitions of overtraining phenomena (74, 84), but an equally pressing concern is the post hoc or circular nature of many of the concepts employed to address overtraining (74). For example, Silva (77) argued that the training stressor and the athlete's adaptive processes determine his or her response to training. However, the roles of the stressor and the adaptive processes are only revealed by the athlete's performance gains or decay. Similarly, Hackney, Pearman, and Nowacki (28) stated that "the ultimate goal of the athletic training process is to improve performance. Overtraining results in the opposite effect, a performance decrease. If the training is inappropriate, performance will not be enhanced and may actually be decreased. The process of overtraining falls into this last category." Clearly, these definitions offer little help in understanding this complex phenomenon. Unfortunately, we cannot yet escape from the post hoc definition of overtraining and so in this presentation we will adopt the definition employed by Stone et al. (84):

> *Overtraining with respect to physical training may be defined as a plateauing and/or a decrease in performance that results from a failure to tolerate or adapt to the training load. If the training load is increased for a few days or weeks, producing a decrease in performance, and subsequent rest or detraining brings performance to higher than previous levels, it has been termed overreaching or supercompensation training.*

The essential issue here is that the athlete exposed to intense training fails to regenerate during the following rest or reduced training period.

As the evidence presented in other chapters of this volume attests, there is an increasing, though still incomplete, understanding of the physiological markers of overtraining (28). The psychosocial signs of the maladaptive response to intense training are also generating increasing consensus. These signs and symptoms of overtraining and burnout are drawn from Hackney et al. (28) and are presented in table 16.1. Even with this increasing specification of the overtraining phenomenon, our ability to predict who will experience a negative response to intense training is poorly developed (32, 40, 48).

The inability to forecast an undesirable response to intense training is due in large part to the complexity of the overtraining event. Training regimens across sports or exercise activities are difficult to compare and there is enormous variability in athletes' abilities to respond to the training demand (13). Based on this, Raglin (74) argued that the distinction between adaptive and maladaptive training is an arbitrary one and that eventual performance is dependent on the interaction between the training program and the individual athlete.

Another contributor to our limited ability to understand the response to training increases is the emphasis, at least in the psychological literature, on descriptive and correlational investigations of overtraining phenomena and the lack of attention to well-controlled, prospective studies (though Lehmann et al. (47) argued that this conclusion applies to the broader overtraining literature, as well). Admittedly, this limitation is in response to the obvious practical and ethical problems engendered by attempts to manipulate training loads and induce athletic performance decrements. However, the result of these strictures is that the dominant model for psychological overtraining experiments involves following athletes or teams over the course of a training season, or part of a season, while performance and psychological variables are monitored. These efforts have given

Table 16.1 Psychosocial Signs and Symptoms of Staleness and Burnout

- Apathy

- Lethargy

- Mental exhaustion

- Sleep disturbance

- Weight loss

- Report of muscle soreness

- Gastrointestinal disturbances

- Appetite loss

- Lowered self-esteem

- Mood changes

- Substance abuse

- Changes in values and beliefs

- Emotional isolation

- Increased anxiety

us a rudimentary understanding of the psychological effects of intense training. Let us give you an example.

Morgan's long-term examination of collegiate swimmers' psychological response to intense training remains the most impressive attempt to understand this phenomenon (61, 62, 70, 75). Morgan and his colleagues (61) summarized a ten-year series of investigations with University of Wisconsin athletes. Over this period, investigators administered the Profile of Mood States (POMS) (55), a nonclinical measure of mood, to approximately 200 male and 200 female swimmers and other collegiate athletes on a variety of different schedules to answer a number of questions related to overtraining. The major findings of this 10-year research effort are as follows:

- Global mood disturbance (the sum of tension, depression, anger, fatigue, and confusion scores minus vigor) increases significantly as the training load increases.
- Following decreases in training load, global mood disturbance returns to baseline. That is, there exists a dose-response relationship between training and mood.
- Even brief periods of rest or taper may be sufficient to restore a more positive mood-state.
- Stale athletes, defined by mood and performance deterioration, demonstrate symptoms similar to those seen in clinical depression.
- Similar mood changes are not found in the general population of college students and so may be attributed to athletes' training regimens.
- The dose-response relationship between training load and mood appears to hold across sports.
- There is great individual variability in mood response to intense training.

Morgan's work is largely devoted to the study of naturally occurring training sequences. One study presents us with some additional rigor by experimentally manipulating training loads while monitoring psychological states. Murphy, Fleck, Dudley, and Callister (67) systematically varied training regimens in judo competitors in residence at the United States Olympic Training Center. Eight male and seven female athletes were comprehensively monitored over a ten-week training period designed in conjunction with the national judo coaching staff. After a four-week baseline period consisting of standard resistance, interval, and specific judo training, athletes were exposed to a four-week increased training volume phase. Resistance and interval training volume increased by 50% while judo training remained constant. In the final sport-specific training phase, judo skill training sessions were doubled, but resistance training returned to baseline levels and interval training dropped below baseline levels. However, based on caloric output estimates, this phase produced the highest training volume. Comprehensive psychological and physical performance assessments were conducted at two, four, eight, and ten weeks.

Measures of isokinetic strength and anaerobic endurance showed decrements in performance during the intense training phases indicating that the training had been adequately demanding. On the POMS, the fatigue and anger scales showed the expected negative response to training. Contrary to the Morgan et al. (61) findings, the total mood disturbance score did not show changes across the training phases. However, on two additional measures of anxiety, scores peaked by the tenth week. Murphy et al. (67) speculated that the involvement of elite athletes participating in a structured workout program was responsible for the failure to closely replicate Morgan's dose-response results. Alternatively, the small sample size may have limited generalizability.

While Morgan et al. (61) argued that the monitoring of mood may aid in the prevention of staleness, we are faced with some inconsistency in the pattern of psychological response to training loads and a good deal of individual variability in this response. Unfortunately, very few studies have attempted to examine, prospectively, potential explanations of or contributors to this variability (47). One experiment that did investigate psychological predictors of overtraining was conducted by Gross (27).

Gross (27) hypothesized that psychological factors may not simply be consequences of intense training but may precipitate overtraining. More specifically, she speculated that the personality characteristic of hardiness may help to explain some of the individual variability in response to demanding training loads. Hardiness is defined by three characteristics—an involvement in the activities of one's life, a belief that one can control or influence these activities, and a perception that these activities are challenging and valuable (39). Gross (27) argued that hardy individuals reduce stress by altering their appraisal of the stressor and by employing coping mechanisms to more effectively manage stress.

To evaluate these assumptions, Gross (27) assessed hardiness, using the cognitive hardiness inventory (69), and mood, using the POMS, of 253 junior high, senior high, and collegiate swimmers across their competitive seasons. As daily training distances increased (from an average of 2,000 yd to almost 12,000 yd) across the season, mood deteriorated.

Significant increases were found for fatigue and total mood disturbance, and a significant decrease for vigor, from preseason to midseason and postseason. However, the high hardiness group (those scoring one standard deviation above the mean of the entire sample) had significantly lower total mood disturbance scores than the low hardiness group (those scoring one standard deviation below the mean of the entire sample). On a measure of coping, high hardiness swimmers also demonstrated more adaptive coping behaviors and fewer maladaptive coping responses.

Gross's (27) findings suggest an initial strategy for conceptualizing the formidable individual variability in response to intense training. Her work begins to help us build a context for understanding the mood changes that can occur when athletes are overworked. We argue that knowledge of both the stressor and the individual athlete's abilities and supports for coping with demands are necessary to fully comprehend the athlete's response to intense training. This is similar to the argument that Smith (81) has made in his cognitive-affective model of athletic burnout. While Smith's model may be extended to overtraining, we believe, as do others (40, 47, 53), that a richer, systemically-oriented model of overtraining will add to our understanding of these phenomena. The remainder of this chapter is devoted to building such a model. We begin by developing a narrative that illustrates the workings of the model.

The Stories of Carl and Hunter

Two athletes, Carl and Hunter, compete for a major club swim team. This is a competitive club with generous facilities and an experienced coach. The coach is energetic, well-liked and attempts to stay current on the coaching and sport science literature. The club includes swimmers from ages six through 18 and has an active parent organization and adequate financial resources. Carl and Hunter swim with a group of about a half dozen other 14- to 16-year-old middle-distance swimmers, most of whom are coming off a successful short course winter season including qualifying for junior nationals. While the coach considers both Carl and Hunter very competent swimmers, qualifying for junior nationals did not, in either case, involve what the coach would have considered "breakthrough performances."

As they approach the long course season, the coach has become increasingly concerned over the fact that the performances of these athletes have not improved much beyond their junior national qualifying times. Given their talent level and the coach's judgment that they are capable of better performances, the coach decides to start preparation for the long course season with a much higher intensity and volume training regimen. In June, the swimmers average 8,000 m/d, a 50% increase in training volume, plus dryland work with weights. Distance work is accomplished primarily in the morning workout with high-intensity intervals in the afternoon session. The athletes continue this pace through the first two meets of season with adequate but not overly impressive performances. At the third meet, an important regional competition preceded by two light workout days, both Carl and Hunter make qualifying times for long course junior nationals.

The swimmers then return to their heavy training program. The coach increases daily distance to 10,000 m with very few light days. Intensity is maintained at a high level. At this point, both athletes begin to complain of fatigue, muscle soreness, and boredom. They both report excessive sleep though they still feel tired. Both appear to be irritable and somewhat moody. Neither seems to be eating enough and both are beginning to show some weight loss. Performance in workouts is deteriorating. Though they have already qualified for junior nationals, the coach has them continue to use local and regional meets as part of their high-quality preparation. He gives them little rest though he does attempt to vary the workouts and build in some enjoyable experiences. Workouts are still long and demanding.

Three weeks before junior nationals the coach takes his swimmers into active rest. Workouts remain intense with a focus on high quality but with longer rest periods and one day off each week. Total daily distance drops significantly. Within the first week, Carl begins to show increased energy and positive moods. He stays focused throughout the workouts and manifests a high level of motivation for training and confidence concerning his competitive performance. He seems gregarious on the pool deck and motivates his teammates. He reports a return of regular sleep and appetite patterns. His performance on timed trials begins to improve.

Hunter, on the other hand, shows no such positive adaptation as the taper period progresses. He appears to be increasingly lethargic and struggles to complete sets. Stroke mechanics seem sloppy and he receives a good deal of criticism from the coach. He isolates himself from teammates and reports anxiety and low confidence about the upcoming meet. One day, late in the taper, he winds up in a shoving match with a teammate after being reprimanded for teasing some of the younger swimmers. He continues to show weight loss and irregular sleep patterns. No performance improvements are evident.

Hunter has now fulfilled our definition of overtraining. He has been stressed beyond his ability to adapt. Carl has experienced overreaching or supercompensation and appears to be using the active rest period to solidify his adaptation to the intense training (84). With this, we turn to the specific components of our model.

A Multisystemic Model of Overtraining

The questions that must be addressed by any model of overtraining are, what factors influenced these athletes to respond in this way and what can we do to aid Hunter? Why does he fail to improve despite the rest? Is the season a loss for him? Is his swimming career over? Can he overcome this extended period of staleness? Should we be concerned about his long-term well-being? Could what happened to Hunter have been prevented? And finally, what makes Carl and Hunter respond to intensive training differently?

Unfortunately, the existing research on the possible detrimental effects of intensive training has not provided us with clear answers to such questions. As we discuss below, preventive efforts based on the monitoring of a few simple, cost efficient parameters may be our most effective strategy. The evidence, however, indicates that these prevention efforts are far from foolproof. For those athletes who become stale, remedial interventions may be needed. The challenge for applied investigators working in this area becomes one of developing an overtraining model that will lead to beneficial preventive and corrective interventions.

From our review of the overtraining literature we can draw at least two preliminary conclusions concerning a possible model of overtraining. First, overtraining or staleness can be thought of as an indication of an athlete's ineffective response to a prolonged stressor. The response to a demanding physical stressor shares similarities with other types of stress reactions such as those experienced by individuals who have lived through a natural disaster (94) or who have dealt with a difficult physical illness or demanding medical treatment (90). The experiences of athletes, like those of Carl and Hunter, appear quite consistent with the current literature on stress and adaptation.

Second, sport scientists investigating overtraining have tended to hold a narrow view of the athlete. Specifically, the athlete is considered only in the context of physical training and sport competition. While researchers typically appreciate the athlete as person, the tendency is to ask our research questions without any consideration of the other worlds or systems that the athlete-person inhabits. There are inherent benefits to conducting assessments in this narrow manner; the questions are simpler, mechanistic, unidirectional, and allow us to adopt designs that appear to maximize internal validity. However, when such parsimonious approaches to studying overtraining fail to promote our understanding, considerations of more comprehensive conceptualizations may become necessary. From this perspective, recognizing that athletes enter exercise protocols with personal histories and a variety of life experiences is potentially extremely valuable.

The next section of this chapter provides a broad or *multisystemic perspective* of the athlete and the person attempting to cope with a prolonged physical stressor. This is accomplished by examining training as a potentially stressful event and then considering that training within a transactional stress and adaptation model (22, 42, 44). From this model, a broad view of the athlete will be constructed.

Then, this systemic model of stress and adaptation will be applied to Carl's and Hunter's responses to the intensive training experience.

A Prolonged Stressor: The Training Regimen

As we examine the process of overtraining one must acknowledge that a tremendous amount of the variability in the athlete's response must be due to the planned increase in training. Further, we suspect that for any athlete, one can design a training program that can be appropriately challenging, overwhelming, or not demanding enough. Given the widely held assumption that today's athletes must train at high volume and intensity to produce success, most elite, and many other skilled and recreational athletes will adopt and adhere to intense training regimens. And by definition, such programs should in and of themselves produce considerable physical and psychological changes independent of whether the eventual result is performance improvement or deterioration (48). This proposition is schematically represented in figure 16.2.

Our swimming scenarios illustrate this assumption, but let us support it with this one additional study of the psychological impact of intense training. We had the opportunity, over two summers, to follow groups of adolescent and young adult male weightlifters participating in junior age group weightlifting camps sponsored by the United States Weightlifting Federation (8, 26). Athletes were chosen for these camps based on prior superior performances and all were candidates for the U.S. national junior weightlifting team. During the first summer camp, 30 athletes were exposed to a one-week training regimen that involved two to three daily 2 hr workouts resulting in a tripling of precamp training volume. Workout intensity was also maintained at a level significantly higher than that to which the athletes were typically exposed. Participants completed the POMS, a state anxiety inventory, and a measure of achievement goal orientation prior to the first training session and after the final session. A short form of the POMS (16) was completed daily. Total mood disturbance increased significantly from pre- to postcamp. The short form of the POMS mirrored this pattern though results failed to reach significance. On the goal orientation measure there was a trend for athletes to report a

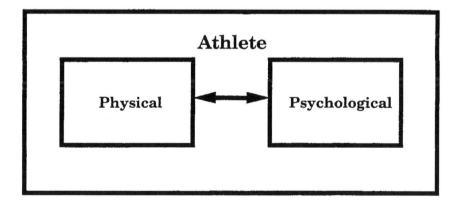

Figure 16.2   The interaction between the physical and psychological aspects of the individual.

decrease in mastery orientation and an increase in ego or outcome focus over the training week. These negative changes in mood and goal orientation were stronger for athletes who reported an injury during the camp. The anxiety measure showed no significant changes. These findings indicate that during this one week of intense training, there was a meaningful deterioration in mood and psychological well-being.

Thirty-one athletes participated in a similar 18-day camp the following summer. Again, mood deteriorated significantly over the course of the camp and this was revealed both by the pre- and postcamp POMS and by the daily brief assessment of mood. This camp also allowed us to assess the psychological skills of these athletes using the Mahoney et al. (52) Psychological Skills Inventory for Sport. Neither demographic measures or psychological skills were related to change in mood over the intense training period. While not entirely consistent with the kind of results reported by Gross (27), these findings suggest to us that with a group of physically gifted athletes, many of whom had been exposed to psychological skills training interventions during previous national training camps, the intense training regimen still produces a meaningful deterioration of mood. That is, the intense training does what it was designed to do. By stressing the athlete, physically and emotionally, the athlete is placed in a position where he or she must adapt to the stressor, and hopefully, extend performance capabilities (23). And this happens to the majority of athletes exposed to intense training loads independent of physical ability or psychological skills.

The Individual: Stress and Adaptation

Many models that attempt to explain the experience of stress emphasize the demand characteristics of the stressor (14, 19). In such stimulus-oriented models, stress is produced by environmental characteristics or events (e.g., the physical training regimen) that increase demands upon or disorganize the individual. The underlying assumption is that the individual has an innate capacity to withstand environmental stressors. When the stressors outweigh the individual's resiliency, that person begins to physically or psychologically deteriorate. Other models can be conceptualized as response-oriented (31). These models classify stress based on how individuals respond to stressors and focus on the individual's level of disorganization or maladaptive functioning.

Across time, research findings have revealed that stimulus and response characteristics are important, but are also insufficient to account for the tremendous amount of individual differences in response to stress (42). Attention, therefore, has shifted to interactional theories of stress adaptation. The emphasis in these theories has been on characteristics of the person as a major mediating mechanism between the stimulus characteristics of the environment and the responses that are invoked.

The dominant theory in this area has been the transactional model of stress and coping (22, 42, 44, 79). Central to this model is the assumption that the individual is influenced by and influences his or her environment. Consistent with Bandura's (2) concept of reciprocal determinism, the individual's behaviors, thoughts, and emotions are both products of situational elements and producers of subsequent situational conditions. A second fundamental assumption in the transactional model is that stress is defined by the individual's cognitive and emotional processes. While situational conditions may appear to be stressful and physiological and psychological components of the response might signal a distressing event, neither is an accurate predictor of the individual's report of stress. Lazarus (43) argued that the quality of a specific stress experience depends not on the demand, but on the individual's perceptual and cognitive processes and behavioral skills or action tendencies. It is these processes that mark a situational condition as stressful.

Lazarus and Folkman's transactional model is depicted in figure 16.3. Readers interested in a more detailed description of the model and supporting research should explore the references cited above, though we will present a brief overview here. If we are to understand Lazarus's model, the first step is a consideration of the *antecedents* of the situational condition. There are two categories of variables that comprise antecedents: person variables and situational factors. Situational factors constitute the demands of the stressor itself. Person variables include the individual's beliefs about the self and world, expectations about upcoming challenges, motivational goals, and self-evaluation. Lazarus viewed person variables as personality traits that, in interaction with situational factors, define how one will construe the situation (79) (see figure 16.3).

The antecedents to the situational condition color the interpretation of the situation and lead to the individual's *appraisal* of the significance of an encounter or event (44, 45). As illustrated in figure 16.3, there are two basic forms of appraisal. Primary appraisal concerns whether something of relevance to the individual's well-being has occurred. Only if the individual has a personal stake in the event, for example a meaningful short- or long-term goal, will there be a stressful response to the interaction. The primary appraisal process results in questions such as these: What is the demand? Is it relevant to the individual? Should that person understand it as a challenge or threat? Lazarus and Smith (46) described three components of primary appraisal. Goal relevance refers to the extent to which a situational encounter impacts personal goals. That is, there are issues in the encounter in which the individual has a personal interest. The second component is goal congruence, or the extent to which an encounter is consistent with the individual's goals. The third component of primary appraisal is ego involvement, which refers to self- and social evaluation based on personal values and morality.

Secondary appraisal concerns the person's perception of coping options. The individual must decide whether any given action can prevent harm, ameliorate it, or produce additional harm or benefit. The fundamental questions under consideration are these: What, if anything, can be done in this situation and how will specific choices affect the individual's well-being? And, are there sufficient resources that can be dedicated to addressing this demand or challenge? According to Lazarus and Smith (46) one aspect of this process is deciding who is accountable or responsible. Secondly, the individual must evaluate his or her potential to cope.

As suggested in figure 16.3, this appraisal process yields immediate reactions or *appraisal outcomes* for the individuals. This response is typically physiological and emotional. The person may also exhibit a tendency to immediate action. Such action tendencies are usually thought of as an urge to act that distinguishes one emotional response from another. It should also be noted that because the appraisal and action processes are actually interactions with the environment, appraisal is an ongoing process. Reappraisal of the situation, therefore, is constantly occurring (79).

The bottom section of figure 16.3 depicts the actual coping processes. Lazarus and Folkman (44) hypothesized that coping consists of cognitive and behavioral efforts to manage specific external or internal demands that are appraised as taxing or exceeding the individual's available resources. Though coping may flow from emotion and be directed at changing the conditions related to the emotion or the

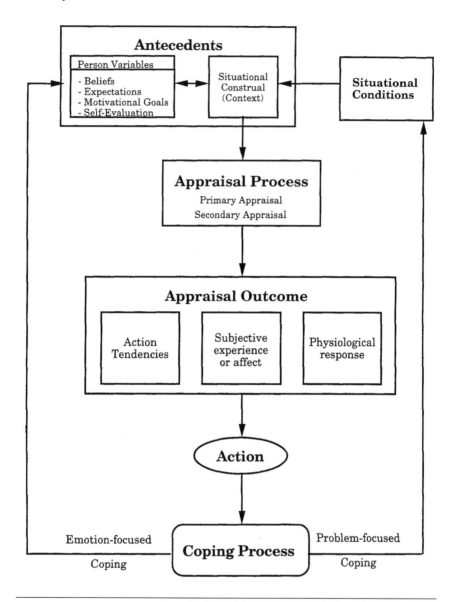

Figure 16.3   The transaction model of stress and adaptation.
Reprinted from Lazarus and Folkman 1984.

emotion itself, coping also affects directly or indirectly subsequent appraisal processes. Folkman and Lazarus (21) described two types of coping. *Problem-focused coping* occurs when the coping process is intended to change the actual conditions of the stressful situation. These are action-centered efforts including strategies of confrontation and playful problem solving. *Emotion-focused coping* strategies involve modification of one's cognitive conception of the person-environment relationship. However, these strategies are not passive ones, rather they involve an internal restructuring of one's plans or a modification of one's perspective or understanding of the stressor. For example, if we successfully avoid thinking about a threat, the fear associated with it may dissipate.

Emotion-focused coping strategies include distancing, escape-avoidance, accepting responsibility or blame, exercising self-control over expression of feelings, or seeking support from others. Research has shown that problem-focused and emotion-focused coping strategies are not mutually exclusive and that an individual may attempt to cope using both strategies (21).

Lazarus and Folkman's (22, 44) transactional model of stress and coping focuses our discussion on the person as the major mediating variable between the stimulus characteristics of the environment and the responses that are invoked. The implication is that an understanding of the experience of stress requires an understanding of the person. In turn, to understand the person we must develop a comprehensive framework for considering the individual actor functioning within complex systems.

The Multisystemic Perspective

We have been arguing that for all athletes, performance is not simply dependent upon what happens in training or competition, but it is also contingent upon events in the larger world. One need only to reflect on the tragic death of the sister of American speed skater Dan Jansen during the 1988 Winter Olympics, or the effects of the fall of communism in Eastern Europe on the athletes of formerly communist countries during the 1992 Olympic Games to understand this. What influences an athlete goes beyond the pool, track, or playing field and includes the support of or struggles with family members, the demands or fulfillment of school or a job, or the ability to depend on a friend. It is not athletic ability alone that allows a 43-year-old cyclist, husband, and employee to make an Olympic team. It is not only the demands of long training sessions that make it difficult for a young child to leave her family to train at an elite level.

The bridge between linear or unidirectional models of causality in sport psychology and a systemic approach to athletic performance can be found in Bandura's (1) notion of reciprocal determinism. Bandura argued that the individual's behavioral skills, cognitive mediators, and environmental context are in a complex, reciprocal interaction. Environmental happenings influence the individual's actions and beliefs; beliefs effect behavior and the environment; and actions and their results influence the environment and subsequent beliefs. This interaction cannot be viewed as a linear sequence of events, rather it must be understood as a set of reciprocally interacting feedback and feed forward mechanisms (95; see figure 16.4).

Reciprocal determinism has been depicted graphically as a triangle with behavioral skills, cognitive mediators, and the environmental context as its three points (2). The bidirectional arrows illustrate the reciprocal, nonlinear influence that exists among the three sets of variables; the influence may start at any point and proceed in any direction, not necessarily following the same path in every situation. An event in the environment may influence an individual's behavioral skills, that in turn may impact the person's cognitive mediational processes. Cognitive mediators may just as easily influence behavioral skills, that bring about some environmental change. The key to reciprocal determinism is to remember that events do not occur in a linear sequence; influence may arise from any set of variables, and may influence one or both of the remaining sets of variables.

Unidirectional Causation

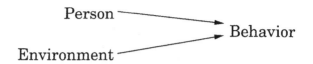

Reciprocal Determinism

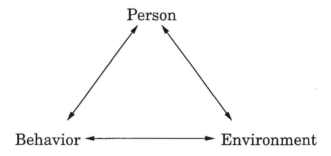

Figure 16.4 Models of unidirectional causation and reciprocal determinism.
Adapted from Bandura, 1978, "The self-system and reciprocal determinism," *American Psychologist* 33: 344-358. © 1978 by the American Psychological Association. Adapted with permission.

When an athlete presents with environmental issues that are restricted to the areas of training and competition, reciprocal determinism fits easily with the application of skill-building intervention strategies. These strategies are directed at enhancing the athlete's ability to manage the challenges of the sport performance. By addressing the performance issues, that may result in changes in the level of the athlete's behavioral skills, these interventions should also influence both cognitive mediators and the environmental context. Hopefully, these changes should produce more successful sport performance outcomes. In situations where the treatment is limited to sport performance, the educational or cognitive-behavioral skill-building model that constitutes the foundation of current sport psychology intervention efforts should be most beneficial (59). As is often the case, however, individuals who present with problem issues rarely fit easily into a neat package. When confronted with an individual who also happens to be an athlete, limiting the scope of inquiry regarding environmental events to the sport context may provide a restricted view of the individual. This restricted view may lead to the application of interventions, that although useful in some aspects, do not address accurately the systemic influences and may result in an unsuccessful counseling or consulting relationship.

In order to avoid the pitfalls of narrow vision, it is necessary to realize that people operate in a variety of contexts, and the sport context is only one of many for the competitive athlete. Further, these contexts and influences are in constant and complex interaction (see figure 16.5). Indeed, the multisystemic model assumes that reciprocity occurs not only within a context but between and among contexts, and that an athlete's failure to perform (or successful performance) may be due to a myriad of interacting influences. Consequently, understanding the athlete's performance requires consideration of both sport and extra-sport contexts.

Perhaps the most productive heuristic for conceptualizing these interrelated influences is the multisystemic approach (30). The multisystemic approach may be seen as a more practical development of Bandura's reciprocal determinism. This approach takes into consideration all areas of an individual's world in order to design and implement interventions. By assessing each of the areas in which the athlete is a participant, the actual influences on sport performance may be addressed. Taking a systemic approach to the issues presented by athletes allows for more appropriately planned intervention strategies which hopefully improve not only sport performance, but also the athlete's functioning in other areas of life.

By adopting a multisystemic view of the individual, presenting problems come to be understood as the athlete's attempts to manage the demands that he or she faces in sport, marriage, family, career, and social relations. For example, increased marital conflict may be the result of the last child leaving home, leaving the parents with the task of learning to readjust to being alone with each other. Alternatively, conflict may come about as a response to a career change that requires increased time commitments to work and corresponding decreases in time commitments to family. Not only do these problems stem from attempts to handle stressors and changes, but they also demonstrate the reciprocal influence sets of variables have on other sets.

To gain a better understanding of the multisystemic perspective, we rely on a brief overview of systems theory suggested by Minuchin (60).

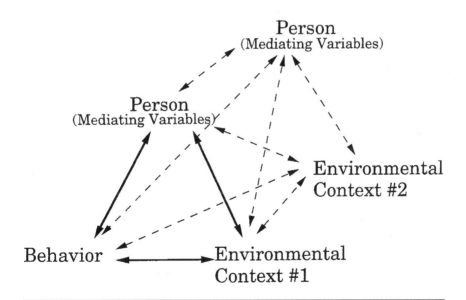

Figure 16.5   The multicontextual model of individual functioning or athletic performance.

Reprinted from Murphy 1995.

• Systems are organized wholes, and elements within the system are necessarily interdependent. These elements are also bound by predictable relationships with one another. Applying this principle to people would be similar to saying that behavior is best understood and predicted when considering as many contextual influences as possible.

• Systems are circular and not linear. To conceptualize this principle, consider two individuals. For example, in training for a triathlon, a wife may spend a great deal of time away from her family. Her husband may feel she is neglecting the family and may try to nag her into spending more time at home. The wife may respond by thinking her husband is trying to keep her from accomplishing her goals and that he is selfish. She becomes angry with him and spends more time training. He nags more, she trains more, and so on. Her actions influenced him, and his actions influenced her, and this feedback loop will continue until someone or something disrupts it.

• Systems strive to maintain homeostasis. The problem with a homeostatic goal is that evolution and change are also inherent in any system. These changes may be brought about by genetic or biological factors such as aging, or environmental factors. They may be the result of psychological changes such as changing confidence levels following a series of successes or failures. All potential stressors, be they developmental, predictable, or unpredictable, bring about evolution and change. Systems, however, attempt to minimize or resist the change. While this resistance can be functional and adaptive, it may also have detrimental effects. The tendency to maintain stability, and therefore minimize helpful change, must be kept in mind when dealing with athletes.

• Complex systems are always composed of subsystems. Consider a family as a complex system. When the family is examined, subsystems can be identified. These subsystems include a spousal subsystem, a parenting subsystem, a sibling subsystem, and others. The subsystems combine to form the whole system, and while each subsystem has its own functions and boundaries, it is clear that there is no mutually exclusive membership in any one subsystem (30).

• Subsystems are separated from larger systems by boundaries, and there are implicit rules and structures that govern interactions across boundaries. These boundaries affect the flow of information to and from systems (30).

To depict the multisystem influences graphically we first return to the very limited picture of the individual athlete presented in figure 16.2. As previously discussed, this athlete is physically and psychologically adjusting to the experience and demands of the world. Remembering the idea of reciprocal determinism, we can predict that the interaction between the physical and mental phenomenology is constant. Physical experiences have a direct and often immediate impact on how the individual feels, thinks, and behaves. Correspondingly, emotional, behavioral, and cognitive feedback also operate in the opposing direction. The individual's thoughts, feelings, and behavior, in turn, impact one's physical performance.

The workings of the multisystemic model are graphically depicted in figure 16.6 as a series of reciprocal relationships among the target individual and others within the system. The overlapping ovals represent the contexts of these reciprocal relationships. The fact that these ovals do overlap makes it clear that influences in one context may readily effect other contexts, or may globally influence all contexts. Beyond the contexts depicted in our figure, or more appropriately, the area in which these contexts are embedded, is the cultural context. Society, culture, politics, and economics all have some effect on the athlete.

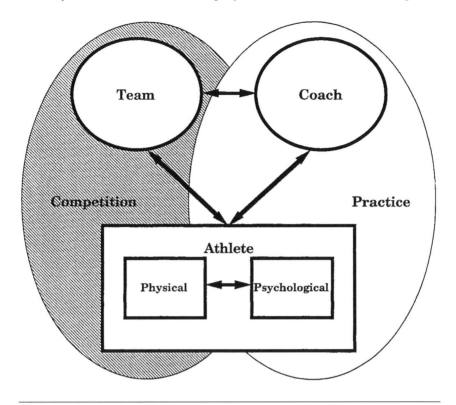

Figure 16.6   A systemic view of the athlete in sport contexts.

Specifically, figure 16.6 represents the sum of the multisystemic influences on the competing athlete. Sport involvement consists of two primary contexts, practice and competition. While these two environments share many characteristics, anyone with experience in sport will understand the striking differences between them. Physiologically and psychologically, an athlete sitting quietly before practice versus before a competition is a very different person. Across these contexts, most athletes are embedded in a system that includes at least one coach, and most likely teammates or a group of other athletes.

The relationship between the athlete and the coach or coaching staff is an important and powerful one for a majority of athletes. Hierarchically, the coach resides in a position of power over the athlete where, typically, the athlete is subordinate to the coach's directives. The typical hierarchical structure of this relationship does not make it unidirectional, but does dictate certain relationship characteristics such as forms of communication, emotional distance, and the scope of behavior. The very conceptualization of the coach's role requires the involvement and input of athletes. And the coach's decisions, actions, and emotions (as well as employment status) are greatly influenced by the athletes under his or her tutelage. It is not uncommon for this to be a very rich relationship where the nature and scope of the reciprocity may be quite varied. The coach can be parent, friend, enemy, supporter, taskmaster, or source of strength, purpose, and motivation. Similarly, the athlete may be childish, rebellious, needy, energized, or compliant toward the coach. On any given day the actions and mood of the athlete and coach are in constant interaction.

The two contexts, practice and competition, sometimes require variations within the coach-athlete relationship. Frequently, athletes are in a subordinate role during training. The training schedule and routine are typically established by the coach. The coach can also exhibit a wide range of behaviors during a training session that would not be acceptable outside of this context. For example, a coach may be shouting encouragement one minute, reprimanding the next, and instructing during the third minute. While a coach's authority may be questioned, the coach can expect his or her directives to be followed, athlete complaints will be temporary, and the hierarchy of the relationship will be maintained. In the competitive setting, the relationship shifts as the coach takes on more of a service and support function, focusing on the athlete's immediate needs, and aiding the athlete to execute well-practiced behaviors or respond constructively to performance difficulties.

Athletes' interaction with their teammates (and sometimes their competitors) can be equally complex and powerful. Depending on the type of sport (e.g., individual versus team) and the role of the athlete within the peer network (e.g., star versus substitute) the characterization of an individual's relationship with teammates can be quite varied. These relationships are also bidirectional as members of a team reliably influence each other. This reciprocal influence within a team of athletes is likely to involve maintenance of a hierarchy, establishment of subsystems, and struggles to maintain homeostasis. As a team moves from practice to competition, the nature of the relationships may evolve. A basketball player may be battling a teammate for a starting position during practice while taking on a supportive role on game day. Or runners may use each other for support and friendship during long practice runs, but stand as competitors on race day. A mistake in practice may be humorous; the same mistake in competition may lead to conflict. Distant relationships in one context may be extremely close in another.

The interactions among teammates, as well as the rules and boundaries within this subsystem, are influenced by, and influence the coach and his or her role within the team. In this way the coach, or coaching staff, not only has reciprocal relationships with each team member, but also a relationship with the team subsystem.

As we have argued throughout, athletes are people with lives that parallel their sport involvement. And as is true in the sport context, athletes are embedded in systems within contexts in their nonsport lives (see figure 16.7). The athlete's nonsport life also contains multiple contexts that influence the person. The two primary environments highlighted here are home and school or work.

Within each of these contexts, the individual is part of a system and, typically, subsystems within the system. At home, the athlete is a member of a family. He or she may be a spouse, parent, roommate, child, or sibling. Subsystems within the home context will have some degree of order and structure. Parental subsystems are often organized as separate entities, although connected to other systems (e.g., siblings). Subsystems among siblings may be organized by age, birth order, and gender. While relationships evolve and change across time, there are tendencies within a system to maintain the status quo. For example, no matter how old or mature an individual may be, his or her mother will maintain key characteristics of the original parent-child relationship.

The work or school context also has a system of relationships that can be described as possessing subsystems, hierarchies, boundaries, and homeostatic functions. At work, the individual will have one or more supervisors who define the structure within that context. The individual will also have coworkers and, possibly, subordinates where subsystems within the context will be formed and main-

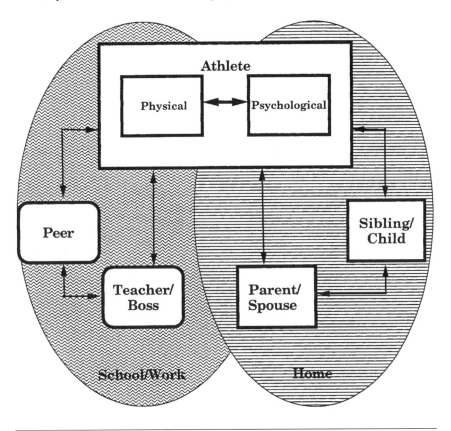

Figure 16.7   A systemic view of the athlete in nonsport contexts.

tained. At school, teachers are part of the classroom system although also separate from the peer subsystem of classmates.

Challenges, demands, and threats may arise within any of these contexts and systems within contexts. As the individual appraises and copes, these challenges can influence not only the individual's sport performance, but also his or her ability to function successfully in the family and at work. These demands may be developmental, predictable, or unpredictable. According to Carter and McGoldrick (11) developmental demands are those events that, although expected during any individual's life span, are often intensely stressful to all members of the family system. These stressful conditions usually occur at transitions in a family and are instances in which the role of a family member in the system changes dramatically or when a family member leaves the system. Some examples of these transition times include the aging of a parent, the birth or death of a family member, a child who leaves for college, or a divorce. Other developmental stressors may also occur in conjunction with an individual's career. For example, a job promotion can have a significant impact on an individual's work context, and the new position may mean that the person must devote a significantly greater amount of time to maintain a successful career. The increased commitment to the work context may require the individual to decrease commitment in other contexts, and this change may negatively impact performance. A career demotion may also produce deteriorating performance in other contexts. Both of these situations carry with them potential financial changes for the individual, and these changes may present additional stressors across many contexts.

There are other predictable demands that are not developmental. An athlete's training season is one example. Periods of intense training require the athlete to reduce the amount of time he or she may allot to family and work. One's spouse must pick up the slack in the family system by assuming more of the caretaking duties, both of children and home, and both partners are left with less time to devote to their own relationship. In the work context, the athlete may be able to maintain previous levels of job performance, but in the more stressful situation of producing a similar quantity and quality of work in a limited time.

There are also demanding situations that are unpredictable, and may for that reason be even more problematic. The competitor has little time to prepare for these capricious stressors, and must deal with them as they come. Such stressful situations might include injuries that effect training or end a competitive season, major health problems experienced by a family member, conflicts that may appear at work or at home. It is not difficult to imagine how any of these stressors would impact not only the immediate context, but also other areas of the athlete's life. Again, understanding the overall impact of any stressor on multiple contexts is central to understanding the issues presented by the athlete. This understanding is necessary to plan useful interventions, rather than providing interventions that act as Band-Aids for any one symptom. Taking a multisystemic approach when dealing with athletes is possible only after understanding complex, multiple contexts.

Before returning to our two swimmers, several important points must be reiterated. Assessment must take into consideration the importance and influence of multiple systems. When gathering information from the client, it is insufficient to focus solely on sport issues. One must remember that valuable information can also be found in other areas of the athlete's life. The same holds true for interventions. Gearing them only toward a sport issue may lead to failure in treatment if the underlying causes of performance problems exist in other contexts. The interventions should be designed to address the conditions that maintain the problem. Just as problems reach out and influence other contexts, so will interventions. Elements in a system are interrelated; when one is changed in some way, the effects of those changes are felt throughout the system.

We now illustrate these assumptions in our two case studies. The application of the multisystemic model to Carl and Hunter should generate a richer understanding of their different responses to the intensive training regimen. The model moves us beyond the concept of "individual differences" and toward potential hypotheses for explaining why the same physical challenge can result in such varied reactions and performance. The model can provide predictors for both antecedents to the young athletes' responses to their training program and the potential resources available for successful coping.

A Multisystemic View of Carl and Hunter

Carl has demonstrated a positive adaptation during the taper period. Table 16.2 presents the key systemic characteristics of Carl's sport and nonsport contexts. Carl has identifiable strengths, signified by the symbol "+," and weaknesses, signified by the symbol "−". There are also characteristics that

Table 16.2 Multi-Contextual Assessment: Carl

Sport Context

Individual

+ Confident   – Easily loses focus

+ Easily motivated   – Refuses role as leader

+ Solid work ethic   – Anxious during meets

+ Tall, physically mature

+ Physically talented, good kinetic sense

+ Dedicated to swimming and hard worker

Coach

+ Greatly values coach's opinion   – Coach not patient when Carl's family

+ Coach understands Carl's lack of focus   conflicts with swim practice
   & compensates (pep talks & flexible
   workouts)

+ Coach believes Carl is very responsive
   to coaching

Teammates

+ Well liked and respected for his swimming

+ Excellent relationship with younger kids
   on team

Nonsport Context

Individual

+ Extroverted   – Confused about future

+ Confident   – Anger control

+ Socially skilled   – Inattentive, impulsive

+ Considerate

Family

+ Mom supports swimming, but is reasonable   – Dad does not support swimming and he

+ Solid with both older brother &   is concerned about academic performance
   younger sister   – Mom and dad argue about Carl's

± Family is #1; everything else #2   swimming

School

+ Well liked by peers and teachers   – "C" average (not acceptable to dad)

+ School not in session   – Last semester was worse yet

Peers

+ Well liked, impressive social skills

+ Physical talents generalize to other sports

+ Receives considerable amount of attention from girls

---

could be considered, depending on the context, either a strength or a weakness, indicated by the symbol "±". At the individual level of the system, Carl is a capable and confident swimmer. Physically, he is talented, with a particularly keen sense of kinetic awareness. He possesses a solid work ethic in the pool and reluctantly understands his leadership role on the team. Competition tends to promote the increased experience of anxiety, but not to a debilitating level. He sometimes thinks and acts impulsively and his attention tends to wander. When asked, Carl reports being uncertain about his future in swimming.

Carl's relationships at the pool make this setting rewarding and comfortable. He views the coach as supportive. Consequently, his behavior towards the coach is attentive and respectful. In response, the coach judges Carl to be an accommodating and eager pupil, dedicated to swimming, and a positive influence on younger swimmers. The coach also seems to understand Carl's impulsivity and moments of inattention. The coach believes that he can tell when Carl's attention wanders and reports that pep talks and playful challenges help Carl focus. Carl, in turn, appreciates attention from the coach and understands the pep talks as the coach's strategy for communicating confidence and praise. Likewise, Carl's relationship with teammates is quite positive. He is well liked and respected for his swimming ability. Younger swimmers look up to him. Parents use Carl as a role model for their young swimmers. Carl, although sometimes reluctant to be identified formally as a leader, thrives on the attention he receives from these athletes. He willingly spends time with them during meets and enjoys encouraging them in the pool.

In nonsport contexts, Carl again demonstrates that he is a confident, but somewhat inattentive individual. Skilled and successful in most social situations, Carl relates to others in a caring and considerate manner. His physical skills are apparent in other sports. He shows anger control difficulties when his mood is down and he reports being uncertain about his future goals.

In the family system, Carl is the middle of three children in an intact family. He has a supportive relationship with his mother and siblings. His mother enjoys swimming and attends most meets and an occasional practice. Commitments to her other children and to her job limit her ability to be involved with the swim team's parent group. Carl knows his mother is interested and is not pressured or embarrassed by her involvement. His siblings occasionally attend meets, his younger sister more regularly than his older brother. Carl reports being very close to his brother, who is not involved in athletics, but provides support when Carl experiences academic difficulties. Carl's father is not supportive of his swimming and is frustrated by Carl's lack of academic success. He also demands, with his wife's encouragement, that the children place the family before everything else. Swimming therefore is a point of contention between Carl and his father, as well as between his father and mother.

Other systems of influence impinging on Carl are school and peer relations. Academics, as suggested before, are a weakness. His grades tend to be average with the previous semester's grades the worst in his high school career. Although his performance is sometimes frustrating to him, it is his father's reactions that he dreads. Indeed, father has threatened to end Carl's swimming. Mother has interceded to postpone this sanction until the fall. Fortunately, school is not in session and, therefore, the current conflict is minimal. Socially, Carl is well liked by peers, younger children, and adults. He receives a considerable amount of attention from girls. His athleticism is often helpful when spending time with others. Within and outside of the school setting, people speak very favorably of him. Even when he struggles, others tend to like, encourage, and value him.

Carl's experience of the intensive training regimen is influenced by these many systems that make up his life. The antecedents to the training demand include his belief in his ability as a swimmer, his supportive relationship with his coach, the nurturing of his mother, the knowledge that others have faith in him, and his tendency to positively evaluate his potential to respond. Therefore, Carl's appraisal of threat may be minimized by the supportive environments at the pool and at home. Similarly, as he evaluates his coping options and then attempts to manage the fatigue, pain, and boredom, his ability to relate to others and actively secure support is more likely to assist him in rebounding emotionally from the stressful training period.

Next, consider Hunter, the swimmer who did not demonstrate a positive adaptation during the taper period. Hunter, like Carl, has many personal strengths, but unlike Carl, Hunter's multisystem picture emphasizes very different, resource-draining aspects of his life. Table 16.3 presents the key systemic characteristics of Hunter's sport and nonsport contexts. At the individual level, Hunter is a smart, dedicated, motivated, and talented swimmer. He holds high standards for his performance while being self-critical. As a result, his confidence in his swimming oscillates, which in turn impacts his mood and his level of frustration. When swimming well, Hunter is a happy, motivated swimmer. When swimming poorly, Hunter tends to appear moody, isolated, and hesitant.

His relationship with others at the pool shows the same mixture of positive and negative features. His coach is aware of Hunter's potential, but expresses confusion about how to manage and mentor him. When Hunter swims well, the coach knows that he can leave Hunter alone and allow the youngster's success to fuel future effort and performance. When Hunter is swimming poorly, the coach knows that his efforts to support Hunter may be rejected. During these periods, Hunter believes that the coach is being patronizing and insincere. He reacts with increased isolation and self-criticism. The coach then retreats and tends to coach Hunter indirectly by making comments to others about him. The result of these interactions is increased tension between coach and athlete. A parallel process occurs with teammates. Hunter has two close friends on the team with whom he spends most of this time. These friends know that it is necessary to leave him alone during bouts of moodiness. Other swimmers in his age group view Hunter as isolated and tend to keep their distance. Hunter has little patience with the younger athletes.

In his nonsport world, Hunter demonstrates his intellectual and academic abilities. Some conflict between high goals and vigorous self-criticism is apparent. Although tall, athletic, and attractive, Hunter evidences a considerable degree of shyness and social anxiety. The confidence that sometimes appears on the pool deck is not evident in social situations.

Hunter's family consists of his mother and himself. He knows very little about his father, who left the family when Hunter was an infant. The father's infrequent contact with the family leaves Hunter confused and upset. His mother, in contrast, is very involved, possibly too involved, with her son's life. She is a leader among the swimmers' parents, attends all swim meets, and frequently attends practice. She tells Hunter that swimming may be his only route to a

Table 16.3 Multi-Contextual Assessment: Hunter

Sport Context

Individual

± Confidence oscillates

+ Motivated

+ Solid work ethic in pool

+ Excellent anxiety management skills

+ Physically talented, good mechanics

+ Dedicated to swimming and hard worker

– Holds high standards

– Self-critical

– Easily frustrated

Coach

+ Coach believes Hunter is talented

+ He respects Hunter's intelligence and work ethic

– Coach confused about how to talk with Hunter; sees Hunter withdraw & reject feedback when he needs it most

– No longer tries to control Hunter's self-criticism

– Finds that the best way to coach Hunter is to provide passive information

Teammates

+ Has two close friends on team

– Little patience with the younger kids on team

– Seen by peers as isolated and not supportive

Nonsport Context

Individual

+ Intelligent, creative

+ Dedicated, motivated

+ Focused

+ Caring

+ Tall, attractive

± Self-critical

– Confused about future

– Tends to obsess

– Not confident

– Shy, socially anxious

Family

+ Close to uncle who was Ivy League swimmer

± Mom very supportive and involved in swim club

± Only child

– No relationship with dad which is bothersome

– Mom believes swimming scholarship is necessary

(continued)

Table 16.3 *continued*

School

+ Top student at academically challenging school

+ Abilities and work product respected by teachers

+ Teachers respect his academic abilities and his ability to challenge himself

+ A handful of close peers who are also academically successful

Peers

± Although not disliked, has only a small circle of friends   – Shy around girls

± Existing friendships are supportive, however most appear to be other isolated adolescents

---

college education. She constantly reminds him that swimming scholarships are available to those who perform well. Hunter does have an uncle who is a former collegiate swimmer. This uncle advises Hunter to stay focused on academics and expresses a more realistic picture of college swimming. Although the uncle's perspective is both helpful and valued by Hunter, the two see each other infrequently.

Hunter is an academic star. He is one of the top students at an academically challenging public school. He is respected and encouraged by his teachers. He is liked and respected by his academically oriented peers. His tendency to be self-critical works well for him at school where it fuels his motivation to learn.

Peer relations outside of school tend to vary. Although not disliked by others, Hunter seems to isolate himself from most of his peers. He has a solid circle of friends who are supportive and loyal. Most of these adolescents are also relatively isolated. He spends most of his free time with these friends engaged in shared interests, or he reads. Girls find him attractive, but shy.

Hunter's experience of the intensive training regimen is also influenced by the multiple systems that impact his life. The antecedents to this demanding training situation include his oscillating belief about his ability as a swimmer, his frustrating relationship with his coach, the enmeshed, yet supportive, demands of his mother, his sense of isolation, and his tendency to critically evaluate his potential in the pool. Hunter's appraisal of threat may be perceived as significant, particularly due to the period of poor swimming performances before the taper period. As he evaluates his coping options and then attempts to deal with negative effects of intensive training, his inability to accept the support of others can be damaging. With this, we turn to issues of prevention and treatment.

Prevention and Treatment

Prevention

We have outlined a comprehensive, systemic model of the psychosocial stressors that we hypothesize influence overtraining. Now the question is, does the paradigm help us to understand the intense training experience and to develop strategies for dealing with negative responses to it? We have argued that the overtraining event is a complex interaction of stress and coping involving physical training variables, biological response systems, individual cognitive, behavioral, and emotional characteristics and skills of the athlete, and a rich variety of environmental stresses and supports. These include interactions with coaches and sport administrators, teammates, peers, funding sources, family members, and other loved ones, employers, teachers and schools, as well as events occurring in the larger community, the culture, and the economy. Being prepared to assess and intervene on these systemic levels is, obviously, a daunting task, but one that should lead to an arsenal of strategies for the prevention and treatment of overtraining.

Prevention efforts have served valuable, though often unappreciated roles, in health and mental health service delivery (33). By preventing disease and dysfunction, one generally produces quality of life benefits at an economic advantage to later treatment programs. This seems particularly relevant for overtraining, where we have yet to develop any simple test to alert us to the impending failure (40); the presence of serious symptoms may indicate that we are already too late to reverse the negative effects, and the cost of an athlete's staleness or burnout is so immediately painful (84).

Without doubt the primary strategy for preventing overtraining is not psychological but requires the design of appropriate and effective training regimens (48). The psychoeducational contributions to this assertion are the necessity of educating coaches and athletes on current scientific information on overtraining and aiding them to develop clear communication on these issues (48). While many coaches and athletes are knowledgeable about overtraining, the "more is better" myth can override the best preparation. More importantly, an athlete who fully understands the coach's plan for the training program, or who participates in the development of that plan, is likely to assume some ownership of the program, and consequently, should be more likely to comply with the training regimen, discuss training problems with the coach (54), and retain a more positive mood (see Blaney (6) for a discussion of the benefits of perceived control on mood). Both athletes and coaches should also be exposed to basic time management (50) and problem-solving skills training (18) to aid in adherence to the training regimen and to minimize other life stressors.

The simplest prevention recommendations involve monitoring physiological states to predict eventual overtraining (28, 32, 48). Unfortunately, and in large measure because of the complex, interacting factors of the systemic model we have proposed, no simple monitoring tool has successfully accomplished this task. At best, physiological and biochemical markers of overtraining have shown only inconsistent or weak trends (32).

Morgan et al. (61), Raglin (74), and others have recommended the regular monitoring of psychological states during training, usually in conjunction with the recording of physiological variables. Morgan's work, reviewed earlier, makes an impressive argument for the monitoring of mood as a concurrent sign of overtraining. More interestingly from a prevention standpoint is the possibility that relatively simple subjective ratings of fatigue and mood may help to predict performance decrements.

As an example, Hooper et al. (32) studied 14 elite Australian swimmers engaged in a 6-month preparation for the national team trials. A comprehensive battery of physiological and biochemical measures was given on five occasions, early in the season, at midseason, late in the season, during the taper, and postcompetition. As part of their training logs, swimmers also recorded their self-ratings of quality of sleep, fatigue, stress, and muscle soreness on a 7-point scale. Hooper et al. classified swimmers as stale if they failed to show performance improvements, had high fatigue ratings, and reported no illnesses. Three of the 14 swimmers were classified as stale and these athletes did not differ in their training programs from their nonstale counterparts. Hooper et al. reported that early, mid- and late-season batteries of subjective ratings of sleep, fatigue, stress, and muscle soreness all served as good predictors of competitive staleness. This grouping of subjective variables generated predictions not significantly different than the predictions made from a larger battery of biological and self-report variables. While this study must be evaluated with consideration being given to the small sample size and some apparent overlap in variables being predicted and predictor variables, the Hooper group's findings are not unlike those reported by Lehmann et al. (47), in which a simple complaint index correlated highly with the best biological predictors of overtraining.

Subjective ratings of mood, fatigue, stress, and soreness may be our most cost-effective prevention strategies. But consistent with the position advocated here, with the position taken by Stone et al. (84) that overtraining is a loss of positive adaptive ability, and with the position advanced by Lehmann et al. (48) that overtraining is an imbalance between stress and stress tolerance, our systemic stress management model allows us to argue that the athlete's task is a complex stress management challenge. The most effective prevention efforts ought to inoculate the athlete against these stressors, that is, help the athlete to become a skillful self-regulator of his or her own stress levels.

Ideally, the intensive training program ought to be designed uniquely for each athlete based not only on the athletic issues but also on a comprehensive assessment of the interacting components of the model. We understand that in all but the rarest of cases, coaches and athletic administrators will have neither the time nor the resources for such efforts. Where social scientists are available, or referrals to sport psychologists or other counselors can be made, we believe that there is empirical evidence to indicate that this is a productive path (59, 95). Consequently, we will briefly review strategies for psychological skill development in athletes and make reference to interventions for other systemic problems. These strategies include goal setting, imagery, arousal management, and cognitive restructuring.

Goal Setting.   Goal setting strategies, first developed in the field of industrial and organizational psychology (49), are employed to help the athlete organize his or her efforts for positive task performance. Typically, goal setting interventions in sport are designed to instruct and encourage the athlete to adopt specific and measurable goals for both practice and competitive environments. Both proximal and distal goals are generated, often in a hierarchical manner, so that the athlete has a planned path to the end goal. Recommendations typically emphasize the use of positive rather than negative goals, challenging but realistic goals, and goals that require the athlete to focus on task mastery rather than outcome (91). While the research on the benefits of goal setting is far from conclusive, evidence does suggest that under certain conditions goal setting may enhance the motivation for involvement and promote more positive self-evaluation of training and competitive performance. Such changes may facilitate task persistence and eventual performance improvements under the stressful conditions generated by intense training (9).

Imagery.   Athletes often report using imagery in the process of athletic task rehearsal and they typically judge this imagery to be a valuable preparatory strategy (51). While some investigations have not supported the use of imaginal rehearsal in sport (20, 65), imagery has been used successfully to improve the performances of ski racers (85, 86), basketball players (58), gymnasts (82), dart throwers (97), golfers (99), volleyball players (76), and swimmers (96).

There is convincing evidence of the beneficial consequences of imagining successful athletic performances (96) and the detrimental effects of imagining athletic failure (99). Suinn (88) suggested that skilled athletes with experience in imagery use may benefit from the application of coping-oriented imagery. Coping imagery involves the presentation of a model who manages performance difficulties and corrects performance problems, a potentially valuable strategy given the demands of intense training.

Arousal Management.   Athletes across sports and at a wide variety of skill levels typically report arousal and other somatic and emotional changes during training and prior to and during competition (24, 51). Though one must acknowledge a good deal of situational and individual specificity (53), moderate arousal is often necessary for optimal performance and extremely low or overly high levels of arousal appear to be debilitating for superior response to competitive demands (41, 53). Interventions to reduce the debilitating arousal an athlete experiences are common (72, 87). These efforts to attenuate over-arousal mirror anxiety reduction and stress management efforts prevalent in clinical psychology (80). Progressive relaxation training, biofeedback, and stress inoculation training (56, 98) have been employed effectively to lower athletes' arousal and anxiety (17, 68, 80, 100).

Efforts to increase arousal, or to psych-up athletes, have relied on athletes' ability to ready themselves psychologically or employ arousing contextual demands. These psyching-up efforts are buttressed by the assumption that mental preparation heightens arousal, thereby preparing the athlete to meet performance demands (92). The ability to regulate arousal, to increase or decrease arousal levels as appropriate, should benefit athletes struggling with difficult training regimens or monotonous, repetitive workouts.

Level of arousal is surely not the only consequential variable mediating the impact of anxiety on performance (53). Athletes' perceptions of the anxiety-arousal context, their previous experience with arousal changes, and their perceived ability to manage or control arousal may play pivotal roles in the arousal-performance relationship. The cognitive contributions to this relationship are often dealt with through interventions based on cognitive restructuring.

Cognitive Restructuring.   Kirschenbaum et al. (36, 37) have proposed a self-regulatory model of athletic skill development. Kirschenbaum argued that an athlete's performance is, at least in part, a test of that individual's skill in self-directed cognition and action. The individual must successfully execute the physical skill, monitor performance, evaluate that performance against some standard or goal, and then alter the execution of the physical skills. Bandura (3) has extended this argument to suggest that both the individual's initiation of and persistence on a task is dependent on his or her sense of self-efficacy, one's belief in one's own ability to perform a task successfully.

Targets of interventions central to these processes include the athlete's self-monitoring (34) and self-instructional behavior (57), and specific attempts to influence the athlete's core cognitive constructs (e.g., the athlete's world view or cognitive belief system; cf., Beck (4)) through cognitive restructuring interventions. Since the athlete's perception and understanding of the intense training experience and his or her response to it are crucial to persistence and competitive performance (81), cognitive self-regulation strategies are potentially promising interventions for maintenance of intense training.

Treatment

Any discussion of the treatment of overtraining has to begin with the admonition for the earliest possible intervention and the obvious assertion that the prevention strategies listed above are applicable here as well. Consequently, the strategies discussed under treatment should also serve as beneficial prevention procedures. Indeed, psychologists legitimize such an assertion by labeling early intervention as secondary prevention (33). The most obvious treatment for the negative response to intense training is the interruption of training and rest (40). The concept of rest should also be extended to the nonsport stressors that the athlete is experiencing. From the systemic stress management position we adopt, the stale athlete must first reduce the overriding stresses of the training program and extrasport systems before new or additional resources can be brought to bear on the problem.

The development of short-term goals should aid the athlete to refocus on productive activities. Communication training with coaches and teammates and significant others may be helpful in soliciting social support (54). Cognitive restructuring can help to develop a more positive motivational orientation to the post-overtraining experience (91, 95). Individual therapy (10, 25, 89), family therapy (29), marital therapy (12), and career counseling (15, 66) may all serve to help the athlete rebound.

Treatment Issues: Carl and Hunter

Let us return one final time to our swimmers. Carl is experiencing a common set of athletic stressors and is responding to this developmental challenge in a productive way. While we could raise issues concerning his focus and impulsivity, competitive anxiety, school performance, and relationship with his father, it is unlikely we would do so unless he requested such attention. Given the communicative relationship between Carl and the coach, one might hope that the coach would encourage Carl to deal with these issues.

Hunter requires rest but also a respite from the stress his enmeshed relationship with his mother brings to the situation. He is also experiencing a classical clinical triad of high self-standards, active self-criticism, moodiness and isolation, a pattern often associated with mood disorders. Hunter would benefit from attention to these issues and the emotional conflicts they bring on. By employing an emotion-focused or avoidance-based coping style, Hunter is not currently confronting the functional problems he faces. Counseling, by helping him to self-regulate more effectively, develop goals for his return to swimming, and deal in a more problem-focused manner with his mother, coach, and team-mates, may be beneficial.

Summary

In another of the ubiquitous pre-Olympic pieces appearing in our newspapers, Harmut Buschbacker, the coach of the U.S. women's rowing team, speaking of his work with his athletes, remarked that it was his "responsibility to come up with a program that won't kill them but makes them better" (7). We certainly agree with his statement and extend the assumption to sport scientists as well. We believe that psychological approaches to prevention and treatment of overtraining can help athletes and coaches to achieve that objective.

Acknowledgments

This work was supported in part by a Tennessee Centers of Excellence grant to the University of Memphis, Department of Psychology. Our thanks to Tim Steenburgh and Dora Compton for their help with this manuscript.

References

1. Bandura, A. 1977. *Social learning theory*. Englewood Cliffs, NJ: Prentice Hall.
2. Bandura, A. 1978. The self system and reciprocal determinism. *American Psychologist* 33: 344-358.
3. Bandura, A. 1989. Human agency in social cognitive theory. *American Psychologist* 44: 1175-1184.
4. Beck, A.T. 1976. *Cognitive therapy and the emotional disorders*. Madison, CT: International Universities Press.
5. Becker, D. 1996. 42-year-old Bostisaurus joins cyclists. *USA Today*, 7 June, sec. C, p. 8.
6. Blaney, P.H. 1977. Contemporary theories of depression: critique and comparison. *Journal of Abnormal Psychology* 36: 203-223.
7. Brown, B. 1996. We're the ones to beat: extremely demanding regimen produces confident crew. *USA Today*, 11 January, sec. C, p. 3.
8. Bryant, V., A.W. Meyers, J.P. Whelan. 1990. Overtraining and mood change in elite junior Olympic weightlifters. Paper presented at the *Association for the Advancement of Applied Sport Psychology*, September, San Antonio.
9. Burton, D. 1993. Goal setting in sport. In *Handbook of research on sport psychology*, eds. R.N. Singer, M. Murphey, L.K. Tennant, 467-491. New York: Macmillan.

10. Carr, C.M., S.M. Murphy. 1995. Alcohol and drugs in sport. In *Sport psychology interventions*, ed. S. M. Murphy, 283-306. Champaign, IL: Human Kinetics.

11. Carter, B., M. McGoldrick, eds. 1988. *The changing family life cycle*. New York: Gardner Press.

12. Coppel, D.B. 1995. Relationship issues in sport: a marital therapy model. In *Sport psychology interventions*, ed. S. M. Murphy, 193-204. Champaign, IL: Human Kinetics.

13. Costill, D.L., M.G. Flynn, J.P. Kirwan, J.A. Houmard, J.B., Mitchell, R. Thomas, S.H. Park. 1988. Effects of repeated days of intensified training on muscle glycogen and swimming performance. *Medicine and Science in Sports and Exercise* 20: 249-254.

14. Cox, T. 1978. *Stress*. Baltimore: University Park Press.

15. Danish, S.J., A. Petitpas, B. Hale. 1995. Psychological interventions: a life development model. In *Sport psychology interventions*, ed. S. M. Murphy, 19-38. Champaign, IL: Human Kinetics.

16. Dean, J., J.P. Whelan, A.W. Meyers. 1990. An incredibly quick way to assess mood states: the incredibly short POMS. Paper presented at the *Association for the Advancement of Applied Sport Psychology*, September, San Antonio.

17. DeWitt, D.J. 1980. Cognitive and biofeedback training for stress reduction with university athletes. *Journal of Sport Psychology* 2: 288-294.

18. D'Zurilla, T.J., M.R. Goldfried. 1971. Problem solving and behavior modification. *Journal of Abnormal Psychology* 78: 107-126.

19. Elliott, G.R., C. Eisdorfer. 1982. *Stress and human health*. New York: Springer.

20. Epstein, M.L. 1980. The relationship of mental imagery and mental rehearsal to performance on a motor task. *Journal of Sports Psychology* 2: 211-220.

21. Folkman, S., R.S. Lazarus. 1980. An analysis of coping in middle-aged community sample. *Journal of Health and Social Behavior* 21: 219-239.

22. Folkman, S., R.S. Lazarus. 1988. The relationship between coping and emotion: Implications for theory and research. *Social Science and Medicine* 26: 309-317.

23. Fry, A.C., W.J. Kraemer, F. van Borselen, N.T. Lynch, N.T. Triplett, L.P. Koziris, S.J. Fleck. 1994. Catecholamine responses to short-term high intensity resistance exercise overtraining. *Journal of Applied Physiology* 77: 941-946.

24. Gould, D., T. Horn, J. Spreemann. 1983. Sources of stress in junior elite wrestlers. *Journal of Sport Psychology* 5: 159-171.

25. Greenspan, M., M.B. Anderson. 1995. Providing psychological services to student athletes: a developmental psychology model. In *Sport psychology interventions,* ed. S. M. Murphy, 177-192. Champaign, IL: Human Kinetics.

26. Grieve, F., F.B. Whelan, A.W. Meyers, J.P. Whelan. 1992. The psychological effects of overtraining on elite Olympic weightlifters. Paper presented at the meeting of the *American Psychological Association*, August, Washington, D. C.

27. Gross, J.D. 1994. Hardiness and mood disturbances in swimmers while overtraining. *Journal of Sport and Exercise Psychology* 16: 135-149.

28. Hackney, A.C., S.N. Pearman, J.M. Nowacki. 1990. Physiological profiles of overtrained and stale athletes: a review. *Journal of Applied Sport Psychology* 1: 21-33.

29. Hellstedt, J.C. 1995. Invisible players: a family systems model. In *Sport psychology interventions,* ed. S. M. Murphy, 117-146. Champaign, IL: Human Kinetics.

30. Henggeler, S.W., C.M. Borduin. 1990. *Family therapy and beyond: a multisystemic approach to treating the behavior problems of children and adolescents.* Pacific Grove, CA: Brooks/Cole.
31. Henry, J.P., P.M. Stephens. 1977. *Stress, health and the social environment: a sociobiological approach to medicine.* New York: Springer.
32. Hooper, S.L., L.T. Mackinnon, A. Howard, R.D. Gordon, A.W. Bachman. 1995. Markers for monitoring overtraining and recovery. *Medicine and Science in Sports and Exercise* 27: 106-112.
33. Iscoe, I., B.L. Bloom, C.D. Spielberger, eds. 1977. *Community psychology in transition.* Washington, D. C.: Hemisphere.
34. Johnston-O'Conner, E. J., D.S. Kirschenbaum. 1986. Something succeeds like success: positive self-monitoring in golf. *Cognitive Therapy and Research* 10: 123-136.
35. Kibler, W.B., T.J. Chandler. 1994. Sport specific conditioning. *American Journal of Sports Medicine* 22: 424-432.
36. Kirschenbaum, D.S., R.M. Bale. 1980. Cognitive behavioral skills in golf: brain power golf. In *Psychology in sports: methods and application,* ed. R. M. Suinn, 334-343. Minneapolis: Burgess International.
37. Kirschenbaum, D.S., D.A. Wittrock. 1984. Cognitive behavioral interventions in sport: a self-regulatory perspective. In *Psychological foundations of sport,* eds. J. M. Silva, R.S. Weinberg, 81-98. Champaign, IL: Human Kinetics.
38. Klein, M., J. Greist, A. Gurman, R. Neimeyer, D. Lesser, N. Bushnell, R. Smith. 1985. Comparative outcome study of group psychotherapy versus exercise treatments for depression. *International Journal of Mental Health* 13: 148-177.
39. Kobasa, S.C., S.R. Maddi, M.C. Puccetti, M.A. Zola. 1985. Effectiveness of hardiness, exercise and social support as resources against illness. *Journal of Psychosomatic Research* 29: 525-533.
40. Kuiper, H., H.A. Keizer. 1988. Overtraining in elite athletes. *Sports Medicine* 6: 79-92.
41. Landers, D.M. 1980. The arousal-performance relationship revisited. *Research Quarterly for Exercise and Sport* 51: 77-90.
42. Lazarus, R.S. 1976. *Patterns of adjustment.* New York: McGraw-Hill.
43. Lazarus, R.S. 1991. Cognition and motivation in emotion. *American Psychologist* 46: 352-367.
44. Lazarus, R.S., Folkman, S. 1984. *Stress, appraisal and coping.* New York: Springer.
45. Lazarus, R.S., R. Launier. 1978. Stress-related transactions between person and environment. In *Perspectives in interactional psychology,* eds. L.A. Pervin, M. Lewis, 287-327. New York: Plenum Press.
46. Lazarus, R.S., C.A. Smith. 1988. Knowledge and appraisal in the cognition-emotion relationship. *Cognition and Emotion* 2: 281-300.
47. Lehmann, M., H.H. Dickhuth, G. Gendrisch, W. Lazar, M. Thum, R. Kaminski, J.F. Aramendi, E. Peterke, W. Wieland, J. Keul. 1991. Training-overtraining: a prospective, experimental study with experienced middle- and long-distance runners. *International Journal of Sports Medicine* 12: 444-452.
48. Lehmann, M., C. Foster, J. Keul. 1993. Overtraining in endurance athletes: a brief review. *Medicine and Science in Sports and Exercise* 25: 854-862.
49. Locke, E.A., G.P. Latham. 1990. *A theory of goal setting and task performance.* Englewood Cliffs, NJ: Prentice Hall.

50. Maher, C.A. 1981. Time management training for school psychologists. *Professional Psychology* 12: 613-620.
51. Mahoney, M.J., M. Avener. 1977. Psychology of the elite athlete: an exploratory study. *Cognitive Therapy and Research* 1: 135-141.
52. Mahoney, M.J., T.J. Gabriel, T.S. Perkins. 1987. Psychological skills and exceptional athletic performance. *Sport Psychologist* 1: 181-199.
53. Mahoney, M.J., A.W. Meyers. 1989. Anxiety and athletic performance: traditional and cognitive developmental perspectives. In *Anxiety and sports*, eds. C. Spielberger, D. Hackfort, 77-94. New York: Hemisphere.
54. McCann, S. 1995. Overtraining and burnout. In *Sport psychology interventions*, ed. S. Murphy, 347-368. Champaign, IL: Human Kinetics.
55. McNair, D.M., M. Lorr, L.F. Droppleman. 1971. *Profile of mood states manual*. San Diego, CA: Educational and Industrial Testing Services.
56. Meichenbaum, D. 1977. *Cognitive-behavior modification*. New York: Plenum Press.
57. Meyers, A.W., R. Schleser, C.J. Cooke, C. Cuvillier. 1979. Cognitive contributions to the development of gymnastic skills. *Cognitive Therapy and Research* 3: 75-85.
58. Meyers, A.W., R. Schleser, T.M. Okwumabua. 1982. A cognitive behavioral intervention for improving basketball performance. *Research Quarterly for Exercise and Sport* 53: 344-347.
59. Meyers, A.W., J.P. Whelan, S. Murphy. 1995. Cognitive behavioral strategies in athletic performance enhancement. In *Progress in behavior modification*, eds. M. Hersen, R. M. Eisler, P. M. Miller, 137-164. Pacific Grove, CA: Brooks/Cole.
60. Minuchin, P.P. 1985. Families and individual development: provocations from the field of family therapy. *Child Development* 56: 289-302.
61. Morgan, W.P., D.R. Brown, J.S. Raglin, P.J. O'Connor, K.A. Ellickson. 1987. Psychological monitoring of overtraining and staleness. *British Journal of Sports Medicine* 21: 107-114.
62. Morgan, W.P., D.L. Costill, M.G. Flynn, J.S. Raglin, P.J. O'Connor. 1988. Mood disturbance following increased training in swimmers. *Medicine and Science in Sports and Exercise* 20: 408-414.
63. Morgan, W.P., S.E. Goldston. 1987. *Exercise and mental health*. New York: Hemisphere.
64. Mulling, C., A.W. Meyers, M. Summerville, R. Neimeyer. 1986. Aerobic exercise and depression: is it worth the effort? Paper presented at the *Association for the Advancement of Applied Sport Psychology*, October, Jekyll Island, GA.
65. Mumford, B., C. Hall. 1985. The effects of internal and external imagery on performing figures in figure skating. *Canadian Journal of Applied Sport Science* 10: 171-177.
66. Murphy, S.M. 1995. Transitions in competitive sport: maximizing individual potential. In *Sport psychology interventions,* ed. S. M. Murphy, 331-346. Champaign, IL: Human Kinetics.
67. Murphy, S.M., S.J. Fleck, G. Dudley, R. Callister. 1990. Psychological and performance concomitants of increased volume training in elite athletes. *Journal of Applied Sport Psychology* 2: 34-50.
68. Murphy, S.M., R.L. Woolfolk. 1987. The effects of cognitive interventions on competitive anxiety and performance on a fine motor skill accuracy task. *International Journal of Sport Psychology* 18: 152-166.

69. Nowack, K.M. 1990. Initial development of an inventory to assess stress and health risk. *American Journal of Health Promotion* 4: 173-180.
70. O'Connor, P.J., W.P. Morgan, J.S. Raglin. 1991. Psychobiological effects of 3 days of increased training in female and male swimmers. *Medicine and Science in Sports and Exercise* 23: 1055-1061.
71. Old man of cycling riding to Olympics. 1996. *The Memphis Commercial Appeal*, 7 June, sec. D, p. 2.
72. Orlick, T. 1986. *Psyching for sport: mental training for sport*. Champaign, IL: Leisure Press.
73. Raglin, J. 1990. Exercise and mental health: beneficial and detrimental effects. *Sports Medicine* 9: 323-329.
74. Raglin, J. 1993. Overtraining and staleness: psychometric monitoring of endurance athletes. In *Handbook of research on sport psychology*, eds. R.N. Singer, M. Murphey, L.K. Tennant, 840-850. New York: Macmillan.
75. Raglin, J.S., W.P. Morgan, P.J. O'Connor. 1991. Changes in mood states during training in female and male college swimmers. *International Journal of Sports Medicine* 12: 849-853.
76. Shick, J. 1970. Effects of mental practice on selected volleyball skills for college women. *Research Quarterly* 41: 88-94.
77. Silva, J.M. III. 1990. An analysis of the training stress syndrome in competitive athletics. *Journal of Applied Sport Psychology* 2: 5-20.
78. Singer, R.N., M. Murphey, L.K. Tennant, eds. 1993. *Handbook of research on sport psychology*. New York: Macmillan.
79. Smith, C.A., R.S. Lazarus. 1990. Emotion and adaptation. In *Handbook of personality: theory and research*, ed. L. A. Pervin. New York: Guilford.
80. Smith, R.E. 1985. A component analysis of athletic stress. In *Competitive sports for children and youths: proceedings of Olympic Scientific Congress*, eds. M. Weiss, D. Gould, 107-112. Champaign, IL: Human Kinetics.
81. Smith, R.E. 1986. Toward a cognitive-affective model of athletic burnout. *Journal of Sport Psychology* 8: 36-50.
82. Start, K.B., A. Richardson. 1964. Imagery and mental practice. *British Journal of Educational Psychology* 34: 280-284.
83. Steptoe, A., M. Kearsley, N. Walters. 1993. Acute mood responses to maximal and submaximal exercise in active and inactive men. *Psychology and Health* 8: 89-99.
84. Stone, M.H., R.E. Keith, J.T. Kearney, S.J. Fleck, G.D. Wilson, N.T. Triplett. 1991. Overtraining: a review of the signs, symptoms and possible causes. *Journal of Applied Sports Science Research* 5: 35-50.
85. Suinn, R.M. 1972. Behavioral rehearsal training for ski racers. *Behavior Therapy* 3: 519-520.
86. Suinn, R.M. 1977. Behavioral methods at the Winter Olympic Games. *Behavior Therapy* 8: 283-284.
87. Suinn, R.M. 1983. *The seven steps to peak performance: manual for mental training for athletes*. Fort Collins, CO: Colorado State University.
88. Suinn, R.M. 1993. Imagery. In *Handbook of research on sport psychology*, eds. R. N. Singer, M. Murphey, L. K. Tennant, 492-510. New York: Macmillan.
89. Swoap, R.A., S.M. Murphy. 1995. Eating disorders and weight management in athletes. In *Sport psychology interventions*, ed. S.M. Murphy, 307-330. Champaign, IL: Human Kinetics.

90. Taylor, S.E., L.G. Aspinwall. 1993. Coping with chronic illness. In *Handbook of stress: theoretical and clinical aspects*, 2d ed., eds. L. Goldberger, S. Breznitz, 511-531. New York: Free Press.
91. Weinberg, R.S., D. Gould. 1995. *Foundations of sport and exercise psychology*. Champaign, IL: Human Kinetics.
92. Weinberg, R., D. Gould, A. Jackson. 1980. Cognition and motor performance: effect of psyching-up strategies on three motor tasks. *Cognitive Therapy and Research* 4: 239-245.
93. Weinstein, W. S., A.W. Meyers. 1983. Running as a treatment for depression: is it worth it? *Journal of Sport Psychology* 5: 288-301.
94. Weisaeth, L. 1993. Disasters: psychological and psychiatric aspects. In *Handbook of stress: theoretical and clinical aspects,* 2d ed., eds. L. Goldberger, S. Breznitz, 591-616. New York: Free Press.
95. Whelan J.P., M.J. Mahoney, A.W. Meyers. 1991. Performance enhancement in sport: a cognitive behavioral domain. *Behavior Therapy* 22: 307-327.
96. White, K.D., R. Ashton, S. Lewis. 1979. Learning a complex skill: effects of mental practice, physical practice, and imagery ability. *International Journal of Sport Psychology* 10: 71-78.
97. Wichman, H., P. Lizotte. 1983. Effects of mental practice and locus of control on performance of dart throwing. *Perceptual and Motor Skills* 56: 807-812.
98. Woolfolk, R.L., P.M. Lehrer, eds. 1984. *Principles and practice of stress management*. New York: Guilford.
99. Woolfolk, R.L., M.W. Parrish, S.M. Murphy. 1985. The effects of positive and negative imagery on motor skill performance. *Cognitive Therapy and Research* 9: 335-341.
100. Ziegler, S.G., J. Klinzing, K. Williamson. 1982. The effects of two stress management training programs on cardiorespiratory efficiency. *Journal of Sport Psychology* 4: 280-289.

Summary, Conclusions, and Future Directions

CHAPTER 17

Future Research Needs and Directions

Michael G. Flynn, PhD

Introduction

The task of summarizing a conference and a text book and determining the future research needs and directions within this field of study is at the same time an honor and a daunting challenge. It is difficult to predict the course of future research in this subject area; however, the preceding chapters provide the history of past research and help us to determine the appropriate directions for ensuing investigations. It is inevitable that a chapter recommending future research needs and directions will reflect the research interests and bias of the author. However, as you reviewed the previous chapters you probably realized that despite rigorous efforts by these and other researchers, many gaps in our knowledge still exist. Therefore, since it is impossible to address all of the research areas that need additional attention, this chapter will focus on general problems and possible solutions related to overtraining and overreaching research.

While the effects of extreme training first attracted the attention of researchers in the 1950s and 1960s (7, 24), focused research in the area of overtraining and overreaching is in its infancy. A renewed emphasis on overtraining research occurred in the mid-1980s. For example, in 1986 and again in 1987, "overtraining and recovery from training" were identified by the United States Olympic Committee Sports Medicine Council (32) as significant problems facing elite athletes. This edict coincided with funding opportunities from USOC and, therefore, spawned an interest in overtraining research in North America. At the same time, while the specific impetus is difficult to identify, similar research initiatives were begun in many other countries.

Research in the area of overtraining and overreaching provides us with an exciting opportunity to foster the efforts of athletes and coaches; however, it is apparent that sport scientists have a great deal of ground to cover before they will be able to provide substantive assistance. In the meantime, common sense may still be one of the most effective tools in the prevention of overtraining. Ideally, sport scientists and competitive athletes will work together to develop a system for monitoring athletes to allow the development of training programs that will maximize performance and at the same time limit the risk of overtraining.

It has become apparent from the preceding chapters that there is no consensus on an effective marker or index of overtraining. Naturally, a simple marker such as heart rate (8) or blood pressure changes (17) would be most desirable, but many of these simple markers have been found to be not consistently predictive (9, 15). The problem of identifying markers of overtraining is exacerbated by the tremendous interindividual differences in tolerance to training loads. For example, in his introduction to the symposium Dr. Peter Snell discussed the plight of an elite triathlete who performed 19 competitive triathlons in a single year! Furthermore, it is evident that we are not in agreement about the treatment of overtrained athletes. For example, it has been frequently suggested, from observation and anecdotal reports, that weeks or months of complete cessation of training may be required to recover from overtraining (18); however, since athletes and coaches are often fearful of rest, efforts need to be made to refine the treatment programs. In the meantime, the only "treatment" is prevention. While it is tenable to suggest that the athletic community has taken notice, and that many athletes and coaches are practicing "safe training," the sports science community has a considerable distance to travel before we can provide substantive information and useful recommendations to athletes.

State of Research on Overtraining and Overreaching

It is important to ask the "where are we now?" question regarding research on overtraining. The preceding chapters provide an excellent framework of research in several areas of overtraining; however, these chapters also make it clear that there are numerous gaps in our knowledge. For example, despite frenetic research activity over the past ten years, there is at present no single marker or index of overtraining for endurance athletes. This may be due to our inability to distinguish between the responses related to acute exercise and intensified training, and those associated with overtraining (see chapter 3). These difficulties may be exacerbated by the models that have frequently been adopted to study these problems (9, 20, 30). That is to say, since institutional review boards would be rightfully quick to question studies designed to induce overtraining in athletes, we are often forced to find ways to imitate overtraining (e.g., brief periods of intensified training) and probably fall woefully short of studying what we intend to study. Unfortunately, it is apparent that in some areas of study (e.g., overtraining in anaerobic or power athletes) there is too little information to even disagree upon at this time.

Terminology

Building a consensus on terminology is a simple but critical element to foster research on overtraining. A number of articles have been published in which the authors have provided their definitions of acute and chronic responses to hard training (4, 18, 25). The terminology used in the preceding chapters, while not universally accepted, may help to prevent confusion regarding future studies in this area. The preface to this volume should be reviewed thoroughly by the reader to further clarify the terminology that has been employed in the preceding pages. At the very least, this book provides a reference for those who choose to adopt these definitions. However, there is also the potential for some confusion. For example, many investigators have used the term overtraining synonymously with overload, i.e., overtraining provides the stimulus for physiological adaptations and overtraining syndrome or staleness may result from chronic overtraining (25). In this text, overtraining has been used as a synonym for overtraining syndrome and staleness.

While used infrequently, a distinction has sometimes been made between sympathetic and parasympathetic overtraining (18, 19). In theory, sympathetic overtraining is associated with hyperexcitability and restlessness and the parasympathetic form with sympathetic insufficiency, and reduced peripheral catecholamine sensitivity. The examination of autonomic balance and adrenergic sensitivity is one of the more promising new areas of research in overtraining (see chapter 2). This terminology, however, has largely been ignored in the literature from many North American and other laboratories.

The use of this terminology by some researchers, combined with an inability or refusal of other researchers to differentiate between sympathetic and parasympathetic forms of overtraining, has led to considerable confusion. It is possible that overtraining presents differently at various stages of overtraining and that athletes may, for example, develop symptoms consistent with sympathetic overtraining and parasympathetic overtraining at different stages. A Band-Aid solution would be to drop the sympathetic and parasympathetic terminology; however, in the short term it would be more reasonable to suggest that researchers using this terminology clearly distinguish between the two forms of overtraining. Over the long term, we are in need of research studies that will help to determine whether these two forms of overtraining are merely observed at different stages of overtraining (18), are elicited under specific training conditions, or by athletes performing selected activities.

We must endeavor to reach a point where terminology is consistent and does not add to the confusion surrounding an already difficult area of study. Investigators should be mindful that differences in terminology exist when conducting research in the future and be willing to use synonyms, or clearly define the terms they are using in their research.

Prevention

As mentioned above, in the absence of a definitive marker or index of overtraining the most effective weapon against overtraining is prevention. Common sense dictates that athletes simply train less, and a number of studies have been completed that demonstrate the potential for competing well with less training (14, 16, 23). Costill et al. (6), for example, divided a collegiate swim team into either a once daily (1.5 h/d) or twice daily (3 h/d) workout group (see figure 17.1). The swimmers remained in these groups for six weeks of early season training and then completed a 2-week taper consisting of about 1,000 m/d. Somewhat surprisingly, the competitive performance of the swimmers was similar between groups

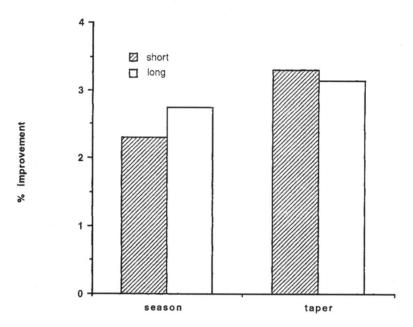

Figure 17.1   The percent improvements in performance in a group of swimmers who trained either once daily (short) or twice daily (long) compared to the previous season (season), and before and after a three week taper at season's end (taper). There were no significant group effects reported.
Adapted from Costill et al. 1991.

following the taper, and the seasonal improvements were not different between the groups. While this study has limitations, it is supported by other reports of athletes performing well on less training (14, 16, 23).

Dr. Keizer pointed out that well-designed training programs are simply "playing with fatigue" (see chapter 8). Additionally, athletes and coaches may be reluctant to reduce training loads. Therefore, another potential tool for the prevention of overtraining is the addition of different training modes to an athlete's sport-specific training (i.e., cross-training). Numerous potential applications for cross-training exist within the overtraining model; however, the most frequently studied is the addition of another training mode to normal run training (10-12, 21, 27). We found no differences in performance or commonly assessed markers of overtraining between well-trained runners who were asked to double their training volume for 10 days using either all running, or a combination of running and cycling (10, 27).

Our earliest studies (10, 27) were an extreme training model designed to examine the differential effects of adding hard training in a sport-specific or non-specific mode. However, the most appropriate application of cross-training to the overtraining model may be supplanting sport-specific training with cross-training, or adding small doses of cross-training in lieu of increased sport-specific training volume or intensity (11, 12, 26). Foster et al. (12), for example, reported that recreational runners performed similarly after the addition of either swim training or run training (10% weekly increase for 8 weeks) to their normal training volume. We found that adding either three additional cycling or three additional running workouts (2 high intensity, 1 moderate intensity) per week to the normal training volume of well-trained runners for six weeks, resulted in significant but similar improvements in 5 km time trial performance (11) (see figure 17.2). Finally, we found that the 5 km time trial performance of well-trained male and female runners was maintained after 30 d of training exclusively in the water with a flotation device (5). We find these results to be particularly intriguing and it leads us to suggest that less stressful training alternatives may be employed without adversely affecting performance. Replication of these studies with elite athletes is required to determine their efficacy and the role they may play in the prevention of overtraining.

The overwhelming majority of investigations in the area are focused on the athletes and their responses to hard training or to brief intensified periods of training (1, 3, 9, 15, 17, 25, 27, 30, 31). While this approach would appear to be logical, it ignores a plausible precipitating influence of overtraining—the coach.

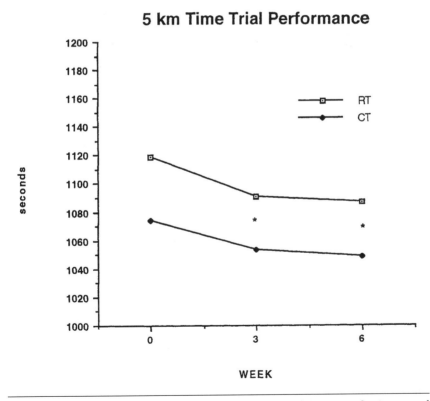

Figure 17.2   The 5 km time trial performance (sec) of runners who increased their training volume and intensity for six weeks by either interval running (RT), or interval cycling (CT). * = significant time effect, i.e., 5 km performance was improved for both groups at weeks 3, and 6 compared to week 0 (11).
Reprinted from Flynn et al. 1994.

That is to say, pairing a highly motivated athlete with an overzealous coach could be a dangerous combination, one that is likely to lead to overtraining. We have all encountered coaches who apparently have the idea that they are paid by the kilometer. The sport science community seems to be in general agreement that coaches might actually achieve greater success with their athletes if they prescribed less training. On the other hand, it would be preposterous to argue that high levels of performance can be achieved without strenuous training. The solution may lie in better communication, improved education of coaches, and the continued development of better working relationships between sport scientists and athletic bodies. This is another area where improved monitoring of athletes might be both an effective preventive measure and an educational tool. It would also be possible to initiate studies that identify bad coaches, e.g., those who have large numbers of ill or injured athletes. However, this approach could foster mistrust and would likely hinder the development of improved relations between coaches and sport scientists.

Implications Beyond Sporting Life

Using the traditional definition (4), overtraining is described as a symptom complex that may include mood disturbances, increased incidence of infection, generalized lethargy, endocrine dysfunction, weight loss, sleeplessness, and poor training or competitive performance. It is natural for those in sport science to focus on the performance aspect of overtraining, but it is important to remember the far-reaching ramifications of this syndrome and the potential for it to impact many other aspects of the athlete's life (see chapter 16). For example, a significant number of top athletes are also university students and the impact of overtraining upon this facet of their lives has largely been ignored by researchers.

Perhaps an example will allow further illustration of this point. While a number of studies have been completed that have examined the effects of hard training on endocrine parameters such as testosterone (1, 9, 10, 13, 28, 29), few investigators have considered the impact that chronically lowered testosterone levels might have on male reproductive function. Ayers et al. (2) reported in a cross-sectional study that 20 marathon runners had reduced testosterone levels compared to controls, but that the values for runners were generally within the normal range. While 18 of 20 runners also had normal sperm counts, two of the runners with the lowest testosterone levels were oligospermatic. We found that increasing the training volume of well-trained recreational runners by 40% for two weeks and 80% for an additional two weeks, did not influence indices of fertility when compared to controls (13). However, two of our runners became oligospermatic during the increased training period (see figure 17.3). When the training volume of these runners was reduced (two weeks at 50% of normal training), their sperm counts were also found to return to the normal range.

Other symptoms of overtraining have the potential to significantly impact the lives of afflicted athletes. Imagine an overtrained athlete (depressed, lethargic, ill) trying to study for exams, or interacting with family, friends, or roommates. These understudied areas of overtraining have the potential to provide far-reaching implications beyond the sporting life of these athletes.

Total Sperm Count

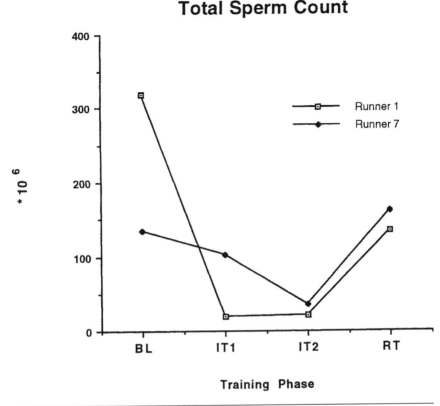

Figure 17.3 Total sperm counts ($*10^6$) of two well-trained runners during an eight week study period. BL = sample obtained after two weeks of normal or baseline training. IT1 = sample obtained after two weeks of training at 140% of baseline training. IT2 = sample obtained after two additional weeks of training at 180% of baseline training. RT = sample obtained after two weeks of recovery training at 50% of baseline training.
Based on data from Hall et al. 1994.

Standardized Methodology

In the early stages of research in a given field there are often differences in methodology that lead to confusion, misinterpretation, and many blind alleys. A number of these issues were raised in David Rowbottom's chapter, e.g., standardized sampling times, correction for plasma volume, allowing sufficient rest between training, and monitoring of athletes. One clear example where differences in methodology have led to misinterpretation is in the measurement of free testosterone. Free testosterone can be measured directly using solid phase radioimmunoassay (3, 9, 10), or it can be derived by measuring testosterone and sex hormone-binding globulin (1, 3). Banfi et al. (3) found that the molar ratio of free testosterone to cortisol was considerably different when direct versus derived measures were used, and would result in considerably different interpretations.

It is inevitable that mistakes will occur as we try to sort out what works best in this field of study. However, we should endeavor to carefully consider other factors that could influence our measures that could inadvertently be attributed to overtraining. For the present, close attention should be paid to time of day for sampling (preferably standardize to 0600-0800 h to allow for sufficient recovery from previous training bouts), consider circa-annual changes, age and gender-related differences, and since intensified training is known to increase plasma volume (9), report both uncorrected values and values corrected for changes in plasma volume in subsequent studies.

Susceptibility

Is it possible that certain groups of athletes are more susceptible to overtraining simply because of the specific demands of their sport? Does the occurrence of musculoskeletal injury when training volume is rapidly increased, "protect" athletes from overtraining? An example that speaks to both of these questions is a comparison of swimming and running training. It is not uncommon for swimmers to complete 15,000 m/d of training during a peak microcycle, that would be calorically equivalent to running approximately 45-60 km/d (22). It is quite possible that after a short time at this daily distance that most runners would succumb to a stress fracture or other musculoskeletal injury, while a number of swimmers might be able tolerate this training volume for an extended period. This should not be interpreted to imply that runners are invulnerable to overtraining, only that athletes participating in sports that do not result in substantial musculoskeletal stress may be more susceptible. It is important to identify which athletes are more likely to be affected by overtraining, so that our efforts may be focused where they might have the greatest impact.

Effective Research Models

A number of researchers have attempted to study overtraining without inducing overtraining (10, 17, 20), by using brief periods of intensified training and determining which physiological markers respond to this training. It is difficult indeed to develop models for overtraining since the desired outcome is so debilitating. An attractive alternative is to monitor groups of athletes during a competitive season or during selected microcycles of training (9, 15), but this model can be complicated by a number of factors. First, research models of this type are largely dependent on what the coach prescribes or upon what the coach and researchers can agree to prescribe. Additionally, since the team goals are oriented toward successful competition, there is no guarantee that some or all of these athletes will become overtrained during the study period. While team monitoring has been quite effective, a third alternative has been largely under used. This alternative would require the identification of athletes who are diagnosed as overtrained. This was rather successfully employed in a study by Barron et al. (4), that resulted in a number of subsequent investigations examining possible hypothalamic-hypophyseal-adrenal axis dysfunction in overtraining. The most effective application of this model might involve a collaborative effort among primary care sport medicine programs that would be likely to encounter overtrained athletes.

Conclusions and New Directions

It is evident that much work needs to be done before we are truly helping athletes. However, some great strides have been made in determining the profile of the overtrained athlete. In addition, excellent progress has been made at some levels of sport in the development of monitoring programs for athletes. While the search for the mechanisms underlying overtraining is important, provision of simple preventive tools to the coach and athlete are most likely to be employed, and will reach the greatest number of athletes.

There are far too many new directions in this area of research to risk making a list and ignoring an important new area of research. In general terms, however, future researchers should consider the following: finding ways to make rest more appealing, determining how much rest is required for recovery after the symptoms of overtraining appear, assessing gender differences with particular attention to gender-specific markers of overtraining, determining differences between aerobic and anaerobic overtraining, examining susceptibility to overtraining in senior athletes and adolescent and preadolescent athletes, examining preventive and alternative training modes such as cross-training and deep water running. These are but a few areas of research that require attention if we are to protect our athletes from the potentially devastating effects of overtraining. Research is the key to the long-term solution of these and other problems facing the athletic community. In the meantime our only defense may be the flimsy shield currently provided by common sense and observation.

References

1. Adlercreutz, H., M. Härkönen, K. Kuoppasalmi, I. Huhtaniemi, H. Tikkanen, K. Remes, A. Dessypris, J. Karvonen. 1986. Effect of training on plasma anabolic and catabolic steroid hormones and their response during exercise. *International Journal of Sports Medicine* 7:S27-S28.
2. Ayers, J.W.T., Y. Komesu, T. Romani, R. Ansbacher. 1985. Anthropometric, hormonal, and psychologic correlates of semen quality in endurance-trained male athletes. *Fertility and Sterility* 43: 917-921.
3. Banfi, G., M. Marinelli, G.S. Roi, V. Agape. 1993. Usefulness of free testosterone/cortisol ratio during a season of elite speed skating athletes. *International Journal of Sports Medicine* 14: 373-379.
4. Barron, G.L., T.D. Noakes, W. Levy, C. Smith, R.P. Millar. 1985. Hypothalamic dysfunction in overtrained athletes. *Journal of Clinical Endocrinology and Metabolism* 60: 803-806.
5. Bushman, B.A., M.G. Flynn, F.F. Andres, C.P. Lambert, M.S. Taylor, W.A. Braun. 1997. Effect of four weeks of deep water run training on running performance. *Medicine and Science in Sports and Exercise* 29: 694-699.

6. Costill, D.L., R. Thomas, R.A. Robergs, D. Pascoe, C. Lambert, S. Barr, W.J. Fink. 1991. Adaptations to swim training: influence of training volume. *Medicine and Science in Sports and Exercise* 23: 371-377.

7. Counsilman, J.E. 1968. *The science of swimming.* Englewood Cliffs, NJ: Prentice Hall.

8. Dressendorfer, R.H., C.E. Wade, J.H. Scaff Jr. 1985. Increased morning heart rate in runners: a valid sign of overtraining? *Physician and Sportsmedicine* 13: 77-86.

9. Flynn, M.G., F.X. Pizza, J.B. Boone Jr., F.F. Andres, T.A. Michaud, J.R. Rodriguez-Zayas. 1994. Indices of training stress during competitive running and swimming seasons. *International Journal of Sports Medicine* 15: 21-26.

10. Flynn, M.G., F.X. Pizza, P.G. Brolinson. 1996. Hormonal responses to excessive training: influence of cross training. *International Journal of Sports Medicine,* in press.

11. Flynn, M.G., K.K. Carroll, H.L. Hall, B.A. Kooiker, C.A. Weideman, C.M. Kasper, P.G. Brolinson. 1994. Cross training: indices of training stress and performance. *Medicine and Science in Sports and Exercise* 26: S153.

12. Foster, C., L.L. Hector, R. Welsh, M. Schraeger, M.A. Green, A.C. Snyder. 1995. Effects of specific versus cross-training on running performance. *European Journal of Applied Physiology* 70: 367-372.

13. Hall, H.L., M.G. Flynn, K.K. Carroll, P.G. Brolinson, S. Shapiro, B.A. Kooiker. 1994. The effects of excessive training and detraining on testicular function. *Medicine and Science in Sports and Exercise* 26: S181.

14. Hickson, R.C., M.A. Rosenkoetter. 1981. Reduced training frequencies and maintenance of increased aerobic power. *Medicine and Science in Sports and Exercise* 13: 13-16.

15. Hooper, S.L., L.T. Mackinnon, A. Howard, R.D. Gordon, A.W. Bachmann. 1995. Markers for monitoring overtraining and recovery in elite swimmers. *Medicine and Science in Sports and Exercise* 27: 106-112.

16. Houmard, J.A. 1991. Impact of reduced training on performance in endurance athletes. *Sports Medicine* 12: 380-393.

17. Kirwan, J.P., D.L. Costill, M.G. Flynn , J.B. Mitchell, W.J. Fink, P.D. Neufer, J.A. Houmard. 1987. Physiological responses to successive days of intense training in competitive swimmers. *Medicine and Science in Sports and Exercise* 20: 255-259.

18. Kuipers, H., H.A. Keizer. 1988. Overtraining and elite athletes: review and directions for the future. *Sports Medicine* 6: 79-92.

19. Lehmann, M., C. Foster, J. Keul. 1993. Overtraining in endurance athletes: a brief review. *Medicine and Science in Sports and Exercise* 25: 854-862.

20. Lehmann, M., U. Gastmann, K.G. Petersen, N. Bachl, A. Seidel, A.N. Khalaf, S. Fischer, J.Keul. 1992. Training-overtraining: performance, and hormone levels, after a defined increase in training volume versus intensity in experienced middle- and long-distance runners. *British Journal of Sports Medicine* 26: 233-242.

21. Loy, S.F., J.J. Hoffman, G.J. Holland. 1995. Benefits and practical use of cross training in sports. *Sports Medicine* 19: 1-8.

22. McArdle, W.D., F.I. Katch, V.L. Katch. 1996. *Exercise physiology: energy, nutrition and human performance,* 4th ed. Baltimore: Williams & Wilkins.

23. McConnell, G.K., D.L. Costill, J.J. Widrick, M.S. Hickey, H. Tanaka, P.B. Gastrin. 1993. Reduced training volume and intensity maintain aerobic ca-

pacity but not performance in distance runners. *International Journal of Sports Medicine* 14: 33-37.

24. Michael, L. 1961. Overtraining in athletes. *Journal of Sports Medicine and Physical Fitness* 1: 99-104.

25. Morgan, W.P., D.R. Brown, J.S. Raglin, P.J. O'Connor, K.A. Ellickson. 1987. Physiological monitoring of overtraining and staleness. *British Journal of Sports Medicine* 21: 107-114.

26. Mutton, D.L., S.F. Loy, D.M. Rogers, G.J. Holland, W.J. Vincent, M. Heng. 1993. Effect of run vs. combined cycle/run training on VO_2max and running performance. *Medicine and Science in Sports and Exercise* 25: 1393-1397.

27. Pizza, F.X., M.G. Flynn, R.D. Starling, P.G. Brolinson, J. Sigg, E.R. Kubitz, R.L. Davenport. 1995. Run training vs. cross training: influence of increased training on running economy, foot impact shock, and run performance. *International Journal of Sports Medicine* 26: 180-184.

28. Urhausen, A., T. Kullmer, W. Kindermann. 1987. A 7-week follow-up study of the behavior of testosterone and cortisol during the competition period in rowers. *European Journal of Applied Physiology* 56: 528-533.

29. Urhausen, A., H. Gabriel, W. Kindermann. 1995. Blood hormones as markers of training stress and overtraining. *Sports Medicine* 20: 251-276.

30. Verde, T.J., S.G. Thomas, R.W. Moore, P. Shek, R. J. Shephard. 1992. Immune responses and increased training of the elite athlete. *Journal of Applied Physiology* 73: 1494-1499.

31. Vervoorn, C. L., J.M. Vermulst, A.M. Boelens-Quist, H.P.F. Koppeschaar, W.B.M. Erich, J.H.H. Thijssen, W.R. deVries. 1992. Seasonal changes in performance and free testosterone/cortisol ratio of elite female rowers. *European Journal of Applied Physiology* 64: 14-21.

32. United States Olympic Committee RFP. 1987. Research Committee of the U.S.O.C. Sports Medicine Council, October.

Index

A

abdominal muscle strain, 181
absolute intensity, 109
Achilles tendinitis, 182
acid-base buffering systems, 77
ACTH. *See* adrenocorticotropic hormone (ACTH)
acute phase reactants, 234
adaptational responses
    and bodily energy stores, 289, 302-303
    cardiovascular, 131-137
    to endurance training, 10-11, 108
        versus resistance training, 108
        use of immunomodulators for, 260-263
    hematologic, 131, 137-139
    and improved performance, 10-12
    musculoskeletal, 171-173, 185-186
        and injury potential, 173-179
    to resistance training, 90-92, 108
    to simultaneous resistance/endurance
        training, 78
    to stress
        and immunosuppression, 244, 257,
            260-263, 266
        individual differences in, 344-347,
            356-359
        and psychosocial issues, 342-354, 356-
            359
adaptive potential, 69-70
ADCC. *See* antibody-dependent cytotoxicity
    (ADCC)
adenine nucleotides, 315
adrenal activity, effects of stress on, 78
adrenal chromaffin cells, 119
adrenocortical dysfunction, and immunosup-
    pression, 257
adrenocorticotropic hormone (ACTH), 150
    actions of, 150-151
    adrenal sensitivity to, 36-39
    effect on thyroid function, 148
    secretion of, 150
        exercise-induced changes in, 100, 117,
            151-152, 155, 257
aerobic training. *See* endurance athletes;
        endurance training; *specific*
        sport
age
    and immune response, 56, 199-200
    and overtraining markers, 56-57, 76, 381
    and stress fractures, 178

alanine, 280, 290, 315
albumin, 311
amenorrhea, 55, 153-154, 156-158, 283-284
American football players, 71
amino acids. *See also specific acid*
    concentration
        alterations in, 31-33, 309-331
        effects of exercise on, 309-310, 327
        and monitoring for overtraining, 54, 57
    dietary considerations, 276, 280, 285, 290-
        291, 314-315, 323
    functions of, 309
ammonia, 54, 93, 314, 323
anabolic-catabolic balance, 36, 289-290
anaerobic training. *See* resistance training;
        specific sport; strength/power
        athletes
androgens, 152-153, 236
anemia, dilutional, 137-138
antibody-dependent cytotoxicity (ADCC), 229
antibody synthesis, and immune suppression,
    236
antiviral medicines, 248, 265. *See also*
        immunomodulators; *specific*
        drug
appetite, 276-277, 285
appraisal outcomes, 345
appraisal process, 345, 357, 359
arousal management, 362
aspartate, 290
aspirin, 210
athletes
    education of, about overtraining, 360
    endurance. *See* endurance athletes; *specific*
        sport
    female. *See* female athletes
    interaction with teammates, 352, 356-357
    male. *See* male athletes
    in nonsport contexts
        implications of overtraining for, 378
        systemic view of, 352-353, 356-359
    nutritional guidelines for, 278-282
    relationship between coach and, 351-352,
        356-357
    strength/power. *See specific sport*; strength/
        power athletes
    young, skeletal immaturity in, and stress
        fractures, 178
autonomic balance, 29-33

List of Contributors

Jacqueline R. Berning, PhD, RD
Biology Department
University of Colorado—Colorado
   Springs
1420 Austin Bluff Parkway
P.O. Box 7150
Colorado Springs, CO 80933-7150
USA

T. Jeff Chandler, EdD
Fitness and Sports Medicine Center
Lexington Clinic
1221 S. Broadway
Lexington, KY 40504
USA

Pamela S. Douglas, MD
Beth Israel Hospital
Harvard Medical School
Harvard-Thorndike Laboratory
330 Brookline Avenue
Boston, MA 02215
USA

Michael G. Flynn, PhD
Department of HKLS
Purdue University
West Lafayette, IN 47906
USA

Carl Foster, PhD
Milwaukee Heart Institute
960 N. 12th Street
P.O. Box 342
Milwaukee, WI 53201-0342
USA

Andrew C. Fry, PhD
Exercise and Sport Science
   Laboratories
The University of Memphis
Memphis, TN 38152
USA

Uwe Gastmann, MD, PhD
University Medical Hospital Ulm
Department of Sports Medicine
Steinhövelstraße 9
D-89075 Ulm
GERMANY

Elena P. Gotovtseva, PhD
Division of Immunology and Clinical
   Allergy
University of Texas—Houston
   Medical School
6431 Fannin 4.044 MFB
Houston, TX 77030
USA

Kevin A. Jacobs, MA
Sport and Exercise Science Section
School of Physical Activity &
   Educational Services
The Ohio State University
337 W. Seventeenth Avenue
Columbus, OH 43210
USA

David Keast, PhD
Department of Microbiology and
   Department of Human Movement
University of Western Australia
Perth, Western Australia 6907
AUSTRALIA

Hans A. Keizer, MD, PhD
Department of Movement Sciences
University of Limburg
P.O. Box 616
6200 MD Maastricht
THE NETHERLANDS

W. Ben Kibler, MD
Fitness and Sports Medicine Center
Lexington Clinic
1221 S. Broadway
Lexington, KY 40504
USA

William J. Kraemer, PhD
Department of Kinesiology
Laboratory for Sports Medicine
21 Recreation Building
The Pennsylvania State University
University Park, PA 16802-5700
USA

Richard B. Kreider, PhD
Exercise & Sport Sciences Laboratory
Department of Human Movement
   Sciences & Education
The University of Memphis
Memphis, TN 38152
USA

Nicole Leenders, MS
Sport and Exercise Science Section
School of Physical Activity &
   Educational Services
The Ohio State University
337 W. Seventeenth Avenue
Columbus, OH 43210
USA

Manfred Lehmann, MD, PhD
University Medical Hospital Ulm
Department of Sports Medicine
Steinhövelstraße 9
D-89075 Ulm
GERMANY

Yufei Liu, MD, PhD
University Medical Hospital Ulm
Department of Sports Medicine
Steinhövelstraße 9
D-89075 Ulm
GERMANY

Werner Lormes, PhD
University Medical Hospital Ulm
Department of Sports Medicine
Steinhövelstraße 9
D-89075 Ulm
GERMANY

Laurel Traeger Mackinnon, PhD
Department of Human Movement
   Studies
The University of Queensland
Brisbane, Queensland 4072
AUSTRALIA

Andrew W. Meyers, PhD
Center for Applied Psychological
   Research
Department of Psychology
University of Memphis
Memphis, TN 38152
USA

Alan R. Morton, PhD
Department of Microbiology and
   Department of Human Movement
University of Western Australia
Perth, Western Australia 6907
AUSTRALIA

Nikolaus Netzer, MD, PhD
University Medical Hospital Ulm
Department of Sports Medicine
Steinhövelstraße 9
D-89075 Ulm
GERMANY

David C. Nieman, DrPH
Department of Health and Exercise
   Science
Appalachian State University
Boone, NC 28608
USA

Bradley C. Nindl, MS
Department of Kinesiology
Noll Physiological Research Center
Center for Sport Medicine
The Pennsylvania State University
University Park, PA 16802-5700
USA

Alexandra Opitz-Gress, MD
University Medical Hospital Ulm
Department of Sports Medicine
Steinhövelstraße 9
D-89075 Ulm
GERMANY

Mary L. O'Toole, PhD
University of Tennessee—Campbell
  Clinic
Department of Orthopaedic Surgery
956 Court Avenue, Room A302
Memphis, TN 38163
USA

David G. Rowbottom, PhD
School of Movement Studies
Queensland University of Technology
Kelvin Grove, Queensland 4059
AUSTRALIA

W. Michael Sherman, PhD
Sport and Exercise Science
School of Physical Activity &
  Educational Services
The Ohio State University
1760 Neil Avenue
Columbus, OH 43210
USA

Jürgen M. Steinacker, MD, PhD
University Medical Hospital Ulm
Department of Sports Medicine
Steinhövelstraße 9
D-89O75 Ulm
GERMANY

Michael H. Stone, PhD
Department of Health and Exercise
  Science
Appalachian State University
Boone, NC 28608
USA

Ida D. Surkina, PhD
National Center for Mental Health
Russian Ministry of Health
Moscow
RUSSIA

Peter N. Uchakin, PhD
Life Sciences Research Laboratories
NASA-Johnson Space Center
Houston, TX 77058
USA

James P. Whelan, PhD
Center for Applied Psychological
  Research
Department of Psychology
University of Memphis
Memphis, TN 38152
USA

About the Editors

Richard B. Kreider, PhD, serves as associate professor and assistant department chair in the Department of Human Movement Sciences and Education at the University of Memphis. His primary research focus has been on optimizing human performance by studying nutritional considerations, physiology of ultraendurance exercise, and overtraining. Dr. Kreider has published more than 100 research articles in scientific journals. He is a Fellow of the American College of Sports Medicine and the research digest editor for the *International Journal of Sport Nutrition.*

Andrew C. Fry, PhD, is an assistant professor in the Department of Human Movement Sciences and Education at the University of Memphis. Dr. Fry's primary research interests lie in resistance exercise, skeletal muscle adaptations to resistance exercise, and overtraining in resistance athletes. He has published more than 75 research articles and also has served as an editorial assistant for the *Journal of Applied Sport Science Research* from 1985 to 1988. Dr. Fry is the supervising editor for the Sport Science and Medical Committee series in *Weightlifting USA,* supervising editor for research summaries in *Strength & Conditioning,* and associate editor of the *Journal of Strength and Conditioning Research.*

Mary O'Toole, PhD, serves as associate professor in the College of Medicine, Department of Orthopaedic Surgery, at the University of Tennessee—Memphis. Her main area of expertise is in the medical and physiological aspects of endurance and ultraendurance exercise; she has studied Ironman triathletes for more than 10 years. Dr. O'Toole has published numerous book chapters, peer-reviewed articles, and abstracts, and has lectured internationally on various aspects of endurance exercise. A Fellow of the American College of Sports Medicine, she has also organized and chaired ACSM symposia on ultraendurance exercise.

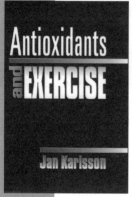